Instrument Rating

Test prep 2002

Study and Prepare for the Instrument Rating, Instrument Flight Instructor (CFII), Instrument Ground Instructor, and Foreign Pilot: Airplane and Helicopter FAA Knowledge Tests

- Effective June 11, 2001
- All FAA Instrument Rating Questions included
- Organized by subject
- Answers, Explanations, References and additional Study Material included for each chapter
- Includes the official FAA Computerized Testing Supplement
- Plus . . . helpful tips and instructions for the FAA Knowledge Test

Calc. 500 ft./min @ 90 knots

Aviation Supplies & Academics, Inc.
Newcastle, Washington

Instrument Rating Test Prep
2002 Edition

Aviation Supplies & Academics, Inc.
7005 132nd Place SE
Newcastle, Washington 98059-3153
(425) 235-1500
www.asa2fly.com

© 2001 ASA, Inc.

FAA Questions herein are from United States government sources and contain current information as of: June 11, 2001.

None of the material in this publication supersedes any documents, procedures or regulations issued by the Federal Aviation Administration.

ASA assumes no responsibility for any errors or omissions. Neither is any liability assumed for damages resulting from the use of the information contained herein.

Important: This Test Prep should be sold with and used in conjunction with *Computerized Testing Supplement for Instrument Rating* (FAA-CT-8080-3D).

ASA reprints the FAA test figures and legends contained within this government document, and it is also sold separately and available from aviation retailers nationwide. Order #ASA-CT-8080-3D.

ASA-TP-I-02
ISBN 1-56027-431-X

Printed in the United States of America
02 01   5 4 3 2 1

About the Contributors

Charles L. Robertson
Associate Professor, UND Aerospace
University of North Dakota

Charles Robertson as flight instructor, associate professor and manager of training at UND Aerospace, contributes a vital and substantial combination of pilot and educator to ASA's reviewing team. After graduating with education degrees from Florida State University in 1967, and Ball State University in 1975, he began his twenty-year career in the United States Air Force as Chief of avionics branch, 58th Military Airlift Squadron, and went on to flight instruction, training for aircraft systems, and airport managing, while gaining many thousands of hours flying international passenger and cargo, aerial refueling and airlift missions. As Division Chief in 1988, Robertson directed the USAF Strategic Air Command's "Alpha Alert Force" and coordinated its daily flight training operations. He holds the CFI Airplane Land, Multi-Engine, Single-Engine and Instrument, the ATP Airplane Land and Multi-Engine, Commercial Pilot, Advanced and Instrument Ground Instructor licenses.

Jackie Spanitz
Director of Curriculum Development
Aviation Supplies & Academics, Inc.

Jackie Spanitz earned a bachelor of science (B.S.) degree with Western Michigan University (WMU), in Aviation Technology and Operations—Pilot option. She completed her masters program with Embry-Riddle Aeronautical University, earning her degree in Aeronautical Science, specializing in Management. As Director of Curriculum Development for ASA, Jackie oversees new and existing product development, ranging from textbooks and flight computers to flight simulation software products, and integration of these products into new and existing curricula. She also provides technical support, research for product development, and project management. Jackie Spanitz holds the CFI Airplane, Land, Single-Engine and Instrument, Commercial Airplane, Land, Single-Engine, Multi-Engine and Instrument, and Advanced and Instrument Ground Instructor certificates; she is the author of *Guide to the Biennial Flight Review*, *Private Pilot Syllabus*, *Instrument Rating Syllabus*, and *Commercial Pilot Syllabus*, and is the technical editor for ASA's Test Prep series.

About ASA: Aviation Supplies & Academics, Inc. (ASA) is an industry leader in the development and sale of aviation supplies and publications for pilots, flight instructors, flight engineers, and aviation maintenance technicians. We manufacture and publish more than 200 products for the aviation industry. Aviators are invited to call 1-800-ASA-2-FLY for a free copy of our catalog. Visit ASA on the web: **www.asa2fly.com**

Contents

Instructions

Preface v
Update Information vi
Description of the Tests vii
   Process for Taking a Knowledge Test viii
      Computer Testing Designees ix
   Use of Test Aids and Materials x
   Cheating or Other Unauthorized Conduct xi
   Validity of Airman Test Reports xi
      Retesting Procedures xi
Eligibility Requirements for the
   Instrument Rating xii
Certificate and Logbook Endorsements xiii
Test-Taking Tips xiv
Suggested Materials for the
   Instrument Rating xv
ASA Test Prep Layout xvi

Chapter 1 **Weather**

The Earth's Atmosphere 1–3
High Altitude Weather 1–4
Temperature 1–4
Wind 1–6
Moisture and Precipitation 1–7
Stable and Unstable Air 1–9
Clouds 1–11
Common IFR Producers 1–12
Air Masses and Fronts 1–14
Turbulence 1–16
Thunderstorms 1–17
Microbursts 1–20
Icing 1–21
Wind Shear 1–23

Chapter 2 **Weather Services**

Aviation Routine Weather Report (METAR) 2–3
Terminal Aerodrome Forecast (TAF) 2–5
Aviation Area Forecast (FA) 2–6
Pilot Reports (UA) and Radar 2–7
Winds and Temperatures Aloft Forecast (FD) 2–9
Inflight Weather Advisories (WA, WS, WST) 2–12
Transcribed Weather Broadcast (TWEB) 2–14
Enroute Flight Advisory Service (EFAS) 2–14
Convective Outlook (AC) 2–15
Surface Analysis Chart 2–16
Weather Depiction Chart 2–16
Radar Summary Chart 2–18
Constant Pressure Chart 2–21
Tropopause Data Chart 2–22
Significant Weather Prognostics 2–23
Severe Weather Outlook Chart 2–27

Chapter 3 **Flight Instruments**

Airspeeds 3–3
Altitudes 3–6
Altimeter Settings 3–9
Vertical Speed Indicator 3–12
Attitude Indicator 3–13
Turn Coordinator 3–16
Heading Indicator 3–24
Magnetic Compass 3–25
Slaved Gyro 3–28
Instrument Errors 3–29
Fundamental Skills 3–33
Attitude Instrument Flying 3–35
Unusual Attitude Recoveries 3–43

Continued

Chapter 4 Navigation

NAVAID Classes	4–3
DME	4–4
VOR	4–7
HSI	4–18
ADF	4–24
RMI	4–29
RNAV and LORAN	4–32
ILS	4–33
MLS and GPS	4–38

Chapter 5 Regulations and Procedures

Requirements for Instrument Rating	5–3
Instrument Currency Requirements	5–6
Equipment Requirements	5–10
Inspection Requirements	5–15
Oxygen Requirements	5–16
Logbook Requirements	5–18
Preflight Requirements	5–19
Airspace	5–20
Cloud Clearance and Visibility Requirements	5–24
Aircraft Accident/Incident Reporting and NOTAMs	5–26
Spatial Disorientation	5–27
Optical Illusions	5–29
Cockpit Lighting and Scanning	5–31
Altitude and Course Requirements	5–32
Communication Reports	5–34
Lost Communication Requirements	5–36

Chapter 6 Departure

Flight Plan Requirements	6–3
Clearance Requirements	6–17
ATC Clearance/Separations	6–18
Departure Procedures (DPs)	6–21
Services Available to Pilots	6–25
Pilot/Controller Roles and Responsibilities	6–28
VFR-On-Top	6–30

Chapter 7 En Route

Instrument Altitudes	7–3
Enroute Low Altitude Chart	7–5
Holding	7–12
Estimated Time Enroute (ETE)	7–22
Fuel Consumption	7–28

Chapter 8 Arrival and Approach

STARs	8–3
Communications During Arrival	8–5
Instrument Approach Terms and Abbreviations	8–8
Instrument Approach Procedures Chart (IAP)	8–12
Radar Approaches	8–29
Visual and Contact Approaches	8–30
Timed Approaches from Holding	8–32
Missed Approach	8–34
RVR	8–34
Inoperative Components	8–36
Hazards on Approach	8–38
Airport Lighting and Marking Aids	8–40
Closing the Flight Plan	8–45

Cross-References:

A: Answer, Subject Matter Knowledge Code, Category & Page Number .... A–1 through A–8

B: Subject Matter Knowledge Code & Question Number .... B–1 through B–8

Preface

Welcome to ASA's Test Prep Series. ASA's test books have been helping pilots prepare for the FAA Knowledge Tests since 1984 with great success. We are confident that with proper use of this book, you will score very well on any of the instrument rating tests.

All of the questions in the FAA Instrument Rating Test Question Bank are included here, and have been arranged into chapters based on subject matter. Topical study, in which similar material is covered under a common subject heading, promotes better understanding, aids recall, and thus provides a more efficient study guide. We suggest you begin by reading the book cover-to-cover. Then go back to the beginning and place emphasis on those questions most likely to be included in your test (identified by the aircraft category above each question). For example: a pilot preparing for the Instrument Airplane test (or Flight Instructor—Instrument, Airplane) would focus on the questions marked "ALL" and "AIR," and a pilot preparing for the Instrument Helicopter test (or Flight Instructor—Instrument, Helicopter) would focus on the questions marked "ALL" and "RTC." Those people preparing for the Instrument Ground Instructor need to study all the questions.

It is important to answer every question assigned on your FAA Knowledge Test. If in their ongoing review, the FAA authors decide a question has no correct answer, is no longer applicable, or is otherwise defective, your answer will be marked correct no matter which one you chose. However, you will not be given the automatic credit unless you have marked an answer. Unlike some other exams you may have taken, there is no penalty for "guessing" in this instance.

The FAA does not supply the correct answers to questions reproduced in this book, is not responsible for answers contained herein, and will not reveal what they consider the correct answers to be. The question and answer choices are duplicated directly from the FAA Question Bank; however, the FAA presents the questions in a different numerical sequence, and they also change the sequence of the A, B, C answer choices on the FAA website (http://afs600.faa.gov). They do this to discourage applicants from learning the test material by rote memory. The ASA test preps include all the questions the FAA will issue at the test centers. A clear explanation is given directly below each question. Be careful to fully understand the intent of each question and corresponding answer while studying, rather than memorize the A, B, C question. If your study leads you to question an answer choice, we recommend you seek the assistance of a local ground or flight instructor. If you still believe the answer needs review, please forward your questions, recommendations, or concerns to:

Aviation Supplies & Academics, Inc.
7005 132nd Place SE
Newcastle, WA 98059-3153

Voice 425.235.1500
Fax 425.235.0128
www.asa2fly.com

Technical Editor:
Jackie Spanitz, ASA

Editor:
Jennie Trerise, ASA

Update Information

Free Test Updates for the One-Year Lifecycle of Test Prep Books

The FAA releases a new test database each spring, and makes amendments to this database approximately twice a year. However, a small number of questions may be withheld from the public for a period of time while the FAA gathers statistics and validates these questions. This means the questions are not available to the public via the internet-posted databases, but they are being issued at the FAA testing centers. In each of these cases, ASA has worded the question to the best of our knowledge, basing it on current figure books, regulations, and procedures, as well as the type of question asked in previous tests.

The questions described above make up a very small percentage of the overall database and are identified by the symbol ^ (printed after the explanation and prior to the subject matter knowledge code — *see* Page xvi, "ASA Test Prep Layout"). You can feel confident that you will be prepared for your FAA Knowledge Exam by using the ASA test prep products. ASA publishes test books each July and stays abreast of all changes to the tests, as well as the new questions that have been validated, and posts these changes on the ASA website as a Test Update. Visit the ASA website before taking your test to be certain you have all the current information:

www.asa2fly.com

Description of the Tests

All test questions are the objective, multiple-choice type, with three answer choices. Each question can be answered by the selection of a single response. Each test question is independent of other questions; that is, a correct response to one does not depend upon, or influence the correct response to another.

A significant number of the questions are "category-specific" and appear only on the airplane test or the helicopter test. The 20-question "added rating" tests are composed mostly of these category-specific questions. A 20-question "added rating" test is administered to an **instrument instructor applicant** (CFII) who already holds an instrument instructor rating in one category (airplane or helicopter) and wishes to meet the knowledge requirements for the other category. The category-specific questions pertain to such knowledge areas as recency of experience and weather minimums.

If you are pursuing a powered lift instrument rating, you may take either the airplane or the helicopter knowledge test. You are not required to take an additional knowledge test when you already hold an instrument rating.

Tests developed from the instrument rating knowledge bank of questions:

The Instrument Rating—Airplane and Helicopter have 60 questions each and 2.5 hours are allowed.

Test Code

- IRA — Instrument Rating—Airplane: Focus on the questions marked ALL and AIR
- IRH — Instrument Rating—Rotorcraft/Helicopter: Focus on the questions marked ALL and RTC

The Instrument Flight Instructor—Airplane and Helicopter, the Ground Instructor—Instrument, and the Instrument Rating—Foreign Pilot tests have 50 questions each and 2.5 hours are allowed.

Test Code

- IFP — Instrument Rating—Foreign Pilot: Focus on the questions marked ALL and AIR or RTC
- FII — Instrument Flight Instructor—Airplane: Focus on the questions marked ALL and AIR
- FIH — Instrument Flight Instructor—Rotorcraft/Helicopter: Focus on the questions marked ALL and RTC
- IGI — Ground Instructor—Instrument: Focus on all the questions in the book

    Note: Instrument ground instructor (IGI) applicants should be prepared to answer any question that appears in the instrument question bank as they are expected to teach all instrument ratings.

All added-rating tests have 20 questions each and 1.0 hour is allowed. The test is for instrument instructors only. Instrument rating candidates must take the 60-question test.

Test Code

- AIF — Instrument Flight Instructor—Airplane (Added Rating): Focus on the questions marked AIR
- HIF — Instrument Flight Instructor—Rotorcraft/Helicopter (Added Rating): Focus on the questions marked RTC

A score of 70 percent must be attained to successfully pass each test.

Instructions

Process for Taking a Knowledge Test

The Federal Aviation Administration (FAA) has available hundreds of computer testing centers worldwide. These testing centers offer the full range of airman knowledge tests including military competence, instrument foreign pilot, and pilot examiner predesignated tests. Refer to the list of computer testing designees (CTDs) at the end of this section.

The first step in taking a knowledge test is the registration process. You may either call the testing centers' 1-800 numbers or simply take the test on a walk-in basis. If you choose to use the 1-800 number to register, you will need to select a testing center, schedule a test date, and make financial arrangements for test payment. You may register for tests several weeks in advance, and you may cancel your appointment according to the CTD's cancellation policy. If you do not follow the CTD's cancellation policies, you could be subject to a cancellation fee.

The next step in taking a knowledge test is providing proper identification. Testing center personnel will not begin the test until your identification is verified.

Authorization requirements should be determined before contacting or going to the computer testing center. Testing center personnel cannot begin the test until provided with the proper documents. A limited number of tests require no authorization. In the instrument rating test area an authorization is not required for Instrument Flight Instructor—Airplane, Instrument Flight Instructor—Helicopter, Instrument Rating—Foreign Pilot, and Ground Instructor—Instrument.

Acceptable forms of authorization:
- A certificate of graduation or a statement of accomplishment certifying the satisfactory completion of the ground school portion of a course from an FAA-certificated pilot school.
- A certificate of graduation or a statement of accomplishment certifying the satisfactory completion of the ground school portion of a course from an agency such as a high school, college, adult education program, U.S. Armed Force, ROTC Flight Training School, or Civil Air Patrol.
- A written statement or logbook endorsement from an authorized instructor certifying that you have accomplished a ground training or home study course required for the rating sought and you are prepared for the knowledge test.
- Failed Airman Test Report, passing Airman Test Report, or expired Airman Test Report (pass or fail), provided that you still have the original Airman Test Report in your possession.

Before you take the actual test, you will have the option to take a sample test. The actual test is time limited; however, you should have sufficient time to complete and review your test.

Upon completion of the knowledge test, you will receive your Airman Test Report, with the testing center's embossed seal, which reflects your score.

The Airman Test Report lists the subject matter knowledge codes for questions answered incorrectly. The total number of subject matter knowledge codes shown on the Airman Test Report is not necessarily an indication of the total number of questions answered incorrectly. Study these knowledge areas to improve your understanding of the subject matter. *See* the *Subject Matter Knowledge Code/ Question Number Cross-Reference* in the back of this book for a complete list of which questions apply to each subject matter knowledge code.

Your instructor is required to provide instruction on each of the knowledge areas listed on your Airman Test Report, and complete an endorsement of this instruction. The Airman Test Report must be presented to the examiner prior to taking the practical test. During the oral portion of the practical test, the examiner is required to evaluate the noted areas of deficiency.

Should you require a duplicate Airman Test Report due to loss or destruction of the original, send a signed request accompanied by a check or money order for the amount of $1 payable to the FAA. Your request should be sent to the Federal Aviation Administration, Airmen Certification Branch, AFS-760, P.O. Box 25082, Oklahoma City, OK 73125.

Computer Testing Designees

The following is a list of the computer testing designees authorized to give FAA knowledge tests. This list should be helpful in case you choose to register for a test or simply want more information. The latest listing of computer testing center locations may be obtained through the FAA website: http://afs600.faa.gov, then select AFS630, Airman Certification, Computer Testing Sites.

Computer Assisted Testing Service (CATS)

1849 Old Bayshore Highway
Burlingame, CA 94010
Applicant inquiry and test registration: 1-800-947-4228
From outside the U.S.: (650) 259-8550

LaserGrade Computer Testing

16209 S.E. McGillivray, Suite L
Vancouver, WA 98683
Applicant inquiry and test registration: 1-800-211-2753 or 1-800-211-2754
From outside the U.S.: (360) 896-9111

International pilots who want to apply for an FAA certificate based on their ICAO foreign certificates should go to the nearest FAA Flight Standards District Office (FSDO). Phone numbers for these offices will be found in the blue pages of the local telephone book. This will allow you to fly a U.S.-registered aircraft while in the U.S. If you hold instrument privileges on your foreign license, you can take a 50-question knowledge test (Instrument Rating—Foreign Pilot), and if you pass it, have instrument privileges added to this FAA private pilot certificate.

If you are outside of the U.S., you will have to go in person to an FAA International Field Office (IFO) and apply for an FAA private pilot certificate. When outside of the U.S. you will only be authorized to fly a U.S.-registered aircraft.

International Field Offices (IFO)

1. Brussels, Belgium (32-2) 508.2721
   FAA C/O American Embassy, PSC 82 Box 002, APO AE 09710
2. Frankfurt, Germany (49-69) 69.705.111
   FAA C/O IFO EA-33 Unit 7580, APO AE 09050
3. London, England (44-181) 754.88.19
   FAA C/O American Embassy, PSC 801 Box 63, FPO AE 09498-4063
4. Singapore (65) 543-1466
   FAA C/O American Embassy, PSC 470 AP 96507-0001

These special certificates will not allow you to fly for hire in the U.S. To qualify for a "clean" FAA commercial pilot certificate or higher, you must meet the full certification requirements of 14 CFR Part 61, for the level of certificate you are requesting. Your current, logged flying time will count towards the required experience. However, all required training, knowledge, and practical tests must be completed.

Instructions

Use of Test Aids and Materials

Airman knowledge tests require applicants to analyze the relationship between variables needed to solve aviation problems, in addition to testing for accuracy of a mathematical calculation. The intent is that all applicants are tested on concepts rather than rote calculation ability. It is permissible to use certain calculating devices when taking airman knowledge tests, provided they are used within the following guidelines. The term "calculating devices" is interchangeable with such items as calculators, computers, or any similar devices designed for aviation-related activities.

1. Guidelines for use of test aids and materials. The applicant may use test aids and materials within the guidelines listed below, if actual test questions or answers are not revealed.

    a. Applicants may use test aids, such as scales, straightedges, protractors, plotters, navigation computers, log sheets, and all models of aviation-oriented calculating devices that are directly related to the test. In addition, applicants may use any test materials provided with the test.

    b. Manufacturer's permanently inscribed instructions on the front and back of such aids listed in 1(a), e.g., formulas, conversions, regulations, signals, weather data, holding pattern diagrams, frequencies, weight and balance formulas, and air traffic control procedures are permissible.

    c. The test proctor may provide calculating devices to applicants and deny them use of their personal calculating devices if the applicant's device does not have a screen that indicates all memory has been erased. The test proctor must be able to determine the calculating device's erasure capability. The use of calculating devices incorporating permanent or continuous type memory circuits without erasure capability is prohibited.

    d. The use of magnetic cards, magnetic tapes, modules, computer chips, or any other device upon which prewritten programs or information related to the test can be stored and retrieved is prohibited. Printouts of data will be surrendered at the completion of the test if the calculating device used incorporates this design feature.

    e. The use of any booklet or manual containing instructions related to the use of the applicant's calculating device is not permitted.

    f. Dictionaries are not allowed in the testing area.

    g. The test proctor makes the final determination relating to test materials and personal possessions that the applicant may take into the testing area.

2. Guidelines for dyslexic applicant's use of test aids and materials. A dyslexic applicant may request approval from the local Flight Standards District Office (FSDO) to take an airman knowledge test using one of the three options listed in preferential order:

    a. Option One. Use current testing facilities and procedures whenever possible.

    b. Option Two. Applicants may use Franklin Speaking Wordmaster® to facilitate the testing process. The Wordmaster® is a self-contained electronic thesaurus that audibly pronounces typed in words and presents them on a display screen. It has a built-in headphone jack for private listening. The headphone feature will be used during testing to avoid disturbing others.

    c. Option Three. Applicants who do not choose to use the first or second option may request a test proctor to assist in reading specific words or terms from the test questions and supplement material. In the interest of preventing compromise of the testing process, the test proctor should be someone who is non-aviation oriented. The test proctor will provide reading assistance only, with no explanation of words or terms. The Airman Testing Standards Branch, AFS-630, will assist in the selection of a test site and test proctor.

Cheating or Other Unauthorized Conduct

Computer testing centers must follow strict security procedures to avoid test compromise. These procedures are established by the FAA and are covered in FAA Order 8080.6, Conduct of Airman Knowledge Tests. The FAA has directed testing centers to terminate a test at any time a test proctor suspects a cheating incident has occurred. An FAA investigation will then be conducted. If the investigation determines that cheating or unauthorized conduct has occurred, then any airman certificate or rating that you hold may be revoked, and you will be prohibited for 1 year from applying for or taking any test for a certificate or rating under 14 CFR Part 61.

Validity of Airman Test Reports

Airman Test Reports are valid for the 24-calendar month period preceding the month you complete the practical test. If the Airman Test Report expires before completion of the practical test, you must retake the knowledge test.

Retesting Procedures

If you receive a grade lower than 70 percent and wish to retest, you must present the following to testing center personnel.

- failed Airman Test Report; and
- a written endorsement from an authorized instructor certifying that additional instruction has been given, and the instructor finds you competent to pass the test.

If you decide to retake the test in anticipation of a better score, you may retake the test after 30 days from the date your last test was taken. The FAA will not allow you to retake a passed test before the 30-day period taken will reflect the official score.

Eligibility Requirements for the Instrument Rating

To be eligible for an instrument rating, a pilot must:

1. Hold at least a current Private Pilot Certificate with a current medical certificate.

2. Be able to read, speak, write, and understand English.

3. Score at least 70 percent on the FAA Knowledge Test on the appropriate subjects.

4. Pass an oral and flight check on the subjects and maneuvers outlined in the Instrument Practical Test Standards (#ASA-8081-4C).

5. Have a total of 50 hours cross-country flight time as pilot-in-command, of which at least 10 hours must be in airplanes for an instrument-airplane rating (each flight must have a landing at least 50 NM from the departure point).

6. Have 40 hours of simulated or actual instrument time (not more than 30 hours may be instruction in a ground trainer if following a 14 CFR Part 142 program, not more than 20 hours if not following a Part 142 program).

7. Have 15 hours instrument flight instruction by an authorized flight instructor.

8. Must have received ground instruction or logged home-study pertaining to IFR regulations, procedures, various methods of navigation, weather report procurement and use, and the *Aeronautical Information Manual*, among others. A complete list of requirements can be found in 14 CFR §61.65.

Certificate and Logbook Endorsements

Instructions

When you go to take your FAA Knowledge Test, you will be required to show proper identification and have certification of your preparation for the examination, signed by an appropriately certified Flight or Ground Instructor. Ground Schools will have issued the endorsements as you complete the course. If you choose a home-study for your Knowledge Test, you can either get an endorsement from your instructor or submit your home-study materials to an FAA Office for review and approval prior to taking the test.

Instrument Rating Endorsement

Endorsement for aeronautical knowledge: 14 CFR §61.65(b)

I certify that Mr./Ms. _____ has received the ground instruction required for the instrument (airplane/helicopter) rating by 14 CFR §61.65 (b)(1) through (10). I have determined he/she is prepared for the _____ knowledge test.

Signed _____ Date _____

CFI Number _____ Expires _____

Flight Instructor Instrument Rating Endorsement

Endorsement for aeronautical knowledge: 14 CFR §61.185 (a) and (b)

I certify that Mr./Ms. _____ has completed the course of instruction required by 14 CFR §61.185, and has logged the ground instruction required by 14 CFR §61.65 (b)(1) through (10). I have determined he/she is prepared for the _____ knowledge test.

Signed _____ Date _____

CFI Number _____ Expires _____

Ground Instructor Instrument Rating Endorsement

Endorsement for aeronautical knowledge: 14 CFR §61.213

I certify that Mr./Ms. _____ is competent to pass the knowledge exam based on the material necessary in 14 CFR §61.213.

Signed _____ Date _____

CFI Number _____ Expires _____

Test-Taking Tips

Follow these time-proven tips, which will help you develop a skillful, smooth approach to test-taking:

1. In order to maintain the integrity of each test, the FAA may rearrange the answer stems to appear in a different order on your test than you see in this book. For this reason, be careful to fully understand the intent of each question and corresponding answer while studying, rather than memorize the A, B, C answer choice.

2. Take with you to the testing center a sign-off from an instructor, photo I.D., the testing fee, calculator, flight computer (ASA's E6-B or CX-2 Pathfinder), plotter, magnifying glass, and a sharp pointer, such as a safety pin.

3. Your first action when you sit down should be to write on the scratch paper the weight and balance and any other formulas and information you can remember from your study. Remember, some of the formulas may be on your E6-B.

4. Answer each question in accordance with the latest regulations and guidance publications.

5. Read each question carefully before looking at the possible answers. You should clearly understand the problem before attempting to solve it.

6. After formulating an answer, determine which answer choice corresponds the closest with your answer. The answer chosen should completely resolve the problem.

7. From the answer choices given, it may appear that there is more than one possible answer. However, there is only one answer that is correct and complete. The other answers are either incomplete, erroneous, or represents popular misconceptions.

8. If a certain question is difficult for you, it is best to mark it for REVIEW and proceed to the other questions. After you answer the less difficult questions, return to those which you marked for review and answer them. Be sure to untag these questions once you've answered them. The review marking procedure will be explained to you prior to starting the test. Although the computer should alert you to unanswered questions, make sure every question has an answer recorded. This procedure will enable you to use the available time to the maximum advantage.

9. Perform each math calculation twice to confirm your answer. If adding or subtracting a column of numbers, reverse your direction the second time to reduce the possibility of error.

10. When solving a calculation problem, select the answer nearest to your solution. The problem has been checked with various types of calculators; therefore, if you have solved it correctly, your answer will be closer to the correct answer than any of the other choices.

11. Remember that information is provided in the FAA Legends and FAA Figures.

12. Remember to answer every question, even the ones with no completely correct answer, to ensure the FAA gives you credit for a bad question.

13. Take your time and be thorough but relaxed. Take a minute off every half-hour or so to relax the brain and the body. Get a drink of water halfway through the test.

14. The same questions are used for the Instrument Rating (airplane/helicopter) as for the CFII (Instrument Flight Instructor) and IGI (Instrument Ground Instructor) Knowledge Tests. For those persons planning to acquire all three of these ratings within a two-year span, we recommend preparing for all three exams at the same time—this will save study time. However, the FAA does require all three tests be taken separately; one test session cannot be applied to all three tests.

Suggested Materials for the Instrument Rating

The following are some of the publications and products recommended for the Instrument Rating. All are reprinted by ASA and available from authorized ASA dealers and distributors.

ASA-CAP-2	*The Complete Advanced Pilot* by Bob Gardner
ASA-AC00-6A	*Aviation Weather*
ASA-AC00-45E	*Aviation Weather Services*
ASA-8083-21	*Rotorcraft Flying Handbook*
ASA-FR-AM-BK	*Federal Aviation Regulations and Aeronautical Information Manual* (combined)
ASA-8083-3	*Airplane Flying Handbook*
ASA-PM-3	*Instrument Flying* (textbook)
ASA-PM-S-I	*Instrument Rating Syllabus*
ASA-8083-15	*Instrument Flying Handbook*
ASA-8081-4C	*Instrument Practical Test Standards* (Airplane & Helicopter)
ASA-8081-9B	*Instrument Flight Instructor Practical Test Standards* (Airplane & Helicopter)
ASA-OEG-I4	*Instrument Oral Exam Guide*
ASA-AC61-23C	*Pilot's Handbook of Aeronautical Knowledge*
ASA-8083-1	*Aircraft Weight and Balance Handbook*
ASA-TP-I-02	*Instrument Rating Test Prep*
ASA-TW-I-02	Prepware™ Exam Software for Instrument Rating
ASA-CP-IFR	Instrument Plotter
ASA-H2G	Jiffyhood (Instrument Training Hood)
ASA-AP-KT-NOS	Approach Binder Kit with Dividers & Sheet Protectors
ASA-AP-KT-7RNG	Approach Binder Kit with Dividers & Sheet Protectors
ASA-CW-8	Book-Style Chart Wallet
ASA-CW-10	Accordion-Fold Chart Wallet
ASA-IP-5.0	Instrument Pilot Trainer Software

ASA Test Prep Layout

The FAA questions have been sorted into chapters according to subject matter. Within each chapter, the questions have been further classified and all similar questions grouped together with a concise discussion of the material covered in each group. This discussion material of "Chapter text" is printed in a larger font and spans the entire width of the page. Immediately following the FAA Question is ASA's Explanation in *italics*. The last line of the Explanation contains the Subject Matter Knowledge Code and further reference (if applicable). *See* the EXAMPLE below.

Figures referenced by the Chapter text only are numbered with the appropriate chapter number, i.e., "Figure 1-1" is Chapter 1's first chapter-text figure.

Some FAA Questions refer to Figures or Legends immediately following the question number, i.e., "4201. (Refer to Figure 14.)." These are FAA Figures and Legends which can be found in the separate booklet: *Computerized Testing Supplement* (CT-8080-XX). This supplement is bundled with the Test Prep and is the exact material you will have access to when you take your computerized test. We provide it separately, so you will become accustomed to referring to the FAA Figures and Legends as you would during the test.

Figures referenced by the Explanation and pertinent to the understanding of that particular question are labeled by their corresponding Question number. For example: the caption "Questions 4245 and 4248" means the figure accompanies the Explanations for both Question 4245 and 4248.

Answers to each question are found at the bottom of each page, and in the Cross-Reference at the back of this book.

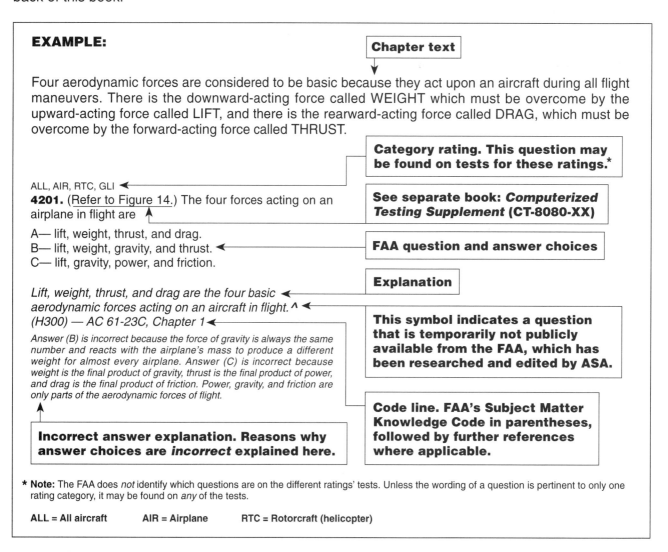

Chapter 1
Weather

The Earth's Atmosphere   *1–3*
High Altitude Weather   *1–4*
Temperature   *1–4*
Wind   *1–6*
Moisture and Precipitation   *1–7*
Stable and Unstable Air   *1–9*
Clouds   *1–11*
Common IFR Producers   *1–12*
Air Masses and Fronts   *1–14*
Turbulence   *1–16*
Thunderstorms   *1–17*
Microbursts   *1–20*
Icing   *1–21*
Wind Shear   *1–23*

Chapter 1 **Weather**

The Earth's Atmosphere

We classify the atmosphere into layers, or spheres, by characteristics exhibited in these layers. The troposphere is the layer from the surface to an average altitude of about 7 miles (37,000 feet). It is characterized by an overall decrease of temperature with increasing altitude. The height of the troposphere varies with latitude and season. It slopes from about 20,000 feet over the poles to about 65,000 feet over the Equator; and it is higher in summer than in winter.

At the top of the troposphere is the tropopause, a very thin layer marking the boundary between the troposphere and the layer above. It is characterized by an abrupt change in temperature lapse rate.

Above the tropopause is the stratosphere. This layer is typified by relatively small changes in temperature with height except for a warming trend near the top.

ALL
4097. A characteristic of the stratosphere is

A—an overall decrease of temperature with an increase in altitude.
B—a relatively even base altitude of approximately 35,000 feet.
C—relatively small changes in temperature with an increase in altitude.

Above the tropopause is the stratosphere. This layer is typified by relatively small changes in temperature with height except for a warming trend near the top. (I20) — AC 00-6A, page 2

Answer (A) is incorrect because temperature increases (not decreases) with an increase in altitude. Answer (B) is incorrect because the stratosphere fluctuates in altitude, as the base is higher at the equator compared to the poles.

ALL
4154. The average height of the troposphere in the middle latitudes is

A—20,000 feet.
B—25,000 feet.
C—37,000 feet.

The height of the troposphere varies with latitude and seasons. It slopes from about 20,000 feet over the poles, to an average of 37,000 feet over the mid-latitudes, to about 65,000 feet over the Equator, and it is higher in summer than in winter. (I20) — AC 00-6A, page 2

ALL
4227. Which feature is associated with the tropopause?

A—Absence of wind and turbulent conditions.
B—Absolute upper limit of cloud formation.
C—Abrupt change in temperature lapse rate.

Temperature over the tropical tropopause increases with height, but temperatures over the polar tropopause remain almost constant. An abrupt change in temperature lapse rate characterizes the tropopause. (I20) — AC 00-6A, page 136

Answer (A) is incorrect because the winds are usually very strong in the tropopause. Answer (B) is incorrect because clouds can form above the tropopause.

Answers

4097 [C]   4154 [C]   4227 [C]

Chapter 1 **Weather**

High Altitude Weather

The jet stream is a river of high speed winds (50 knots or more) associated the tropopause. The location of the jet stream changes seasonally. In the winter, the jet stream moves south and increases in velocity. During the summer, the jet stream moves north and slows.

ALL
4155. A jet stream is defined as wind of

A—30 knots or greater.
B—40 knots or greater.
C—50 knots or greater.

A jetstream occurs in an area of intensified temperature gradients characteristic of the break in the tropopause. The concentrated winds, by arbitrary definition, must be 50 knots or greater to classify as a jetstream. (I32) — AC 00-6A, page 136

ALL
4168. The strength and location of the jet stream is normally

A—stronger and farther north in the winter.
B—weaker and farther north in the summer.
C—stronger and farther north in the summer.

In mid-latitudes, wind speed in the jetstream averages considerably stronger in winter than in summer. Also the jet shifts farther south in winter than in summer. (I32) — AC 00-6A, page 137

Temperature

The major source of all weather is the sun. Changes or variations of weather patterns are caused by the unequal heating of the Earth's surface. In aviation, surface and aloft temperature is measured in degrees Celsius (°C).

Standard temperature is 15°C at sea level. To calculate International Standard Atmosphere (ISA), use the average lapse rate of 2°C per 1,000 feet.

ALL
4096. The primary cause of all changes in the Earth's weather is

A—variation of solar energy received by the Earth's regions.
B—changes in air pressure over the Earth's surface.
C—movement of the air masses.

Every physical process of weather is accompanied by or is the result of a heat exchange. Differences in solar energy create temperature variations. These temperature variations create forces that drive the atmosphere in its endless motion. (I21) — AC 00-6A, page 7

Answer (B) is incorrect because changes in air pressure are due to temperature variations. Answer (C) is incorrect because movement of air masses is a result of varying temperatures and pressures.

ALL
4095. How much colder than standard temperature is the forecast temperature at 9,000 feet, as indicated in the following excerpt from the Winds and Temperature Aloft Forecast?

FT	6000	9000
	0737-04	1043-10

A—3°C.
B—10°C.
C—7°C.

According to the winds and temperatures aloft forecast, the temperature is -10°C at 9,000 feet. Using the average lapse rate of 2°C per 1,000 feet, the temperature change from sea level to 9,000 feet is 18°C. Standard sea level temperature is 15°C. Subtract 18°C from 15°C to get -3°C. Compared to the winds and temperatures aloft forecast for 9,000 feet, the difference is 7°C (10 – 3). (I21) — AC 00-45E, Chapter 4

Answer (A) is incorrect because 3°C is the standard temperature at 9,000 feet, which is not what the question is asking for. Answer (B) is incorrect because 10°C is the given temperature at 9,000 feet, which is not what the question is asking for.

Answers

4155 [C]   4168 [B]   4096 [A]   4095 [C]

Chapter 1 **Weather**

ALL
4113. If the air temperature is +8°C at an elevation of 1,350 feet and a standard (average) temperature lapse rate exists, what will be the approximate freezing level?

A—3,350 feet MSL.
B—5,350 feet MSL.
C—9,350 feet MSL.

Temperature normally decreases with increasing altitude throughout the troposphere. This decrease of temperature with altitude is defined as lapse rate. The average decrease of temperature (average lapse rate) in the troposphere is 2°C per 1,000 feet. An 8°C loss is necessary to reach 0°C, or freezing, in this situation. At 2°/1,000 feet the amount of altitude gain necessary would be:

1. *8°C ÷ 2 = 4 or 4,000 ft*

2.   1,350   *ft MSL (altitude at +8°C)*

   + 4,000   *ft (altitude gain necessary to reach 0°C)*

     5,350   *ft MSL (approximate freezing level)*

(I21) — AC 00-6A, page 9

ALL
4094. A common type of ground or surface based temperature inversion is that which is produced by

A—warm air being lifted rapidly aloft in the vicinity of mountainous terrain.
B—the movement of colder air over warm air, or the movement of warm air under cold air.
C—ground radiation on clear, cool nights when the wind is light.

An increase in temperature with altitude is defined as an inversion. An inversion often develops near the ground on clear, cool nights when wind is light. The ground radiates and cools much faster than the overlying air. Air in contact with the ground becomes cold while the temperature a few hundred feet above changes very little. Thus, temperature increases with height. (I21) — AC 00-6A, page 9

Answer (A) is incorrect because when warm air is lifted, an unstable situation occurs, and a temperature inversion requires stable conditions. Answer (B) is incorrect because warm air over cold air constitutes an inversion (not cold air over warm air).

ALL
4112. The most frequent type of ground- or surface-based temperature inversion is that produced by

A—radiation on a clear, relatively still night.
B—warm air being lifted rapidly aloft in the vicinity of mountainous terrain.
C—the movement of colder air under warm air, or the movement of warm air over cold air.

An inversion often develops near the ground on clear, cool nights when wind is light. The ground radiates and cools much faster than the overlying air. Air in contact with the ground becomes cold while the temperature a few hundred feet above changes very little. Thus, temperature increases with height. (I21) — AC 00-6A, page 9

Answer (B) is incorrect because it describes orographic lifting. Answer (C) is incorrect because it describes fronts.

ALL
4114. What feature is associated with a temperature inversion?

A—A stable layer of air.
B—An unstable layer of air.
C—Air mass thunderstorms.

A temperature inversion occurs when the temperature increases with altitude. A stable layer of air is characterized by warmer air lying above colder air. With an inversion, the layer is stable and convection is suppressed. (I21) — AC 00-6A, page 52

Answer (B) is incorrect because unstable air is characterized by a decrease in temperature with an increase in altitude. Answer (C) is incorrect because air mass thunderstorms are characteristic of unstable conditions.

ALL
4125. A temperature inversion will normally form only

A—in stable air.
B—in unstable air.
C—when a stratiform layer merges with a cumuliform mass.

If the temperature increases with altitude through a layer (an inversion), the layer is stable and convection is suppressed. Air may be unstable beneath the inversion. (I21) — AC 00-6A, page 9

Answer (B) is incorrect because unstable air has warmer air below colder air. Answer (C) is incorrect because when a stratiform layer merges with a cumuliform mass it is associated with a cold front occlusion.

Answers

4113 [B]   4094 [C]   4112 [A]   4114 [A]   4125 [A]

Chapter 1 **Weather**

ALL
4200. Which weather conditions should be expected beneath a low-level temperature inversion layer when the relative humidity is high?

A—Smooth air and poor visibility due to fog, haze, or low clouds.
B—Light wind shear and poor visibility due to haze and light rain.
C—Turbulent air and poor visibility due to fog, low stratus-type clouds, and showery precipitation.

A ground-based inversion favors poor visibility by trapping fog, smoke, and other restrictions into low levels of the atmosphere. Wind just above the inversion may be relatively strong. A wind shear zone develops between the calm and the stronger winds above. Eddies in the shear zone cause airspeed fluctuations as an aircraft climbs or descends through the inversion. (I21) — AC 00-6A, pages 10 and 88

Answer (B) is incorrect because wind shear may be expected within (not beneath) a low-level temperature inversion. Answer (C) is incorrect because inversions cause steady precipitation and create a stable layer of air, thus making it smooth (not turbulent).

Wind

The rules in the Northern Hemisphere are:

1. Air circulates in a clockwise direction around a high pressure system;
2. Air circulates in a counterclockwise direction around a low pressure system;
3. The closer the isobars are together, the stronger the wind speed;
4. Due to surface friction (up to about 2,000 feet AGL), surface winds do not exactly parallel the isobars, but move outward from the center of the high toward lower pressure.
5. Coriolis force is at a right angle to wind direction and directly proportional to wind speed. The force deflects air to the right in the Northern Hemisphere.

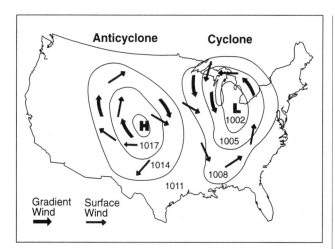

Figure 1-1. Gradient and surface wind

ALL
4105. What causes surface winds to flow across the isobars at an angle rather than parallel to the isobars?

A—Coriolis force.
B—Surface friction.
C—The greater density of the air at the surface.

Friction between the wind and the surface slows the wind. As frictional force slows the wind speed, Coriolis force decreases. However, friction does not affect pressure gradient force. Pressure gradient and Coriolis forces are no longer in balance. The stronger pressure gradient force turns the wind at an angle across the isobars toward lower pressure until the three forces balance. The angle of surface wind to isobars is about 10° over water, increasing with roughness of terrain. (I23) — AC 00-6A, page 30

Answer (A) is incorrect because as wind decreases, so does the Coriolis force. Answer (C) is incorrect because the density of the air has little effect on the relation to the winds and the isobars.

Answers

4200 [A]   4105 [B]

ALL
4106. Winds at 5,000 feet AGL on a particular flight are southwesterly while most of the surface winds are southerly. This difference in direction is primarily due to

A—a stronger pressure gradient at higher altitudes.
B—friction between the wind and the surface.
C—stronger Coriolis force at the surface.

Surface winds and winds at altitude can differ due to friction. Friction between the wind and the surface slows the wind. (I23) — AC 00-6A, page 30

Answer (A) is incorrect because the pressure gradient is relatively uniform at altitudes. Answer (C) is incorrect because the winds are weaker at the surface, therefore the Coriolis force is weaker.

ALL
4107. What relationship exists between the winds at 2,000 feet above the surface and the surface winds?

A—The winds at 2,000 feet and the surface winds flow in the same direction, but the surface winds are weaker due to friction.
B—The winds at 2,000 feet tend to parallel the isobars while the surface winds cross the isobars at an angle toward lower pressure and are weaker.
C—The surface winds tend to veer to the right of the winds at 2,000 feet and are usually weaker.

Close to the earth, wind direction is modified by the contours over which it passes and wind speed is reduced by friction with the surface. Also, the winds at the surface are at an angle across the isobars due to the stronger pressure gradient. At levels 2,000 feet above the surface, the speed is greater and the direction is usually parallel to the isobars. (I23) — AC 00-6A, page 30

Answer (A) is incorrect because the winds at 2,000 feet and those at the surface flow in different directions due to the Coriolis force being weaker at the surface. Answer (C) is incorrect because surface winds do not veer to the right of the winds at 2,000 feet, the winds at 2,000 feet veer to the right of the surface winds.

ALL
4108. Which force, in the Northern Hemisphere, acts at a right angle to the wind and deflects it to the right until parallel to the isobars?

A—Centrifugal.
B—Pressure gradient.
C—Coriolis.

Coriolis force is at a right angle to wind direction and directly proportional to wind speed. In the Northern Hemisphere, the air is deflected to the right. (I23) — AC 00-6A, page 25

Answer (A) is incorrect because centrifugal force acts outwardly to any moving objective in a curved path. Answer (B) is incorrect because pressure gradient causes the wind to move perpendicular to the isobars, but it is then deflected by Coriolis force.

Moisture and Precipitation

Air contains moisture (water vapor). The water vapor content of air can be expressed in two different ways: relative humidity and dew point.

Relative humidity relates the actual water vapor present in the air to that which could be present. Temperature largely determines the maximum amount of water vapor the air can hold. Warm air can hold more water vapor than cold air can. Air with 100% relative humidity is said to be saturated, and air with less than 100% is unsaturated.

Dew point is the temperature to which air must be cooled to become saturated by the water already present in the air.

When water vapor condenses on large objects, such as leaves, windshields, or airplanes, it will form dew. When it condenses on microscopic particles, such as salt, dust, or combustion by-products (condensation nuclei), it will form clouds or fog.

If the temperature and dew point spread is small and decreasing, condensation is about to occur. If the temperature is above freezing, fog or low clouds will be most likely to develop.

The growth rate of precipitation is enhanced by upward currents. Cloud particles collide and merge into a larger drop in the more rapid growth process. This process produces larger precipitation particles and does so more rapidly than the simple condensation growth process. Upward currents also support larger drops.

Answers

4106 [B]   4107 [B]   4108 [C]

Chapter 1 Weather

If wet snow is encountered at your flight altitude, then the temperature is above freezing at your altitude. Since melting snow has been encountered, the freezing level must be at a higher altitude.

The presence of ice pellets at the surface is evidence that there is freezing rain at a higher altitude. Rain falling through colder air may become super-cooled, freezing on impact as freezing rain; or it may freeze during its descent, falling as ice pellets.

ALL
4104. Clouds, fog, or dew will always form when

A—water vapor condenses.
B—water vapor is present.
C—the temperature and dew point are equal.

When temperature reaches the dew point, water vapor can no longer remain invisible but is forced to condense, becoming visible on the ground as dew, appearing in the air as fog or clouds, or falling to the earth as rain. (I24) — AC 00-6A, page 38

Answer (B) is incorrect because there is almost always water vapor present. Answer (C) is incorrect because even when the temperature and dew point are equal (100% humidity), sufficient condensation nuclei must be present for water vapor to condense.

ALL
4101. To which meteorological condition does the term "dew point" refer?

A—The temperature to which air must be cooled to become saturated.
B—The temperature at which condensation and evaporation are equal.
C—The temperature at which dew will always form.

Dew point is the temperature to which air must be cooled to become saturated by the water vapor already present in the air. Aviation weather reports normally include the air temperature and dew point temperature. Dew point, when related to air temperature, reveals qualitatively how close the air is to saturation. (I24) — AC 00-6A, page 38

Answer (B) is incorrect because it takes higher temperatures for water to evaporate, and lower temperatures for water vapor to condense. Answer (C) is incorrect because the formation of dew depends upon the temperatures of the surface and the relative humidity.

ALL
4103. The amount of water vapor which air can hold largely depends on

A—relative humidity.
B—air temperature.
C—stability of air.

Temperature largely determines the maximum amount of water vapor air can hold. Warm air can hold more water vapor than cool air. (I24) — AC 00-6A, page 38

Answer (A) is incorrect because relative humidity does not determine the amount of water vapor air can hold, but rather measures the existing amount of water vapor compared to the amount that could be held. Answer (C) is incorrect because stability of air pertains to the temperature lapse rate, not moisture.

ALL
4159. What enhances the growth rate of precipitation?

A—Advective action.
B—Upward currents.
C—Cyclonic movement.

Cloud particles collide and merge into a larger drop in the more rapid growth process. This process produces larger precipitation particles and does so more rapidly than the simple condensation growth process. Upward currents enhance the growth rate and also support larger drops. (I24) — AC 00-6A, page 43

ALL
4102. What temperature condition is indicated if wet snow is encountered at your flight altitude?

A—The temperature is above freezing at your altitude.
B—The temperature is below freezing at your altitude.
C—You are flying from a warm air mass into a cold air mass.

Wet snow at your altitude means that the temperature is above freezing since the snow has begun to melt. For snow to form, water vapor must go from the vapor state to the solid state (known as sublimation) with the temperature below freezing. Since melting snow has been encountered, the freezing level must be at a higher altitude. (I24) — AC 00-6A, page 43

Answer (B) is incorrect because wet snow requires a temperature above freezing (not below). Answer (C) is incorrect because the temperature is colder above you, but not necessarily in front of you.

Answers

4104 [A]	4101 [A]	4103 [B]	4159 [B]	4102 [A]

ALL
4099. The presence of ice pellets at the surface is evidence that

A—there are thunderstorms in the area.
B—a cold front has passed.
C—there is freezing rain at a higher altitude.

Rain falling through colder air may become supercooled, freezing on impact as freezing rain; or it may freeze during its descent, falling as ice pellets. Ice pellets always indicate freezing rain at higher altitude. (I24) — AC 00-6A, page 43

Answers (A) and (B) are incorrect because thunderstorms and cold fronts do not necessarily cause ice pellets.

ALL
4161. Which precipitation type normally indicates freezing rain at higher altitudes?

A—Snow.
B—Hail.
C—Ice pellets.

Ice pellets always indicate freezing rain at higher altitudes. (I24) — AC 00-6A, page 43

Answer (A) is incorrect because snow indicates that the air above is already below freezing. Answer (B) is incorrect because hail indicates instability where supercooled water droplets have begun to freeze.

Stable and Unstable Air

Atmospheric stability is defined as the resistance of the atmosphere to vertical motion. A stable atmosphere resists any upward or downward movement. An unstable atmosphere allows an upward or downward disturbance to grow into a vertical (convective) current.

Characteristics of unstable air are: cumuliform clouds, showery precipitation, rough air (turbulence), and good visibility, except in blowing obstructions. Characteristics of stable air are: stratiform clouds and fog, continuous precipitation, smooth air, and fair to poor visibility in haze and smoke.

Determining the stability of the atmosphere requires measuring the difference between the actual existing (ambient) temperature lapse rate of a given parcel of air and the dry adiabatic (3°C per 1,000 feet) lapse rate.

ALL
4121. Stability can be determined from which measurement of the atmosphere?

A—Low-level winds.
B—Ambient lapse rate.
C—Atmospheric pressure.

A change in ambient temperature lapse rate of an air mass can tip the balance between stable or unstable air. The ambient lapse rate is the rate of decrease in temperature with altitude. (I25) — AC 00-6A, page 49

Answer (A) is incorrect because stability cannot be determined from low-level winds. Answer (C) is incorrect because the atmospheric pressure does not affect the stability of the air.

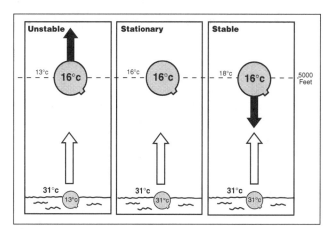

Figure 1-2

Answers
4099 [C]   4161 [C]   4121 [B]

Chapter 1 Weather

ALL
4122. What determines the structure or type of clouds which form as a result of air being forced to ascend?

A—The method by which the air is lifted.
B—The stability of the air before lifting occurs.
C—The amount of condensation nuclei present after lifting occurs.

Cloud type or structure is determined by whether the air is stable or unstable within the layer forced upward. When stable air is forced upward, the air tends to retain horizontal flow and any cloudiness is flat and stratified. When unstable air is forced upward, the disturbance grows, and resulting cloudiness shows "heaped" or cumulus development. (I25) — AC 00-6A, page 50

Answers (A) and (C) are incorrect because the method in which air is lifted and the nuclei present do not determine cloud structure or type.

ALL
4124. Unsaturated air flowing up slope will cool at the rate of approximately (dry adiabatic lapse rate)

A—3°C per 1,000 feet.
B—2°C per 1,000 feet.
C—2.5°C per 1,000 feet.

Unsaturated air moving upward and downward cools and warms at about 3°C (5.4°F) per 1,000 feet. (I25) — AC 00-6A, page 48

Answer (B) is incorrect because 2°C per 1,000 feet is the lapse rate for stable air. Answer (C) is incorrect because 2.5°C per 1,000 feet is the rate of converging temperature and dew point in a convective current of unsaturated air.

ALL
4115. What type of clouds will be formed if very stable moist air is forced up slope?

A—First stratified clouds and then vertical clouds.
B—Vertical clouds with increasing height.
C—Stratified clouds with little vertical development.

When stable air is forced upward, the air tends to retain horizontal flow and any cloudiness is flat and stratified. When unstable air is forced upward, the disturbance grows and any resulting cloudiness shows extensive vertical development. (I25) — AC 00-6A, page 51

Answers (A) and (B) are incorrect because vertical clouds are not characteristic of stable air.

ALL
4118. What type clouds can be expected when an unstable air mass is forced to ascend a mountain slope?

A—Layered clouds with little vertical development.
B—Stratified clouds with considerable associated turbulence.
C—Clouds with extensive vertical development.

When stable air is forced upward, the air tends to retain horizontal flow and any cloudiness is flat and stratified. When unstable air is forced upward, the disturbance grows and any resulting cloudiness shows extensive vertical development. (I25) — AC 00-6A, page 51

Answers (A) and (B) are incorrect because layered and stratified clouds are characteristic of stable air.

ALL
4123. Which of the following combinations of weather producing variables would likely result in cumuliform-type clouds, good visibility, rain showers, and possible clear-type icing in clouds?

A—Unstable, moist air, and no lifting mechanism.
B—Stable, dry air, and orographic lifting.
C—Unstable, moist air, and orographic lifting.

Unstable air favors convection. A cumulus cloud forms in a convective updraft and builds upward. The initial lifting that triggers a cumuliform cloud can be either orographic (topographical, i.e., mountains) or by surface heating. For convective cumuliform clouds to develop, the air must be unstable after saturation. (I25) — AC 00-6A, page 48

Answer (A) is incorrect because without a lifting mechanism, clouds will not have vertical development, resulting in showery rain and icing in clouds. Answer (B) is incorrect because stable conditions result in stratiform clouds, and dry air does not result in showery rain or icing when lifted.

Answers

4122 [B]   4124 [A]   4115 [C]   4118 [C]   4123 [C]

Clouds

Cloud types are divided into four families: high clouds, middle clouds, low clouds, and clouds with extensive vertical development. The first three families are further classified according to the way they are formed. The high cloud family is cirriform and includes cirrus, cirrocumulus, and cirrostratus. They are composed almost entirely of ice crystals. Clouds formed by vertical currents in unstable air are cumulus, meaning "accumulation or heap." They are characterized by their lumpy, billowy appearance. Clouds formed by the cooling of a stable layer are stratus, meaning "stratified or layered." They are characterized by their uniform, sheet-like appearance.

In addition to the above, the prefix nimbo or the suffix nimbus means "raincloud." A heavy, swelling cumulus type cloud which produces precipitation is a "cumulonimbus." The high cloud family is cirriform and composed almost entirely of ice crystals.

Lenticular clouds are associated with a mountain wave. Crests of standing waves may be marked by stationary, lens-shaped clouds known as "standing lenticular" clouds. Mountain waves and lenticular clouds indicate strong turbulence.

ALL
4131. The suffix "nimbus", used in naming clouds, means a

A—cloud with extensive vertical development.
B—rain cloud.
C—dark massive, towering cloud.

The prefix "nimbo-" or the suffix "-nimbus" means raincloud. Thus, stratified clouds from which rain is falling are called nimbostratus. A heavy, swelling cumulus-type cloud which produces precipitation is a cumulonimbus. (I26) — AC 00-6A, page 53

Answer (A) is incorrect because "cumulo" means a cloud with extensive vertical development. Answer (C) is incorrect because cumulonimbus is a dark massive, towering cloud.

ALL
4134. What are the four families of clouds?

A—Stratus, cumulus, nimbus, and cirrus.
B—Clouds formed by updrafts, fronts, cooling layers of air, and precipitation into warm air.
C—High, middle, low, and those with extensive vertical development.

For identification purposes, cloud types are divided into four "families." The families are high clouds, middle clouds, low clouds, and clouds with extensive vertical development. (I26) — AC 00-6A, page 53

ALL
4157. A high cloud is composed mostly of

A—ozone.
B—condensation nuclei.
C—ice crystals.

The high cloud family is cirriform and includes cirrus, cirrocumulus, and cirrostratus. They are composed almost entirely of ice crystals. (I26) — AC 00-6A, page 54

ALL
4133. Which family of clouds is least likely to contribute to structural icing on an aircraft?

A—Low clouds.
B—High clouds.
C—Clouds with extensive vertical development.

Two conditions are necessary for structural icing in flight:

1. *The aircraft must be flying through visible water such as rain or cloud droplets, and*
2. *The temperature, at the point where the moisture strikes the aircraft must be 0°C or colder.*

The high cloud family is composed almost entirely of ice crystals. Because ice crystals are already frozen, they most likely won't stick to an aircraft. (I29) — AC 00-6A, pages 54 and 92

Answers
4131 [B]　　　4134 [C]　　　4157 [C]　　　4133 [B]

Chapter 1 **Weather**

ALL
4129. Which clouds have the greatest turbulence?

A—Towering cumulus.
B—Cumulonimbus.
C—Altocumulus castellanus.

Cumulonimbus clouds are the ultimate manifestation of instability. They are vertically developed clouds of large dimensions with dense boiling tops, often crowned with thick veils of dense cirrus (the anvil). Nearly the entire spectrum of flying hazards are contained in these clouds, including violent turbulence. They should be avoided at all times. (I28) — AC 00-6A, page 61

Answer (A) is incorrect because a towering cumulus has primarily updrafts. Answer (C) is incorrect because altocumulus castellanus are turbulent, but not nearly as turbulent as cumulonimbus clouds.

ALL
4130. Standing lenticular clouds, in mountainous areas, indicate

A—an inversion.
B—unstable air.
C—turbulence.

Standing lenticular altocumulus clouds are formed on the crests of waves created by barriers to the wind flow. The clouds show little movement, hence the name "standing." However, wind can be quite strong blowing through such clouds. The presence of these clouds is a good indication of very strong turbulence and should be avoided. (I28) — AC 00-6A, page 85

ALL
4132. The presence of standing lenticular altocumulus clouds is a good indication of

A—a jet stream.
B—very strong turbulence.
C—heavy icing conditions.

Standing lenticular altocumulus clouds are formed on the crests of waves created by barriers to the wind flow. The clouds show little movement, hence the name "standing." However, wind can be quite strong blowing through such clouds. The presence of these clouds is a good indication of very strong turbulence and they should be avoided. (I28) — AC 00-6A, page 85

ALL
4149. Fair weather cumulus clouds often indicate

A—turbulence at and below the cloud level.
B—poor visibility.
C—smooth flying conditions.

Fair weather cumulus clouds often indicate bumpy turbulence beneath and in the clouds but good visibility. The cloud tops indicate the approximate upper limit of convection; flight above is usually smooth. (I28) — AC 00-6A, page 81

Answer (B) is incorrect because fair weather cumulus clouds produce good visibility. Answer (C) is incorrect because fair weather cumulus clouds produce turbulence.

Common IFR Producers

Fog is a surface-based cloud composed of either water droplets or ice crystals. Fog is the most frequent cause of surface visibility below 3 miles, and is one of the most common and persistent weather hazards encountered in aviation. Fog may form by cooling the air to its dew point or by adding moisture to the air near the ground. A small temperature/dew point spread is essential to the formation of fog. An abundance of condensation nuclei from combustion products makes fog prevalent in industrial areas. Fog is classified by the way it is formed:

1. Radiation fog (ground fog) is formed when terrestrial radiation cools the ground, which in turn cools the air in contact with it. When the air is cooled to its dew point (or within a few degrees), fog will form. This fog will form most readily in warm, moist air over low, flatland areas on clear, calm nights.

2. Advection fog (sea fog) is formed when warm, moist air moves (wind is required) over colder ground or water. An example is an air mass moving inland from the coast in winter.

3. Upslope fog is formed when moist, stable air is cooled to its dew point as it moves (wind is required) up sloping terrain. Cooling will be at the dry adiabatic lapse rate of approximately 3°C per 1,000 feet.

Answers
4129 [B]   4130 [C]   4132 [B]   4149 [A]

ALL
4163. Fog is usually prevalent in industrial areas because of

A—atmospheric stabilization around cities.
B—an abundance of condensation nuclei from combustion products.
C—increased temperatures due to industrial heating.

Abundant condensation nuclei enhance the formation of fog. Thus, fog is prevalent in industrial areas where byproducts of combustion provide a high concentration of these nuclei. (I31) — AC 00-6A, page 126

ALL
4156. Under which condition does advection fog usually form?

A—Moist air moving over colder ground or water.
B—Warm, moist air settling over a cool surface under no-wind conditions.
C—A land breeze blowing a cold air mass over a warm water current.

Advection fog forms when moist air moves over colder ground or water. (I31) — AC 00-6A, page 127

Answer (B) is incorrect because it describes radiation fog. Answer (C) is incorrect because it describes steam fog.

ALL
4164. In which situation is advection fog most likely to form?

A—An air mass moving inland from the coast in winter.
B—A light breeze blowing colder air out to sea.
C—Warm, moist air settling over a warmer surface under no-wind conditions.

Advection fog forms when moist air moves over colder ground or water. During the winter, advection fog over the central and eastern United States results when moist air from the Gulf of Mexico spreads northward over cold ground. The fog may extend as far north as the Great Lakes. (I31) — AC 00-6A, page 127

ALL
4165. In what localities is advection fog most likely to occur?

A—Coastal areas.
B—Mountain slopes.
C—Level inland areas.

Advection fog forms when moist air moves over colder ground or water. It is most common along coastal areas but often develops deep into continental areas. (I31) — AC 00-6A, page 127

ALL
4166. What types of fog depend upon a wind in order to exist?

A—Steam fog and down slope fog.
B—Precipitation-induced fog and ground fog.
C—Advection fog and up slope fog.

Advection fog forms when moist air moves over colder ground or water. Upslope fog forms as a result of moist, stable air being cooled adiabatically as it moves up sloping terrain. (I31) — AC 00-6A, page 127

ALL
4167. What situation is most conducive to the formation of radiation fog?

A—Warm, moist air over low, flatland areas on clear, calm nights.
B—Moist, tropical air moving over cold, offshore water.
C—The movement of cold air over much warmer water.

Conditions favorable for radiation fog are clear skies, little or no wind, and small temperature/dew point spread (high relative humidity). The fog forms almost exclusively at night or near daybreak. (I31) — AC 00-6A, page 126

Answer (B) is incorrect because it describes advection fog. Answer (C) is incorrect because it describes steam fog.

ALL
4169. Which conditions are favorable for the formation of radiation fog?

A—Moist air moving over colder ground or water.
B—Cloudy sky and a light wind moving saturated warm air over a cool surface.
C—Clear sky, little or no wind, small temperature/dew point spread, and over a land surface.

Conditions favorable for radiation fog are clear skies, little or no wind, and small temperature/dew point spread (high relative humidity). The fog forms almost exclusively at night or near daybreak. (I31) — AC 00-6A, page 126

Answers (A) and (B) are incorrect because they both describe advection fog.

Answers

| 4163 | [B] | 4156 | [A] | 4164 | [A] | 4165 | [A] | 4166 | [C] | 4167 | [A] |
| 4169 | [C] | | | | | | | | | | |

Chapter 1 **Weather**

ALL
4162. Which weather condition can be expected when moist air flows from a relatively warm surface to a colder surface?

A—Increased visibility.
B—Convective turbulence due to surface heating.
C—Fog.

Advection fog forms when moist air moves over colder ground or water. (I32) — AC 00-6A, page 127

Air Masses and Fronts

When a body of air (or air mass) comes to rest, or moves slowly over a large geographical area that has fairly uniform properties of temperature and moisture, the air mass acquires the temperature/moisture properties of that area it covers. Characteristics of an unstable air mass include: cumuliform clouds, showery precipitation, rough air (turbulence), and good visibility, except in blowing obstructions. Characteristics of a stable air mass include: stratiform clouds and fog, continuous precipitation, smooth air, and fair to poor visibility in haze and smoke.

As air masses move out of their source regions, they come in contact with other air masses of different properties. The zone between two different air masses is a frontal zone or front. Across this zone, temperature, humidity and wind often change rapidly over short distances. The three principal types of fronts are the cold front, the warm front, and the stationary front. Frontal waves and cyclones (areas of low pressure) usually form on slow-moving cold fronts or stationary fronts.

ALL
4158. An air mass is a body of air that

A—has similar cloud formations associated with it.
B—creates a wind shift as it moves across the Earth's surface.
C—covers an extensive area and has fairly uniform properties of temperature and moisture.

If a body of air (or air mass) comes to rest, or moves slowly over a large geographical area that has fairly uniform temperatures and moisture content, the body of air (or air mass) acquires the temperature/moisture properties of the geographical area it covers. Therefore, it becomes fairly uniform in these properties over an extensive area. (I27) — AC 00-6A, page 63

Unstable Air	Stable Air
Cumuliform clouds	Stratiform clouds and fog
Showery precipitation	Continuous precipitation
Rough air (turbulence)	Smooth air
Good visibility, except in blowing obstructions	Fair to poor visibility in haze and smoke

Questions 4116, 4117, 4119, 4120 and 4128

ALL
4116. The general characteristics of unstable air are

A—good visibility, showery precipitation, and cumuliform-type clouds.
B—good visibility, steady precipitation, and stratiform-type clouds.
C—poor visibility, intermittent precipitation, and cumuliform-type clouds.

The stability of an air mass determines its typical weather characteristics. When one type of air mass overlies another, conditions change with height. Characteristics typical of an unstable and a stable air mass are shown in the figure below. (I27) — AC 00-6A, page 64

Answers (B) and (C) are incorrect because steady precipitation, stratiform clouds, and poor visibility are characteristic of stable air.

ALL
4120. What are some characteristics of unstable air?

A—Nimbostratus clouds and good surface visibility.
B—Turbulence and poor surface visibility.
C—Turbulence and good surface visibility.

The stability of an air mass determines its typical weather characteristics. When one type of air mass overlies another, conditions change with height. Characteristics typi-

Answers
4162 [C]   4158 [C]   4116 [A]   4120 [C]

cal of an unstable and a stable air mass are shown in the figure above. (I27) — AC 00-6A, page 64

Answers (A) and (B) are incorrect because nimbostratus clouds and poor surface visibility are characteristic of stable air.

ALL
4128. Which are characteristics of an unstable cold air mass moving over a warm surface?

A—Cumuliform clouds, turbulence, and poor visibility.
B—Cumuliform clouds, turbulence, and good visibility.
C—Stratiform clouds, smooth air, and poor visibility.

Cool air moving over a warm surface is heated from below, generating instability and increasing the possibility of showers. Stability of an air mass determines its typical weather characteristics. When one type of air mass overlies another, conditions change with height. Characteristics typical of an unstable air mass are cumuliform clouds, turbulence, and good visibility. See the figure to the left. (I27) — AC 00-6A, page 64

Answer (A) is incorrect because poor visibility is a characteristic of stable air. Answer (C) is incorrect because stratiform clouds, smooth air, and poor visibility are characteristics of stable air.

ALL
4117. Which is a characteristic of stable air?

A—Fair weather cumulus clouds.
B—Stratiform clouds.
C—Unlimited visibility.

The stability of an air mass determines its typical weather characteristics. When one type of air mass overlies another, conditions change with height. Characteristics typical of an unstable and a stable air mass are shown in the figure to the left. (I27) — AC 00-6A, page 64

Answers (A) and (C) are incorrect because fair weather cumulus clouds and unlimited visibility are characteristic of unstable air.

ALL
4119. What are the characteristics of stable air?

A—Good visibility, steady precipitation, and stratus-type clouds.
B—Poor visibility, intermittent precipitation, and cumulus-type clouds.
C—Poor visibility, steady precipitation, and stratus-type clouds.

The stability of an air mass determines its typical weather characteristics. When one type of air mass overlies another, conditions change with height. Characteristics typical of an unstable and a stable air mass are shown in the figure above. (I27) — AC 00-6A, page 64

Answers (A) and (B) are incorrect because good visibility, intermittent precipitation, and cumulus type clouds are all characteristics of unstable air.

ALL
4098. Steady precipitation, in contrast to showers, preceding a front is an indication of

A—stratiform clouds with moderate turbulence.
B—cumuliform clouds with little or no turbulence.
C—stratiform clouds with little or no turbulence.

Steady precipitation is a characteristic of stable air, which has little or no turbulence. Stratiform clouds are associated with stable air. (I27) — AC 00-6A, page 72

Answer (A) is incorrect because stratiform clouds are a characteristic of stable air, which has little or no turbulence. Answer (B) is incorrect because precipitation from cumuliform clouds is of a shower type and the clouds are turbulent.

ALL
4127. Frontal waves normally form on

A—slow moving cold fronts or stationary fronts.
B—slow moving warm fronts and strong occluded fronts.
C—rapidly moving cold fronts or warm fronts.

Frontal waves and cyclones (areas of low pressure) usually form on slow-moving cold fronts or stationary fronts. (I27) — AC 00-6A, page 66

ALL
4136. Which weather phenomenon is always associated with the passage of a frontal system?

A—A wind change.
B—An abrupt decrease in pressure.
C—Clouds, either ahead or behind the front.

Wind always changes across a front. Wind discontinuity may be in direction, in speed, or in both. Temperature and humidity also may change. (I27) — AC 00-6A, page 65

Answer (B) is incorrect because pressure will increase abruptly when flying into colder air and will decrease gradually when flying into warmer air. Answer (C) is incorrect because there may be insufficient moisture to produce clouds.

Answers
4128 [B]   4117 [B]   4119 [C]   4098 [C]   4127 [A]   4136 [A]

Turbulence

If severe turbulence is encountered either inside or outside of clouds, the airplane's airspeed should be reduced to maneuvering speed and the pilot should attempt to maintain a level flight attitude because the amount of excess load that can be imposed on the wing will be decreased. An attempt to maintain a constant altitude will greatly increase the stresses that are applied to the aircraft. *See* Figure 1-3 below.

ALL
4160. If you fly into severe turbulence, which flight condition should you attempt to maintain?
A—Constant airspeed (V_A).
B—Level flight attitude.
C—Constant altitude and constant airspeed.

When flying in severe turbulence, maintaining positive aircraft control may be nearly impossible to do. In attempting to maintain a constant altitude, the stresses applied to the aircraft are greatly increased. Stresses will be lessened if the aircraft is held in a constant attitude and allowed to "ride the waves." Wide fluctuations in airspeed will probably have to be tolerated in order to reduce aircraft stresses. (I30) — AC 00-6A, page 123

Answer (A) is incorrect because airspeed should be maintained at or below V_A. Answer (C) is incorrect because a constant attitude (not airspeed or altitude) should be maintained.

Intensity	Aircraft reaction	Reaction inside aircraft	Reporting term definition
Light	Turbulence that momentarily causes slight, erratic changes in altitude and/or attitude (pitch, roll, yaw). Report as *Light Turbulence*;* or Turbulence that causes slight, rapid and somewhat rhythmic bumpiness without appreciable changes in altitude or attitude. Report as *Light Chop*.	Occupants may feel a slight strain against belts or shoulder straps. Unsecured objects may be displaced slightly. Food service may be conducted and little or no difficulty is encountered in walking.	Occasional – less than 1/3 of the time. Intermittent – 1/3 to 2/3 of the time. Continuous – more than 2/3 of the time.
Moderate	Turbulence that is similar to Light Turbulence but of greater intensity. Changes in altitude and/or attitude occur but the aircraft remains in positive control at all times. It usually causes variations in indicated airspeed. Report as *Moderate Turbulence*.* or Turbulence that is similar to Light Chop but of greater intensity. It causes rapid bumps or jolts without appreciable changes in aircraft altitude or attitude. Report as *Moderate Chop*.	Occupants feel definite strains against seat belts or shoulder straps. Unsecured objects are dislodged. Food service and walking are difficult.	NOTE 1. Pilots should report location(s), time (UTC), intensity, whether in or near clouds, altitude, type of aircraft and, when applicable, duration of turbulence. 2. Duration may be based on time between two locations or over a single location. All locations should be readily identifiable.
Severe	Turbulence that causes large, abrupt changes in altitude and/or attitude. It usually causes large variations in indicated airspeed. Aircraft may be momentarily out of control. Report as *Severe Turbulence*.*	Occupants are forced violently against seat belts or shoulder straps. Unsecured objects are tossed about. Food service and walking are impossible.	
Extreme	Turbulence in which the aircraft is violently tossed about and is practically impossible to control. It may cause structural damage. Report as *Extreme Turbulence*.*		

* High level turbulence (normally above 15,000 feet AGL) that is not associated with cumuliform cloudiness, including thunderstorms, should be reported as CAT (clear air turbulence) preceded by the appropriate intensity, or light or moderate chop.

Figure 1-3. Turbulence Reporting Criteria

Answers
4160 [B]

ALL
4916. If severe turbulence is encountered during your IFR flight, the airplane should be slowed to the design maneuvering speed because the

A—maneuverability of the airplane will be increased.
B—amount of excess load that can be imposed on the wing will be decreased.
C—airplane will stall at a lower angle of attack, giving an increased margin of safety.

If during flight, rough air or severe turbulence is encountered, the airspeed should be reduced to maneuvering speed or less, to minimize the stress on the airplane structure. (H303) — AC 61-23C, Chapter 1

Answer (A) is incorrect because the maneuverability will not be increased by slowing down the aircraft. Answer (C) is incorrect because the airplane will always stall at the same critical angle of attack.

ALL
4210. A pilot reporting turbulence that momentarily causes slight, erratic changes in altitude and/or attitude should report it as

A—light turbulence.
B—moderate turbulence.
C—light chop.

Light turbulence is defined as turbulence that momentarily causes slight, erratic changes in attitude and/or altitude. (I67) — AC 00-45E, Chapter 14

Answer (B) is incorrect because moderate turbulence will cause greater changes in attitude and altitude. Answer (C) is incorrect because light chop is slight or moderate rhythmic bumps with negligible changes in attitude or altitude.

Thunderstorms

A thunderstorm is a local storm produced by a cumulonimbus cloud. It is always accompanied by lightning and thunder, usually with strong gusts of wind, heavy rain, and sometimes, with hail. Three conditions necessary for the formation of a thunderstorm are: sufficient water vapor, an unstable lapse rate, and an initial upward boost (lifting). The initial upward boost can be caused by heating from below, frontal lifting, or by mechanical lifting (wind blowing air upslope on a mountain).

There are three stages of a thunderstorm:

1. The **cumulus** stage is characterized by continuous updrafts, and these updrafts create low-pressure areas.
2. The **mature** stage is characterized by updrafts and downdrafts inside the cloud. Precipitation inside the cloud aids in the development of these downdrafts, and the start of rain from the base of the cloud signals the beginning of the mature stage. Thunderstorms reach their greatest intensity during the mature stage.
3. The **dissipating** stage is characterized predominantly by downdrafts.

Thunderstorms that generally produce the most intense hazard to aircraft are called squall-line thunderstorms. These nonfrontal, narrow bands of thunderstorms often develop ahead of a cold front.

Embedded thunderstorms are those that are obscured by massive cloud layers and cannot be seen.

ALL
4148. What are the requirements for the formation of a thunderstorm?

A—A cumulus cloud with sufficient moisture.
B—A cumulus cloud with sufficient moisture and an inverted lapse rate.
C—Sufficient moisture, an unstable lapse rate, and a lifting action.

For a thunderstorm to form, the air must have:

1. *Sufficient water vapor,*
2. *An unstable lapse rate, and*
3. *An initial upward boost (lifting) to start the storm process in motion.*

(I30) — AC 00-6A, page 111

Answers (A) and (B) are incorrect because neither cumulus clouds nor an inverted lapse rate are necessary elements for the formation of a thunderstorm.

Answers
4916 [B]   4210 [A]   4148 [C]

Chapter 1 Weather

ALL
4126. Which weather phenomenon signals the beginning of the mature stage of a thunderstorm?

A—The start of rain at the surface.
B—Growth rate of cloud is maximum.
C—Strong turbulence in the cloud.

Precipitation beginning to fall from the cloud base signals that a downdraft has developed and a cell has entered the mature stage. (I30) — AC 00-6A, page 111

Answer (B) is incorrect because maximum growth rate occurs during the cumulus stage. Answer (C) is incorrect because strong turbulence may occur during any stage of a thunderstorm.

ALL
4143. During the life cycle of a thunderstorm, which stage is characterized predominately by downdrafts?

A—Cumulus.
B—Dissipating.
C—Mature.

A thunderstorm cell during its life cycle progress through three stages: the cumulus, the mature, and the dissipating. Downdrafts characterize the dissipating stage. (I30) — AC 00-6A, page 111

Answer (A) is incorrect because the cumulus stage is characterized by updrafts. Answer (C) is incorrect because the mature stage is characterized by precipitation beginning to fall.

ALL
4147. What is an indication that downdrafts have developed and the thunderstorm cell has entered the mature stage?

A—The anvil top has completed its development.
B—Precipitation begins to fall from the cloud base.
C—A gust front forms.

Precipitation beginning to fall from the cloud base signals that a downdraft has developed and a cell has entered the mature stage. (I30) — AC 00-6A, page 111

Answer (A) is incorrect because the anvil top will normally complete development near the end of the mature stage. Answer (C) is incorrect because a gust front may be formed by any number of cells in a thunderstorm.

ALL
4137. Where do squall lines most often develop?

A—In an occluded front.
B—In a cold air mass.
C—Ahead of a cold front.

A squall line (instability line) is a nonfrontal, narrow band of active thunderstorms. Often it develops ahead of a cold front in moist, unstable air, but it may develop in unstable air far removed from any front. (I30) — AC 00-6A, page 114

ALL
4142. If squalls are reported at your destination, what wind conditions should you anticipate?

A—Sudden increases in wind speed of at least 16 knots, rising to 22 knots or more, lasting for at least 1 minute.
B—Peak gusts of at least 35 knots for a sustained period of 1 minute or longer.
C—Rapid variation in wind direction of at least 20° and changes in speed of at least 10 knots between peaks and lulls.

A squall is a sudden increase in speed of at least 16 knots to a sustained speed of 22 knots or more, lasting for at least one minute. (I57) — AC 00-45E, Chapter 2

Answer (B) is incorrect because peak gusts of at least 35 knots refers to "peak wind." Answer (C) is incorrect because squalls refer to wind speed (not direction) changes.

ALL
4145. Which thunderstorms generally produce the most severe conditions, such as heavy hail and destructive winds?

A—Warm front.
B—Squall line.
C—Air mass.

A squall line often contains severe steady-state thunderstorms and presents the single most intense weather hazard to aircraft. It usually forms rapidly, generally reaching maximum intensity during the late afternoon and the first few hours of darkness. Hail competes with turbulence as the greatest thunderstorm hazard to aircraft. (I30) — AC 00-6A, page 114

Answers

| 4126 | [A] | 4143 | [B] | 4147 | [B] | 4137 | [C] | 4142 | [A] | 4145 | [B] |

ALL
4141. What is indicated by the term "embedded thunderstorms"?

A— Severe thunderstorms are embedded within a squall line.
B— Thunderstorms are predicted to develop in a stable air mass.
C— Thunderstorms are obscured by massive cloud layers and cannot be seen.

Usually, thunderstorms are quite visible to the pilot. However, when a thunderstorm is present but not visible to the pilot due to cloud cover, such as a thick stratus layer, the thunderstorm is said to be "embedded." (I27) — AC 00-6A, page 72

Answer (A) is incorrect because a squall line is a line of thunderstorms that can usually be seen. Answer (B) is incorrect because thunderstorms are virtually non-existent in stable air masses.

ALL
4144. Which weather phenomenon is always associated with a thunderstorm?

A— Lightning.
B— Heavy rain showers.
C— Supercooled raindrops.

A thunderstorm is a local storm produced by a cumulonimbus cloud. It is always accompanied by lightning and thunder, usually with strong gusts of wind, heavy rain, and sometimes, with hail. (I30) — AC 00-6A, page 116

Answer (B) is incorrect because it may hail instead of rain in a thunderstorm. Answer (C) is incorrect because the occurrence of supercooled raindrops depends on the temperature.

ALL
4146. Which procedure is recommended if a pilot should unintentionally penetrate embedded thunderstorm activity?

A— Reverse aircraft heading or proceed toward an area of known VFR conditions.
B— Reduce airspeed to maneuvering speed and maintain a constant altitude.
C— Set power for recommended turbulence penetration airspeed and attempt to maintain a level flight attitude.

Following are some do's and don'ts during thunderstorm penetration:

1. *Do keep your eyes on your instruments. Looking outside the cockpit can increase the danger of temporary blindness from lightning.*

2. *Don't change power settings; maintain settings for reduced airspeed.*

3. *Do maintain a constant attitude; let the aircraft "ride the waves." Maneuvers that try to maintain constant altitude increase stresses on the aircraft.*

4. *Don't turn back once you are in the thunderstorm. A straight course through the storm most likely is the quickest way out of the hazards. In addition, turning maneuvers increase stresses on the aircraft.*

(I30) — AC 00-6A, page 123

Answer (A) is incorrect because turning maneuvers increase the load on the aircraft, and a straight path will be the most direct route out of the thunderstorm. Answer (B) is incorrect because the pilot should attempt to maintain a constant attitude (not airspeed or altitude).

Answers
4141 [C]   4144 [A]   4146 [C]

Chapter 1 **Weather**

Microbursts

Microbursts are small-scale intense downdrafts which, as they get near the ground, spread outward from the center in all directions. Maximum downdrafts at the center of a microburst may be as strong as 6,000 feet per minute. As your aircraft approaches, encounters, and exits a microburst it will experience a sudden increase in performance due to a headwind followed by a strong downdraft at the microburst center and then a severe loss of performance as the downdraft becomes a sudden tailwind. Since the initial headwind may be 45 knots and the tailwind also 45 knots, the total wind shear passing across the microburst would be 90 knots.

An individual microburst will seldom last longer than 15 minutes from the time it strikes the ground until dissipation. The horizontal winds continue to increase during the first 5 minutes with the maximum intensity winds lasting approximately 2-4 minutes.

ALL
4251. What is the expected duration of an individual microburst?

A—Two minutes with maximum winds lasting approximately 1 minute.
B—One microburst may continue for as long as 2 to 4 hours.
C—Seldom longer than 15 minutes from the time the burst strikes the ground until dissipation.

An individual microburst will seldom last longer than 15 minutes from the time it strikes the ground until dissipation. However, there may be multiple microbursts in the area. (J25) — AIM ¶7-1-24

ALL
4252. Maximum downdrafts in a microburst encounter may be as strong as

A—8,000 feet per minute.
B—7,000 feet per minute.
C—6,000 feet per minute.

The downdrafts can be as strong as 6,000 feet per minute in a microburst encounter. (J25) — AIM ¶7-1-24(d)(2)

ALL
4253. An aircraft that encounters a headwind of 45 knots, within a microburst, may expect a total shear across the microburst of

A—40 knots.
B—80 knots.
C—90 knots.

Horizontal winds near the surface can be as strong as 45 knots resulting in a 90-knot shear (headwind to tailwind) across the microburst. These strong horizontal winds occur within a few hundred feet of the ground. (J25) — AIM ¶7-1-24(d)(2)

ALL
4254. (Refer to Figure 13.) If involved in a microburst encounter, in which aircraft positions will the most severe downdraft occur?

A—4 and 5.
B—2 and 3.
C—3 and 4.

The strongest downdraft occurs at point 3. Although there is also downdraft at points 2 and 4, the correct answer is C (3 and 4) because the combination of downdraft and tailwind at 4 has a more negative effect on performance than the combination of downdraft and headwind at 2. (J25) — AIM ¶7-1-24(e)

ALL
4255. (Refer to Figure 13.) When penetrating a microburst, which aircraft will experience an increase in performance without a change in pitch or power?

A—3.
B—2.
C—1.

At position 1 the aircraft is in the outflow, initially experiencing a headwind with little or no downdraft resulting in a temporary performance increase. (J25) — AIM ¶7-1-24(e)

Answers

4251 [C]   4252 [C]   4253 [C]   4254 [C]   4255 [C]

ALL
4256. (Refer to Figure 13.) The aircraft in position 3 will experience which effect in a microburst encounter?

A—Decreasing headwind.
B—Increasing tailwind.
C—Strong downdraft.

Answer C is the best answer. Although all three answer choices are occurring, the strong downdraft is the most severe effect at point 3. (J25) — AIM ¶7-1-24(e)

ALL
4257. (Refer to Figure 13.) What effect will a microburst encounter have upon the aircraft in position 4?

A—Strong tailwind.
B—Strong updraft.
C—Significant performance increase.

In position 4, performance decreases from the strong tailwind and the strong downdraft. (J25) — AIM ¶7-1-24(e)

ALL
4258. (Refer to Figure 13.) How will the aircraft in position 4 be affected by a microburst encounter?

A—Performance increasing with a tailwind and updraft.
B—Performance decreasing with a tailwind and downdraft.
C—Performance decreasing with a headwind and downdraft.

At position 4, the aircraft experiences a severe performance decrease. The wind contours indicate a strong downdraft and increasing tailwind which may cause an aircraft to sink toward ground impact. (J25) — AIM ¶7-1-24(e)

Icing

Structural icing occurs on an aircraft whenever supercooled condensed droplets of water make contact with any part of the aircraft that is also at a temperature below freezing. An in-flight condition necessary for structural icing is visible moisture (clouds or raindrops). Aircraft structural ice will most likely have the highest accumulation in freezing rain which indicates a warmer temperature at a higher altitude.

Frost is described as ice deposits formed by sublimation on a surface when the temperature of the collecting surface is at or below the dew point of the adjacent air and the dew point is below freezing. Frost causes early airflow separation on an airfoil resulting in a loss of lift. Therefore, all frost should be removed from the lifting surfaces of an airplane before flight or it may prevent the airplane from becoming airborne.

Test data indicate that ice, snow, or frost formations having a thickness and surface roughness similar to medium or course sandpaper on the leading edge and upper surface of a wing can reduce wing lift by as much as 30% and increase drag by 40%.

An operational consideration if you fly into rain which freezes on impact is that temperatures are above freezing at some higher altitude. As the rain falls through air that is below freezing, its temperature begins to fall below freezing yet without freezing solid. This is freezing rain. The process requires the temperature of the rain must be above freezing before it becomes supercooled. Therefore, when freezing rain is encountered, it indicates that warmer temperatures are above.

Answers
4256 [C]   4257 [A]   4258 [B]

Chapter 1 Weather

ALL
4100. Which conditions result in the formation of frost?

A—The temperature of the collecting surface is at or below freezing and small droplets of moisture are falling.
B—When dew forms and the temperature is below freezing.
C—Temperature of the collecting surface is below the dewpoint of surrounding air and the dewpoint is colder than freezing.

In order for frost to form, the air temperature must be below the dew point and the dew point of the surrounding air must be colder than freezing. Water vapor will then sublimate directly as ice crystals or frost. (I29) — AC 00-6A, page 41

Answer (A) is incorrect because droplets of moisture are not necessary, and the dew point must also be colder than freezing. Answer (B) is incorrect because dew does not have to form, and the temperature must be below the dew point and the dew point below freezing.

ALL
4151. Why is frost considered hazardous to flight operation?

A—Frost changes the basic aerodynamic shape of the airfoil.
B—Frost decreases control effectiveness.
C—Frost causes early airflow separation resulting in a loss of lift.

Frost does not change the basic aerodynamic shape of the wing, but the roughness of its surface spoils the smooth flow of air, thus causing a slowing of the airflow. This slowing of the air causes early air flow separation over the affected airfoil, resulting in a loss of lift. (I29) — AC 00-6A, page 102

ALL
4152. In which meteorological environment is aircraft structural icing most likely to have the highest rate of accumulation?

A—Cumulonimbus clouds.
B—High humidity and freezing temperature.
C—Freezing rain.

The condition most favorable for very hazardous icing is the presence of many large, supercooled water droplets, also called freezing rain. (I29) — AC 00-6A, page 99

Answer (A) is incorrect because due to temperature variations within them, cumulonimbus clouds (thunderstorms) do not necessarily cause structural icing to accumulate. Answer (B) is incorrect because high humidity alone can not produce structural icing; visible moisture must be present.

ALL
4153. What is an operational consideration if you fly into rain which freezes on impact?

A—You have flown into an area of thunderstorms.
B—Temperatures are above freezing at some higher altitude.
C—You have flown through a cold front.

As the rain falls through air that is below freezing, its temperature begins to fall below freezing yet without freezing solid. This is freezing rain. The process requires that the temperature of the rain must be above freezing before it becomes supercooled. Therefore, when freezing rain is encountered, it indicates that warmer temperatures are above. (I29) — AC 00-6A, page 43

ALL
4171. Test data indicate that ice, snow, or frost having a thickness and roughness similar to medium or coarse sandpaper on the leading edge and upper surface of an airfoil

A—reduce lift by as much as 50 percent and increase drag by as much as 50 percent.
B—increase drag and reduce lift by as much as 25 percent.
C—reduce lift by as much as 30 percent and increase drag by 40 percent.

Test data indicate that ice, snow, or frost formations having a thickness and surface roughness similar to medium or coarse sandpaper on the leading edge and upper surface of a wing can reduce wing lift by as much as 30% and increase drag by 40%. (I29) — AC 120-58

Answers

| 4100 | [C] | 4151 | [C] | 4152 | [C] | 4153 | [B] | 4171 | [C] |

Wind Shear

Wind shear is defined as a change in wind direction and/or speed in a very short distance in the atmosphere. This can occur at any level of the atmosphere and can be detected by the pilot as a sudden change in airspeed.

Low-level (low altitude) wind shear can be expected during strong temperature inversions, on all sides of a thunderstorm and directly below the cell. Low-level wind shear can also be found near frontal activity because winds can be significantly different in the two air masses which meet to form the front.

In warm front conditions, the most critical period is before the front passes. Warm front shear may exist below 5,000 feet for about 6 hours before surface passage of the front. The wind shear associated with a warm front is usually more extreme than that found in cold fronts.

The shear associated with cold fronts is usually found behind the front. If the front is moving at 30 knots or more, the shear zone will be 5,000 feet above the surface 3 hours after frontal passage.

There are two potentially hazardous shear situations:

1. *Loss of Tailwind* — A tailwind may shear to either a calm or headwind component. In this instance, initially the airspeed increases, the aircraft pitches up, and altitude increases. Lower than normal power would be required initially, followed by a further decrease as the shear is encountered, and then an increase as glide slope is regained.

2. *Loss of Headwind* — A headwind may shear to a calm or tailwind component. Initially, the airspeed decreases, the aircraft pitches down, and altitude decreases.

ALL
4138. Where does wind shear occur?

A— Exclusively in thunderstorms.
B— Wherever there is an abrupt decrease in pressure and/or temperature.
C— With either a wind shift or a wind speed gradient at any level in the atmosphere.

Wind shear may be associated with either a wind shift or a wind speed gradient at any level in the atmosphere. (I28) — AC 00-6A, page 86

ALL
4139. What is an important characteristic of wind shear?

A— It is primarily associated with the lateral vortices generated by thunderstorms.
B— It usually exists only in the vicinity of thunderstorms, but may be found near a strong temperature inversion.
C— It may be associated with either a wind shift or a wind speed gradient at any level in the atmosphere.

Wind shear may be associated with either a wind shift or a wind speed gradient at any level in the atmosphere. Low-level wind shear may result from a frontal passage, thunderstorm activity, or low-level temperature inversion. Wind shear may also be found at higher altitudes in association with a frontal passage due to wind shift through the frontal zone. (I28) — AC 00-6A, page 86

ALL
4150. What is an important characteristic of wind shear?

A— It is an atmospheric condition that is associated exclusively with zones of convergence.
B— The Coriolis phenomenon in both high- and low-level air masses is the principal generating force.
C— It is an atmospheric condition that may be associated with a low-level temperature inversion, a jet stream, or a frontal zone.

Wind shear may be associated with either a wind shift or a wind speed gradient at any level in the atmosphere. Low-level wind shear may result from a frontal passage, thunderstorm activity, or low-level temperature inversion. Wind shear may also be found at higher altitudes in association with a frontal passage due to wind shift through the frontal zone. (I28) — AC 00-6A, page 86

Answers
4138 [C]   4139 [C]   4150 [C]

Chapter 1 Weather

ALL
4140. Which is a characteristic of low-level wind shear as it relates to frontal activity?

A—With a warm front, the most critical period is before the front passes the airport.
B—With a cold front, the most critical period is just before the front passes the airport.
C—Turbulence will always exist in wind-shear conditions.

Wind shear occurs with a cold front just after the front passes the airport and for a short period thereafter. If the front is moving 30 knots or more, the frontal surface will usually be 5,000 feet above the airport about three hours after the frontal passage. With a warm front, the most critical period is before the front passes the airport. Warm front wind shear may exist below 5,000 feet for approximately six hours. The problem ceases to exist after the front passes the airport. Data compiled on wind shear indicates that the amount of shear in warm fronts is much greater than that found in cold fronts. (I28) — AC 00-6A, page 88

Answer (B) is incorrect because cold front wind shear occurs as the front passes or just after (not before). Answer (C) is incorrect because turbulence may or may not exist in wind shear conditions.

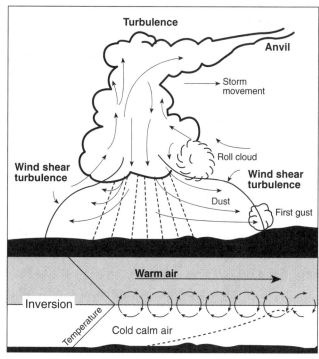

Question 4238

ALL
4238. Hazardous wind shear is commonly encountered near the ground

A—during periods when the wind velocity is stronger than 35 knots.
B—during periods when the wind velocity is stronger than 35 knots and near mountain valleys.
C—during periods of strong temperature inversion and near thunderstorms.

You can be relatively certain of a shear zone in a low-level temperature inversion, if you know that the wind at 2,000 to 4,000 feet is 25 knots or more. Wind shear turbulence is also found near the ground outside thunderstorm clouds. See the figure above. (I23) — AC 00-6A, pages 88, 115, 119

Answer (A) is incorrect because winds stronger than 35 knots are hazardous but not specifically due to wind shear. Answer (B) is incorrect because hazardous wind shear is normally located on the leeward side of mountains (not just mountain valleys).

ALL
4135. Where can wind shear associated with a thunderstorm be found? Choose the most complete answer.

A—In front of the thunderstorm cell (anvil side) and on the right side of the cell.
B—In front of the thunderstorm cell and directly under the cell.
C—On all sides of the thunderstorm cell and directly under the cell.

The winds around a thunderstorm are complex. Wind shear can be found on all sides of a thunderstorm cell and in the downdraft directly under the cell. The wind shift line, or gust front, associated with thunderstorms can precede the actual storm by 15 nautical miles or more. (I30) — AC 00-6A, page 115

Answers
4140 [A]   4238 [C]   4135 [C]

ALL
4720. When passing through an abrupt wind shear which involves a shift from a tailwind to a headwind, what power management would normally be required to maintain a constant indicated airspeed and ILS glide slope?

A—Higher than normal power initially, followed by a further increase as the wind shear is encountered, then a decrease.
B—Lower than normal power initially, followed by a further decrease as the wind shear is encountered, then an increase.
C—Higher than normal power initially, followed by a decrease as the shear is encountered, then an increase.

When on an approach in a tailwind condition that shears into a calm wind or headwind, initially the IAS and pitch will increase and the aircraft will balloon above the glide slope. Power should first be reduced to correct this condition. Power should be further reduced as the aircraft proceeds through the wind shear, then increased. (K04) — AC 00-54A

ALL
4727. While flying a 3° glide slope, a constant tailwind shears to a calm wind. Which conditions should the pilot expect?

A—Airspeed and pitch attitude decrease and there is a tendency to go below glide slope.
B—Airspeed and pitch attitude increase and there is a tendency to go below glide slope.
C—Airspeed and pitch attitude increase and there is a tendency to go above glide slope.

When an aircraft that is flying with a tailwind suddenly encounters calm air, it is the same as encountering a headwind. Initially, the IAS and pitch will increase and the aircraft will balloon above the glide slope. (K04) — AC 00-54A

Answer (A) is incorrect because the airspeed and pitch increase and there is a tendency to go above the glide slope. Answer (B) is incorrect because there is a tendency to go above the glide slope.

ALL
4739. Thrust is managed to maintain IAS, and glide slope is being flown. What characteristics should be observed when a headwind shears to be a constant tailwind?

A—PITCH ATTITUDE: Increases; REQUIRED THRUST: Increased, then reduced; VERTICAL SPEED: Increases; IAS: Increases, then decreases to approach speed.
B—PITCH ATTITUDE: Decreases; REQUIRED THRUST: Increased, then reduced; VERTICAL SPEED: Increases; IAS: Decreases, then increases to approach speed.
C—PITCH ATTITUDE: Increases; REQUIRED THRUST: Reduced, then increased; VERTICAL SPEED: Decreases; IAS: Decreases, then increases to approach speed.

When a headwind shears to a constant tailwind: the nose pitches down, thrust must be increased initially to resume normal approach speed and then reduced as the airspeed stabilizes, vertical speed increases, and indicated airspeed decreases and then increases to approach speed. See the figure below. (K04) — AC 00-54A

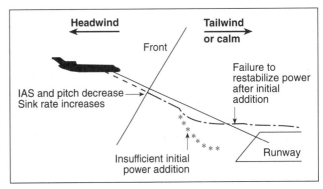

Question 4739

Answers
4720 [B]   4727 [C]   4739 [B]

ALL
4755. While flying a 3° glide slope, a headwind shears to a tailwind. Which conditions should the pilot expect on the glide slope?

A—Airspeed and pitch attitude decrease and there is a tendency to go below glide slope.
B—Airspeed and pitch attitude increase and there is a tendency to go above glide slope.
C—Airspeed and pitch attitude decrease and there is a tendency to remain on the glide slope.

When a headwind shears to a tailwind the airspeed decreases, the nose pitches down, and the aircraft will begin to drop below the glide slope. See the figure below. (K04) — AC 00-54A, ¶6(a)(1)

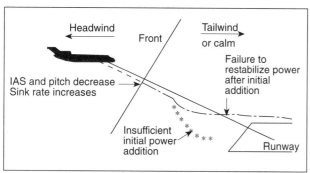

Question 4755

ALL
4917. When a climb or descent through an inversion or wind-shear zone is being performed, the pilot should be alert for which of the following change in airplane performance?

A—A fast rate of climb and a slow rate of descent.
B—A sudden change in airspeed.
C—A sudden surge of thrust.

There are two potentially hazardous shear situations:

1. *A tailwind may shear to either a calm or headwind component. In this instance, initially the airspeed increases, the aircraft pitches up and the altitude increases.*

2. *A headwind may shear to a calm or tailwind component. In this situation, initially the airspeed decreases, the aircraft pitches down, and the altitude decreases.*

(K04) — AC 00-54A

Answer (A) is incorrect because a fast rate of climb and a slow rate of descent is safer than the reverse situation. Answer (C) is incorrect because thrust does not vary with wind shear.

Answers
4755 [A]   4917 [B]

Chapter 2
Weather Services

Aviation Routine Weather Report (METAR)   *2-3*

Terminal Aerodrome Forecast (TAF)   *2-5*

Aviation Area Forecast (FA)   *2-6*

Pilot Reports (UA) and Radar   *2-7*

Winds and Temperatures Aloft Forecast (FD)   *2-9*

Inflight Weather Advisories (WA, WS, WST)   *2-12*

Transcribed Weather Broadcast (TWEB)   *2-14*

Enroute Flight Advisory Service (EFAS)   *2-14*

Convective Outlook (AC)   *2-15*

Surface Analysis Chart   *2-16*

Weather Depiction Chart   *2-16*

Radar Summary Chart   *2-18*

Constant Pressure Chart   *2-21*

Tropopause Data Chart   *2-22*

Significant Weather Prognostics   *2-23*

Severe Weather Outlook Chart   *2-27*

Aviation Routine Weather Report (METAR)

An international weather reporting code is used for weather reports (METAR) and forecasts (TAFs) worldwide. The reports follow the format shown in Figure 2-1.

For aviation purposes, the ceiling is the lowest broken or overcast layer, or vertical visibility into an obscuration.

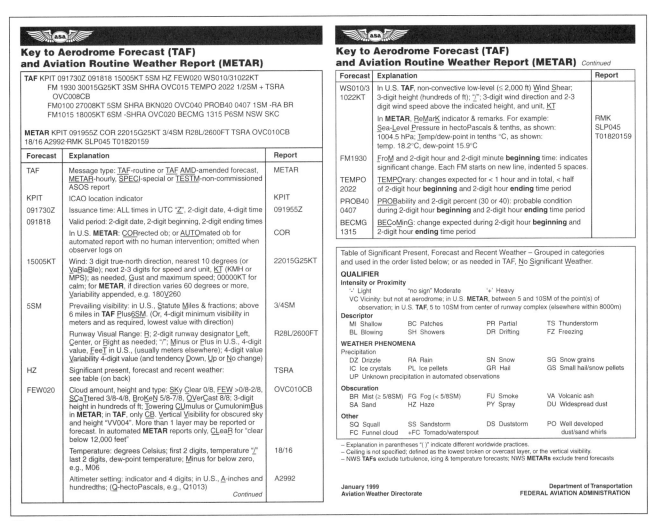

Figure 2-1

ALL
4202. A ceiling is defined as the height of the

A—highest layer of clouds or obscuring phenomena aloft that covers over 6/10 of the sky.
B—lowest layer of clouds that contributed to the overall overcast.
C—lowest layer of clouds or obscuring phenomena aloft that is reported as broken or overcast.

Ceiling is defined as height of the lowest layer of clouds or obscuring phenomena aloft that is reported as broken or overcast and not classified as thin; or, as vertical visibility into surface-based obscuring phenomena that hides all the sky. (I55) — AC 00-45E, Chapter 2

Answer (A) is incorrect because a ceiling is defined as the lowest layer of clouds. Answer (B) is incorrect because a ceiling is specifically the lowest layer of clouds which is broken or overcast.

Answers
4202 [C]

Chapter 2 Weather Services

ALL
4196. The station originating the following weather report has a field elevation of 1,300 feet MSL. From the bottom of the overcast cloud layer, what is its thickness? (tops of OVC are reported at 3800 feet)

SPECI KOKC 2228Z 28024G36KT 3/4SM BKN008 OVC020 28/23 A3000

A—500 feet.
B—1,700 feet.
C—2,500 feet.

Will Rogers airport (KOKC) reports clouds broken at 800 feet (BKN008) and an overcast ceiling at 2,000 feet (OVC020). This means the bottom of the overcast layer is 3,300 feet (1,300 MSL + 2,000 feet AGL). The tops of the overcast are reported at 3,800 feet. Therefore, the overcast layer is 500 feet thick (3,800 – 3,300). (I55) — AC 00-45E, Chapter 2

ALL
4203. The reporting station originating this Aviation Routine Weather Report has a field elevation of 620 feet. If the reported sky cover is one continuous layer, what is its thickness? (tops of OVC are reported at 6500 feet)

METAR KMDW 121856Z AUTO 32005KT 1 1/2SM +RABR OVC007 17/16 A2980

A—5,180 feet.
B—5,800 feet.
C—5,880 feet.

Chicago Midway airport (KMDW) reports an overcast ceiling at 700 feet (OVC007). This means the bottom of the overcast layer is 1,320 feet (620 MSL + 700 feet AGL). The tops of the overcast layer are reported at 6,500 feet. Therefore, the overcast layer is 5,180 feet thick (6,500 – 1,320). (I57) — AC 00-45E, Chapter 2

ALL
4205. What is meant by the entry in the remarks section of METAR surface report for KBNA?

METAR KBNA 211250Z 33018KT 290V260 1/2SM R31/2700FT +SN
BLSNFG VV008 00/M03 A2991 RMK RAE42SNB42

A—The wind is variable from 290° to 360.
B—Heavy blowing snow and fog on runway 31.
C—Rain ended 42 past the hour, snow began 42 past the hour.

RMK — remarks follow
RAE42SNB42 — rain ended 42 past the hour (RAE42), snow began 42 past the hour (SNB42)

(I55) — AC 00-45E, Chapter 2

ALL
4182. What significant sky condition is reported in this METAR observation?

METAR KBNA 1250Z 33018KT 290V360 1/2SM R31/2700FT +SN BLSNFG VV0008 00/M03 A2991 RMK RERAE42SNB42

A—Runway 31 ceiling is 2700 feet.
B—Sky is obscured with vertical visibility of 800 feet.
C—Measured ceiling is 300 feet overcast.

METAR KBNA 1250Z — routine weather report for Nashville Metro airport, time 1250 UTC
33018KT 290V360 — wind from 330° at 18 knots, variable between 290° and 360°
1/2SM — visibility 1/2 statute mile
R31/2700FT — RVR for runway 31 is 2,700 feet
+SN BLSNFG — heavy snow, visibility obstructed by blowing snow and fog
VV008 — sky obscured with vertical visibility of 800 feet
00/M03 — temperature 0°C, dew point -3°C
A2991 — altimeter setting 29.91"
RMK RAE42SNB42 — remarks follow, recent weather events, rain ended 42 past the hour, snow began 42 past the hour

(I56) — AC 00-45E, Chapter 2

Answers

4196 [A] 4203 [A] 4205 [C] 4182 [B]

Terminal Aerodrome Forecast (TAF)

A Terminal Aerodrome Forecast (TAF) is a concise statement of the expected meteorological conditions at an airport during a specified period (usually 24 hours). TAFs use the same code used in the METAR weather reports (*See* Figure 2-1, page 2-3).

TAFs are issued in the following format:

TYPE / LOCATION / ISSUANCE TIME / VALID TIME / FORECAST

Note: the "/" above are for separation purposes and do not appear in the actual TAFs.

ALL
4170. The body of a Terminal Aerodrome Forecast (TAF) covers a geographical proximity within a

A—5 nautical mile radius of the center of an airport.
B—5 statute mile radius from the center of an airport runway complex.
C—5 to 10 statute mile radius from the center of an airport runway complex.

An Aviation Terminal Forecast (TAF) is a concise statement of the expected meteorological conditions within a 5-statute-mile radius from the center of an airport's runway complex during a 24-hour time period. (I57) — AC 00-45E, Chapter 4

ALL
4176. Which primary source should be used to obtain forecast weather information at your destination for the planned ETA?

A—Area Forecast.
B—Radar Summary and Weather Depiction Charts.
C—Terminal Aerodrome Forecast (TAF).

A TAF is a concise statement of the expected meteorological conditions at an airport during a specified period (usually 24 hours). (I57) — AC 00-45E, Chapter 4

Answer (A) is incorrect because the Area Forecast (FA) is a forecast of general weather conditions over an area the size of several states. It is used to determine forecast en route weather and to interpolate conditions at airports which do not have TAFs issued. Answer (B) is incorrect because the Radar Summary and Weather Depiction Charts show current weather, not forecasts.

ALL
4177. A "VRB" wind entry in a Terminal Aerodrome Forecast (TAF) will be indicated when the wind is

A—3 knots or less.
B—6 knots or less.
C—9 knots or less.

A variable wind is encoded as "VRB" when wind direction fluctuates due to convective activity or low wind speeds (3 knots or less). (I57) — AC 00-45E, Chapter 4

ALL
4178. When the visibility is greater than 6 SM on a TAF it is expressed as

A—6PSM.
B—P6SM.
C—6SMP.

Expected visibilities greater than 6 miles are forecast as a Plus 6SM (P6SM). (I57) — AC 00-45E, Chapter 4

ALL
4180. What is the forecast wind at 1800Z in the following TAF?

KMEM 091740Z 1818 00000KT 1/2SM RAFG OVC005=

A—Calm.
B—Unknown.
C—Not recorded.

A calm wind is forecast as "00000KT." (I57) — AC 00-45E, Chapter 4

ALL
4228. From which primary source should you obtain information regarding the weather expected to exist at your destination at your estimated time of arrival?

A—Weather Depiction Chart.
B—Radar Summary and Weather Depiction Chart.
C—Terminal Aerodrome Forecast.

A Terminal Aerodrome Forecast (TAF) is a prediction of weather conditions to be expected for a specific airport rather than a larger area. (I43) — AC 00-45E, Chapter 4

Answers (A) and (B) are incorrect because they are not forecasts, but rather reports of current or past conditions.

Answers

| 4170 [B] | 4176 [C] | 4177 [A] | 4178 [B] | 4180 [A] | 4228 [C] |

Chapter 2 Weather Services

ALL
4204. What is the wind shear forecast in the following TAF?

TAF
KCVG 231051Z 231212 12012KT 4SM -RA BR
  OVC008
WS005/27050KT TEMPO 1719 1/2SM -RA FG
 FM1930 09012KT 1SM -DZ BR VV003 BECMG 2021
  5SM HZ=

A—5 feet AGL from 270° at 50 KT.
B—50 feet AGL from 270° at 50 KT.
C—500 feet AGL from 270° at 50 KT.

WS005/27050KT — low level wind shear at 500 feet, wind 270° at 50 knots.

(I57) — AC 00-45E, Chapter 4

Aviation Area Forecast (FA)

An aviation Area Forecast (FA) is a forecast of general weather conditions over an area the size of several states. It is used to determine forecast enroute weather and to interpolate conditions at airports which do not have TAFs issued. FAs are issued 3 times a day, and are comprised of four sections: a communications and product header section, a precautionary statement section, and two weather sections: a SYNOPSIS section, and a VFR CLOUDS/WX section.

ALL
4179. "WND" in the categorical outlook in the Aviation Area Forecast means that the wind during that period is forecast to be

A—at least 6 knots or stronger.
B—at least 15 knots or stronger.
C—at least 20 knots or stronger.

"WND" indicates that winds, sustained or gusty, are expected to be 20 knots or greater. (I57) — AC 00-45E, Chapter 4

ALL
4201. Area forecasts generally include a forecast period of 18 hours and cover a geographical

A—terminal area.
B—area less than 3000 square miles.
C—area the size of several states.

An Aviation Area Forecast (FA) is a forecast of general weather conditions over an area the size of several states. They are the six areas in the contiguous 48 states, Alaska, Hawaii, and the Gulf of Mexico. (I57) — AC 00-45E, Chapter 4

Answers

4204 [C]   4179 [C]   4201 [C]

Pilot Reports (UA) and Radar

Aircraft in flight are the only means of directly observing cloud tops, icing, and turbulence; therefore, no observation is more timely than one made from the cockpit. While the FAA encourages pilots to report inflight weather, a report of any unforecast weather is required by regulation. A PIREP (UA) is usually transmitted in a prescribed format. *See* Figure 2-2.

Pilots seeking weather avoidance assistance should keep in mind that ATC radar limitations and frequency congestion may limit the controller's capability to provide this service.

ALL
4092. Which is true regarding the use of airborne weather-avoidance radar for the recognition of certain weather conditions?

A—The radarscope provides no assurance of avoiding instrument weather conditions.
B—The avoidance of hail is assured when flying between and just clear of the most intense echoes.
C—The clear area between intense echoes indicates that visual sighting of storms can be maintained when flying between the echoes.

Radar only detects precipitation. Not all clouds contain enough moisture to be detected by weather radar, which makes it possible to fly into instrument weather conditions despite using radar. (I31) — AC 00-45E, Chapter 3

Answer (B) is incorrect because hail can be thrown miles away from thunderstorms, and radar can only detect the actual storm. Answer (C) is incorrect because clouds and fog are not shown on radar, so clear areas on the radar do not guarantee VFR conditions.

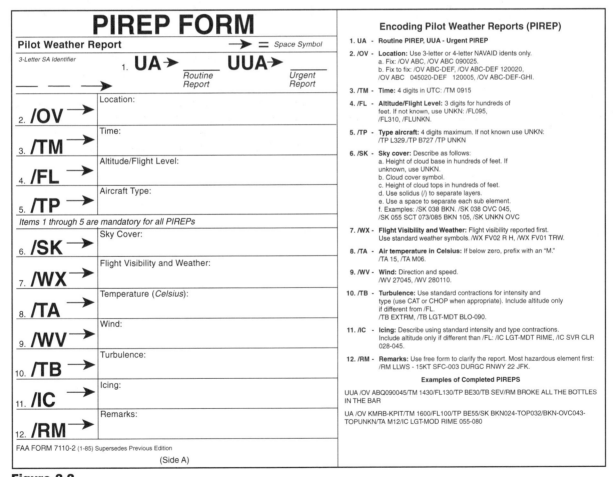

Figure 2-2

Answers
4092 [A]

Chapter 2 Weather Services

ALL
4468. Pilots on IFR flights seeking ATC in-flight weather avoidance assistance should keep in mind that

A—ATC radar limitations and, frequency congestion may limit the controllers capability to provide this service.
B—circumnavigating severe weather can only be accommodated in the en route areas away from terminals because of congestion.
C—ATC Narrow Band Radar does not provide the controller with weather intensity capability.

The controller's primary function is to provide safe separation between aircraft. Any additional service, such as weather avoidance assistance, can only be provided to the extent that it does not degrade the primary function. The separation workload is generally greater than normal when weather disrupts the usual flow of traffic. ATC radar limitations and frequency congestion may also be a factor in limiting the controller's capability to provide additional service. (J25) — AIM ¶7-1-12(b)(4)

Answer (B) is incorrect because controllers can provide in-flight weather avoidance in terminal areas when workload conditions permit. Answer (C) is incorrect because ATC Narrow Band Radar does provide weather intensity capability.

ALL
4198. Which response most closely interprets the following PIREP?

UA/OV OKC 063064/TM 1522/FL080/TP C172/TA -04/ WV 245040/TB LGT/RM IN CLR

A—64 nautical miles on the 63 degree radial from Oklahoma City VOR at 1522 UTC, flight level 8,000 ft. Type of aircraft is a Cessna 172.
B—Reported by a Cessna 172, turbulence and light rime icing in climb to 8,000 ft.
C—63 nautical miles on the 64 degree radial from Oklahoma City, thunderstorm and light rain at 1522 UTC.

UA/OV OKC 063064 — *pilot report, aircraft located 64 nautical miles on the 63° radial from Oklahoma City VOR*
TM 1522 — *time is 1522 UTC*
FL080 — *reporting aircraft is at flight level 8,000 feet MSL*
TP C172 — *reporting aircraft type is Cessna 172*
TA -4 — *temperature -4°C*
WV 245040 — *wind is from 245° at 40 knots*
TB LGT — *light turbulence*
RM IN CLR — *remarks, in the clear*
(I56) — AC 00-45E, Chapter 3

ALL
4220. Interpret this PIREP.

MRB UA/OV MRB/TM1430/FL060/TPC182/SK BKN BL/WX RA/TB MDT

A—Ceiling 6,000 feet intermittently below moderate thundershowers; turbulence increasing westward.
B—FL 60,000, intermittently below clouds; moderate rain, turbulence increasing with the wind.
C—At 6,000 feet; between layers; moderate turbulence; moderate rain.

A pilot reported over (OV) MRB at 1430 Zulu time (TM 1430), at 6,000 feet (FL060) in a Cessna 182 (TPC182), the sky cover is broken between layers (SK BKN BL), the weather included rain (WX RA) and moderate turbulence (TB MDT). (I56) — AC 00-45E, Chapter 3

Answers

4468 [A]   4198 [A]   4220 [C]

Winds and Temperatures Aloft Forecast (FD)

The Winds and Temperatures Aloft Forecast shows wind direction, wind velocity, and the temperature that is forecast to exist at specified levels. No winds are forecast for a level within 1,500 feet of the surface and no temperatures are forecast for the 3,000-foot level or for a level within 2,500 feet of the surface. The wind direction is shown in tens of degrees with reference to true north, and the velocity is shown in knots. Temperatures are in degrees Celsius, and are negative above 24,000 feet.

Each 6-digit group includes wind and temperature. For example, the entry "2045-26" indicates wind from 200° true north at 45 knots, temperature -26°C.

A forecast wind from 160° at 115 knots with a temperature of 34°C at the 30,000-foot level would be encoded "661534," with 50 added to the wind direction and 100 subtracted from the velocity. A wind coded "7699" would indicate wind from 260° at 199 knots or more. "9900" indicates winds light and variable, speed less than 5 knots.

The Observed Winds Aloft Chart shows temperature, wind direction, and speed at selected stations. Arrows with pennants and barbs indicate wind direction and speed. Each pennant is 50 knots, each barb is 10 knots, and each half barb is 5 knots. Wind direction is shown by an arrow drawn to the nearest 10°, with the second digit of the coded direction entered at the outer end of the arrow. Thus, a wind in the northwest quadrant with the digit 3 indicates 330°, and a wind in the southwest quadrant with the digit 3 indicates 230°. *See* Figure 2-3.

ALL
4191. Which values are used for winds aloft forecasts?

A—Magnetic direction and knots.
B—Magnetic direction and MPH.
C—True direction and knots.

The wind direction for all reports and forecasts is always given in relation to true north. The wind speed is always reported and forecast in knots. (I63) — AC 00-45E, Chapter 4

ALL
4172. What wind direction and speed is represented by the entry 9900+00 for 9,000 feet, on an Winds and Temperatures Aloft Forecast (FD)?

A—Light and variable; less than 5 knots.
B—Vortex winds exceeding 200 knots.
C—Light and variable; less than 10 knots.

When the forecast speed is less than 5 knots, the coded group is "9900" and read, LIGHT AND VARIABLE. The "+00" is the outside air temperature for that altitude in degrees Celsius. (I63) — AC 00-45E, Chapter 4

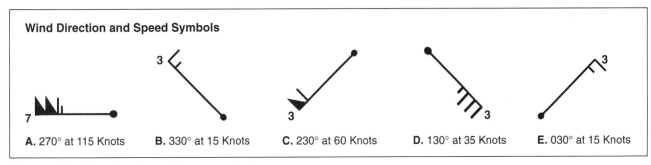

Figure 2-3. Wind direction and speed symbols

Answers
4191 [C]   4172 [A]

Chapter 2 **Weather Services**

ALL
4188. When is the temperature at one of the forecast altitudes omitted at a specific location or station in the Winds and Temperatures Aloft Forecast (FD)?

A—When the temperature is standard for that altitude.
B—For the 3,000-foot altitude (level) or when the level is within 2,500 feet of station elevation.
C—Only when the winds are omitted for that altitude (level).

No temperatures are forecast for the 3,000-foot level or for a level within 2,500 feet of station elevation. (I63) — AC 00-45E, Chapter 4

ALL
4189. When is the wind-group at one of the forecast altitudes omitted at a specific location or station in the Winds and Temperatures Aloft Forecast (FD)? When the wind

A—is less than 5 knots.
B—is less than 10 knots.
C—at the altitude is within 1,500 feet of the station elevation.

No winds are forecast within 1,500 feet of station elevation. (I63) — AC 00-45E, Chapter 4

ALL
4190. Decode the excerpt from the Winds and Temperature Aloft Forecast (FD) for OKC at 39,000 feet.

FT   3000   6000   39000
OKC   830558

A—Wind 130° at 50 knots, temperature -58°C.
B—Wind 330° at 105 knots, temperature -58°C.
C—Wind 330° at 205 knots, temperature -58°C.

Encoded wind speeds 100 to 199 knots have 50 added to the direction code and 100 subtracted from the speed, so 830558 would be: Wind direction 83 – 50 = 330°; wind speed 05 + 100 = 105 knots; temperatures are negative above 24,000 feet, so 58 becomes -58°C. (I63) — AC 00-45E, Chapter 4

ALL
4192. (Refer to Figure 2.) What approximate wind direction, speed, and temperature (relative to ISA) should a pilot expect when planning for a flight over PSB at FL 270?

A—260° magnetic at 93 knots; ISA +7°C.
B—280° true at 113 knots; ISA +3°C.
C—255° true at 93 knots; ISA +6°C.

To find the conditions at PSB at FL270, you must interpolate between 24,000 and 30,000:

Altitude	Code	Wind direction	Wind speed	Temperature
24,000	2368-26	230°	68	-26°C
30,000	781939	280°	119	-39°C
27,000		255°	93	-33°C

To find the temperature relative to ISA, multiply the standard temperature lapse rate (-2°C per 1,000 feet) times the altitude in thousands of feet (in this case 27), then subtract +15°C for standard day conditions:

1. *-2°C x 27 = -54°C*
2. *-54°C – 15 = -39°C (ISA)*

Therefore, -33°C (today) is 6°C warmer than -39°C, so the temperature is ISA + 6°C. (I63) — AC 00-45E, Chapter 4

ALL
4193. (Refer to Figure 2.) What approximate wind direction, speed, and temperature (relative to ISA) should a pilot expect when planning for a flight over ALB at FL 270?

A—270° magnetic at 97 knots; ISA -4°C.
B—260° true at 110 knots; ISA +5°C.
C—275° true at 97 knots; ISA +4°C.

To find the conditions at ALB at FL270, you must interpolate between 24,000 and 30,000:

Altitude	Code	Wind direction	Wind speed	Temperature
24,000	2777-28	270°	77	-28°C
30,000	781842	280°	118	-42°C
27,000		275°	97	-35°C

To find the temperature relative to ISA, multiply the standard temperature lapse rate (-2°C per 1,000 feet) times the altitude in thousands of feet (in this case 27), then subtract +15°C for standard day conditions:

1. *-2°C x 27 = 54°C*
2. *-54°C – 15 = -39°C (ISA)*

Therefore, -35°C (today) is 4°C warmer than -39°C, so the temperature is ISA + 4°C. (I63) — AC 00-45E, Chapter 4

Answers

4188 [B]   4189 [C]   4190 [B]   4192 [C]   4193 [C]

Chapter 2 **Weather Services**

ALL
4194. (Refer to Figure 2.) What approximate wind direction, speed, and temperature (relative to ISA) should a pilot expect when planning for a flight over EMI at FL 270?

A—265° true; 100 knots; ISA +3°C.
B—270° true; 110 knots; ISA +5°C.
C—260° magnetic; 100 knots; ISA -5°C.

To find the conditions at EMI at FL270, you must interpolate between 24,000 and 30,000:

Altitude	Code	Wind direction	Wind speed	Temperature
24,000	2891-30	280°	91	-30°C
30,000	751041	250°	110	-41°C
27,000		265°	100	-36°C

To find the temperature relative to ISA, multiply the standard temperature lapse rate (-2°C per 1,000 feet) times the altitude in thousands of feet (in this case 27), then subtract +15°C for standard day conditions:

1. *-2°C x 27 = 54°C*
2. *-54°C – 15 = -39°C (ISA)*

Therefore, -36°C (today) is 3°C warmer than -39°C, so the temperature is ISA + 3°C. (I63) — AC 00-45E, Chapter 4

ALL
4199. A station is forecasting wind and temperature aloft at FL 390 to be 300° at 200 knots; temperature -54° C. How would this data be encoded in the FD?

A—300054.
B—809954.
C—309954.

Wind speed above 100 knots have 50 added to the direction code (first 2 digits). Wind speeds at 200 knots or greater are coded 99 (middle 2 digits). Temperatures above 24,000 feet are negative so the minus sign is omitted (last 2 digits). The 300° at 200 knots and 54° is a encoded as 809954. (I63) — AC 00-45E, Chapter 4

Answer (A) is incorrect because 300054 would be 300° at 0 knots and -54°C. Answer (C) is incorrect because 309954 would be 300° at 99 knots and -54°C.

ALL
4246. (Refer to Figure 12.) What is the approximate wind direction and velocity at 34,000 feet (see arrow C)?

A—290°/50 knots.
B—330°/50 knots.
C—090°/48 knots.

Arrow C (in Oregon):

Line extends to the northwest with a 9 at the end of it = 290°
Pennant = 50 knots

(I63) — AC 00-45E, Chapter 10

Answers (B) and (C) are incorrect because the 9 at the end of the line indicates the wind is coming from 290° (not 330° or 090°).

ALL
4247. (Refer to Figure 12.) The wind direction and velocity on the Observed Winds Aloft Chart (see arrow A) is indicated from the

A—northeast at 35 knots.
B—northwest at 47 knots.
C—southwest at 35 knots.

Arrow A (in Indiana):

Line extends to the southwest with a 3 at the end of it = 230°

3 full barbs (10 knots each) and 1 half barb (5 knots) = 35 knots

(I63) — AC 00-45E, Chapter 10

Answers (A) and (B) are incorrect because the line extends to the southwest, therefore the wind is from the southwest, and the barbs indicate the wind has a velocity of 35 knots.

ALL
4249. (Refer to Figure 12.) What is the approximate wind direction and velocity at CVG at 34,000 feet (see arrow A)?

A—040°/35 knots.
B—097°/40 knots.
C—230°/35 knots.

Arrow A (in Indiana):

Line extends to the southwest with a 3 at the end of it = 230°

3 full barbs (10 knots each) and 1 half barb (5 knots) = 35 knots

(I63) — AC 00-45E, Chapter 10

Answers (A) and (B) are incorrect because the line extends to the southwest, therefore the wind is from the southwest (not the east or northeast).

Answers

| 4194 | [A] | 4199 | [B] | 4246 | [A] | 4247 | [C] | 4249 | [C] |

Chapter 2 **Weather Services**

ALL
4250. (Refer to Figure 12, arrow B.) What is the approximate wind direction and velocity at BOI (see arrow B)?

A—270°/55 knots.
B—250°/95 knots.
C—080°/95 knots.

Arrow B (in Idaho):

*Line extends to the southwest with a 5 at the end = 250°
Pennant (50 knots), 4 full barbs (10 knots each), 1 half
barb (5 knots) = 95 knots*

(I63) — AC 00-45E, Chapter 10

Inflight Weather Advisories (WA, WS, WST)

Severe Weather Watch bulletins (WW) define areas of possible severe thunderstorms or tornado activity. They are unscheduled and are issued as required.

HIWAS is a continuous broadcast over selected VORs of In-Flight Weather Advisories; i.e. SIGMETs, CONVECTIVE SIGMETs, AIRMETs, Severe Weather Forecast Alerts (AWW), and Center Weather Advisories (CWA).

Inflight Weather Advisories advise pilots en route of the possibility of encountering hazardous flying conditions that may not have been forecast at the time of the preflight weather briefing.

AIRMETs (WA) contain information on weather that may be hazardous to single engine, other light aircraft, and VFR pilots. The items covered are moderate icing or turbulence, sustained winds of 30 knots or more at the surface, widespread areas of IFR conditions, and extensive mountain obscurement.

SIGMETs (WS) advise of weather potentially hazardous to all aircraft. The items covered are severe icing, severe or extreme turbulence, and widespread sandstorms, dust storms or volcanic ash lowering visibility to less than 3 miles.

SIGMETs and AIRMETs are broadcast upon receipt and at 30-minute intervals (H + 15 and H + 45) during the first hour. If the advisory is still in effect after the first hour, an alert notice will be broadcast. Pilots may contact the nearest FSS to ascertain whether the advisory is pertinent to their flights.

CONVECTIVE SIGMETs (WST) cover weather developments such as tornadoes, lines of thunderstorms, and embedded thunderstorms, and they also imply severe or greater turbulence, severe icing, and low-level wind shear. Convective SIGMET Bulletins are issued hourly at H+55. Unscheduled Convective SIGMETs are broadcast upon receipt and at 15-minute intervals for the first hour (H + 15, H + 30, H + 45).

ALL
4186. When are severe weather watch bulletins (WW) issued?

A—Every 12 hours as required.
B—Every 24 hours as required.
C—Unscheduled and issued as required.

A severe weather watch bulletin (WW) defines areas of possible severe thunderstorms or tornado activity. They are unscheduled and are issued as required. (I57) — AC 00-45E, Chapter 4

ALL
4181. SIGMETs are issued as a warning of weather conditions potentially hazardous

A—particularly to light aircraft.
B—to all aircraft.
C—only to light aircraft operations.

A SIGMET advises of weather potentially hazardous to all aircraft other than convective activity. In the continental U.S., items covered are: severe icing, severe or extreme turbulence, duststorms, sandstorms, or volcanic ash lowering visibilities to less than 3 miles. (I57) — AC 00-45E, Chapter 4

Answer (A) is incorrect because SIGMETs apply to all aircraft. Answer (C) is incorrect because AIRMETs (not SIGMETs) apply only to light aircraft operations.

Answers

4250　[B]　　　　4186　[C]　　　　4181　[B]

ALL

4183. Which meteorological condition is issued in the form of a SIGMET (WS)?

A—Widespread sand or dust storms lowering visibility to less than 3 miles.
B—Moderate icing.
C—Sustained winds of 30 knots or greater at the surface.

A SIGMET advises of weather potentially hazardous to all aircraft other than convective activity. In the continental U.S., items covered are: severe icing, severe or extreme turbulence, duststorms, sandstorms, or volcanic ash lowering visibilities to less than 3 miles. (I57) — AC 00-45E, Chapter 4

Answer (B) is incorrect because SIGMETs advise of severe (not moderate) icing. Answer (C) is incorrect because AIRMETs (not SIGMETs) advise of sustained winds of 30 knots or greater at the surface.

ALL

4184. A pilot planning to depart at 1100Z on an IFR flight is particularly concerned about the hazard of icing. What sources reflect the most accurate information on icing conditions (current and forecast) at the time of departure?

A—Low-Level Significant Weather Prognostic Chart, and the Area Forecast.
B—The Area Forecast, and the Freezing Level Chart.
C—Pilot weather reports (PIREPs), AIRMETs, and SIGMETs.

Almost all of the listed reports and forecasts will contain information about icing hazards. SIGMETs and AIRMETs are issued when there is an icing hazard. PIREPs (pilot reports) contain information from pilots actually flying in the area forecast to have icing conditions. (I56) — AC 00-45E, Chapter 4

Answers (A) and (B) are incorrect because Low-Level Sig Weather Prog Charts are not issued for icing hazards. Area Forecasts only show basic icing forecasts, and Freezing Level Charts show the freezing level (not icing).

ALL

4241. The Hazardous Inflight Weather Advisory Service (HIWAS) is a continuous broadcast over selected VORs of

A—SIGMETs, CONVECTIVE SIGMETs, AIRMETs, Severe Weather Forecasts Alerts (AWW), and Center Weather Advisories.
B—SIGMETs, CONVECTIVE SIGMETs, AIRMETs, Wind Shear Advisories, and Severe Weather Forecast Alerts (AWW).
C—Wind Shear Advisories, Radar Weather Reports, SIGMETs, CONVECTIVE SIGMETs, AIRMETs, and Center Weather Advisories (CWA).

HIWAS is a continuous broadcast over selected VORs of In-Flight Weather Advisories; i.e., SIGMETs, CONVECTIVE SIGMETs, AIRMETs, Severe Weather Forecast Alerts (AWW), and Center Weather Advisories (CWA). (I54) — AC 00-45E, Chapter 1

ALL

4187. What is the maximum forecast period for AIRMETs?

A—Two hours.
B—Four hours.
C—Six hours.

AIRMETs are issued with a maximum forecast period of 6 hours. If conditions persist beyond the 6-hour forecast period, the AIRMET must be updated and reissued. (I57) — AC 00-45E, Chapter 4

ALL

4467. AIRMETs are issued on a scheduled basis every

A—15 minutes after the hour only.
B—15 minutes until the AIRMET is canceled.
C—six hours.

AIRMETs are issued on a scheduled basis every six hours, with unscheduled amendments issued as required. (J25) — AIM ¶7-1-5(h)

Answers

4183 [A]   4184 [C]   4241 [A]   4187 [C]   4467 [C]

Transcribed Weather Broadcast (TWEB)

A TWEB is a continuous broadcast of weather information on low/medium frequency NAVAIDs and on selected VORs. The information provided varies depending on the type of equipment available. Generally, the broadcast contains route-oriented data with specially prepared NWS forecasts, in-flight advisories, and winds aloft plus preselected current information, such as weather reports, NOTAMs, and special notices. These broadcasts are made available primarily for preflight and in-flight planning, and as such, should not be considered as a substitute for specialist-provided preflight briefings.

ALL
4185. Which forecast provides specific information concerning expected sky cover, cloud tops, visibility, weather, and obstructions to vision in a route format?

A—DFW FA 131240.
B—MEM TAF 132222.
C—249 TWEB 252317.

The TWEB Route Forecast is similar to the Area Forecast (FA) except more specific information is contained in a route format. Forecast sky cover (height and amount of cloud bases), cloud tops, visibility (including vertical visibility), weather and obstructions to vision are described for a corridor 25 miles either side of the route. (I57) — AC 00-45E, Chapter 4

Answer (A) is incorrect because "FA" indicates it is an Area Forecast. Answer (B) is incorrect because "TAF" indicates it is a Terminal Aerodrome Forecast.

Enroute Flight Advisory Service (EFAS)

EFAS provides enroute aircraft with weather advisories pertinent to the type of flight, route to be flown, and the altitude. The service is normally available throughout the conterminous United States from 6 a.m. to 10 p.m. local time. Aircraft flying at 5,000 feet AGL to 17,500 feet MSL can obtain the service by calling "Flight Watch" on radio frequency 122.0 MHz.

ALL
4504. On what frequency should you obtain En Route Flight Advisory Service below FL 180?

A—122.1T/112.8R.
B—123.6.
C—122.0.

EFAS (Enroute Flight Advisory Service) provides enroute aircraft with weather advisories pertinent to the type of flight, route to be flown, and the altitude. The service is normally available throughout the conterminous United States from 6 a.m. to 10 p.m. local time. Aircraft flying between 5,000 feet AGL and 18,000 feet AGL can obtain the service by calling "Flight Watch" on radio frequency 122.0 Mhz. (J35) — AIM ¶7-1-4(a)

Answer (A) is incorrect because it is an example of communications with an FSS through a VOR. Answer (B) is incorrect because 123.6 is not the EFAS frequency.

Answers
4185 [C]   4504 [C]

Convective Outlook (AC)

A Convective Outlook (AC) describes the prospects for general thunderstorm activity during the following 24 hours. Areas in which there is a high, moderate, or slight risk of severe thunderstorms are included as well as areas where thunderstorms may approach severe limits. Outlooks are transmitted by the National Severe Storm Forecast Center in Kansas City, MO at 0700Z and 1500Z. Pilots should use the Outlook primarily for planning flights later in the day.

ALL
4215. What information is provided by a Convective Outlook (AC)?

A—It describes areas of probable severe icing and severe or extreme turbulence during the next 24 hours.
B—It provides prospects of both general and severe thunderstorm activity during the following 24 hours.
C—It indicates areas of probable convective turbulence and the extent of instability in the upper atmosphere (above 500 MB).

A Convective Outlook (AC) describes prospects for general thunderstorm activity during the following 24 hours. Use the outlook primarily for planning flights later in the day. The charts show conditions as they are forecast to be at the valid time of the chart. (I65) — AC 00-45E, Chapter 12

Answer (A) is incorrect because SIGMETs (not Convective Outlooks) describe both severe turbulence and icing. Answer (C) is incorrect because a Constant Pressure Analysis Chart (not Convective Outlooks) indicates areas of turbulence and instability.

ALL
4226. Which weather forecast describes prospects for an area coverage of both severe and general thunderstorms during the following 24 hours?

A—Terminal Aerodrome Forecast.
B—Convective Outlook.
C—Radar Summary Chart.

A Convective Outlook (AC) describes the prospects for general thunderstorm activity during the following 24 hours. (I57) — AC 00-45E, Chapter 12

Answer (A) is incorrect because Terminal Aerodrome Forecasts are given for a specific airport. Answer (C) is incorrect because Radar Summary Charts are issued hourly and are valid until 35 minutes past the hour.

ALL
4175. What does a Convective Outlook (AC) describe for a following 24 hour period?

A—General thunderstorm activity.
B—A severe weather watch bulletin.
C—When forecast conditions are expected to continue beyond the valid period.

A Convective Outlook (AC) describes the prospects for general thunderstorm activity during the following 24 hours. Areas in which there is a high, moderate, or slight risk of severe thunderstorms are included as well as areas where thunderstorms may approach severe limits (approaching is defined as winds greater than or equal to 35 knots but less than 50 knots and/or hail equal to or greater than 1/2 inch in diameter). (I64) — AC 00-45E, Chapter 12

Answers

4215 [B]   4226 [B]   4175 [A]

Chapter 2 **Weather Services**

Surface Analysis Chart

The Surface Analysis Chart depicts frontal positions, pressure patterns, temperature, dew point, wind, weather, and obstructions to vision as of the valid time of the chart.

ALL
4209. The Surface Analysis Chart depicts

A—actual pressure systems, frontal locations, cloud tops, and precipitation at the time shown on the chart.
B—frontal locations and expected movement, pressure centers, cloud coverage, and obstructions to vision at the time of chart transmission.
C—actual frontal positions, pressure patterns, temperature, dew point, wind, weather, and obstructions to vision at the valid time of the chart.

The Surface Analysis Chart provides a ready means of locating pressure systems and fronts and also gives an overview of winds, temperatures and dew points as of the time of the plotted observations. The chart also shows weather and obstruction to vision symbols. It does not show cloud heights or tops or the expected movement of any item depicted. Expected movement would be contained in a forecast. (I58) — AC 00-45E, Chapter 5

Answer (A) is incorrect because cloud tops are not depicted by the Surface Analysis Chart. Answer (B) is incorrect because current conditions are depicted, not expected movement.

Weather Depiction Chart

The Weather Depiction Chart is computer-prepared from Surface Aviation Observations (SAO). It gives a broad overview of the observed flying category conditions at the valid time of the chart. This chart begins at 01Z each day, is transmitted at three hour intervals, and is valid at the time of the plotted data.

A legend in the lower portion of the chart describes the method of identifying areas of VFR, MVFR, and IFR conditions. Total sky cover is depicted by shading of the station circle as shown in Figure 2-4 on the next page.

Cloud height is entered under the station circle in hundreds of feet above ground level. Weather and obstructions to vision symbols are normally shown to the left of the station circle. Visibility (in statute miles) is entered to the left of any weather symbol. If the visibility is greater than 6 miles it will be omitted. Figure 2-5 on the next page shows examples of plotted data.

ALL
4206. (Refer to Figure 4.) What is the meaning of a bracket (]) plotted to the right of the station circle on a weather depiction chart?

A—The station represents the en route conditions within a 50 mile radius.
B—The station is an automated observation location.
C—The station gives local overview of flying conditions for a six hour period.

Data for the Weather Depiction Chart comes from the observations reported by both manual and automated observation locations. The automated stations are denoted by a bracket (]) plotted to the right of the station circle. (I59) — AC 00-45E, Chapter 6

Answers

4209 [C]   4206 [B]

ALL

4207. (Refer to Figure 4.) The Weather Depiction Chart indicates the heaviest precipitation is occurring in

A—north central Florida.
B—north central Minnesota.
C—central South Dakota.

The shaded area in north central Minnesota indicates IFR conditions with thunderstorms and rain. (I59) — AC 00-45E, Chapter 6

Answer (A) is incorrect because the shaded area in central Florida indicates IFR conditions; however, there is no indication of any precipitation. Answer (C) is incorrect because central South Dakota is not reporting any precipitation, despite the warm front shown across the northern part of the state.

ALL

4208. (Refer to Figure 4.) The Weather Depiction Chart in the area of northwestern Wyoming, indicates

A—overcast with scattered rain showers.
B—1,000-foot ceilings and visibility 3 miles or more.
C—500-foot ceilings and continuous rain, less than 3 miles visibility.

The shaded area in northwestern Wyoming indicates IFR with ceiling less than 1,000 feet and/or visibility less than 3 miles The 2 dots indicate continuous rain, while the completely filled in station circle indicates the skies are overcast. The 5 below the station circle indicates the ceiling is 500 feet. (I59) — AC 00-45E, Chapter 6

Symbol	Total sky cover
○	Sky Clear (less than 1/10)
◔	1/10 to 5/10 inclusive (Scattered)
◑	5/10 to 9/10 inclusive (Broken)
●	10/10 (Overcast) ✓
⊗	Sky obscured or partially obscured

Figure 2-4

Plotted	Interpreted
◯ 8	Few clouds, base 800 feet, visibility more than 6
◑ 12	Broken sky cover, ceiling 1,200 feet, rain shower, visibility more than 6
5 ∞ ●	Thin overcast, visibility 5 in haze
▲◔ 30	Scattered at 3,000 feet, clouds topping ridges, visibility more than 6
2 = ○	Sky clear, visibility 2, ground fog or fog
1/2 ↟ ⊗	Sky partially obscured, visibility 1/2, blowing snow, no cloud layers observed
1/4 ✶ ⊗	Sky partially obscured, visibility 1/4, snow, no cloud layers observed
1 ⚡ ● 12	Overcast, ceiling 1,200 feet, thunderstorm, rain shower, visibility 1
Ⓜ	Data missing

Note: Since a partial and a total obscuration (X) is entered as total sky cover, it can be difficult to determine if a height entry is a cloud layer above a partial obscuration or vertical visibility into a total obscuration. Check the SAO.

Figure 2-5

Answers

4207 [B]   4208 [C]

Chapter 2 **Weather Services**

Radar Summary Chart

The Radar Summary Chart will yield a three-dimensional view of clouds and precipitation when used in conjunction with other charts and reports. Radar detects only drops or ice particles of precipitation size, it does not detect clouds and fog. It is the only chart that shows lines and cells of hazardous thunderstorms.

Areas from which precipitation echoes were received are shown on the Radar Summary Chart, with echo intensity indicated by contours as shown in the figure for Question 4230. Echo heights are displayed in hundreds of feet MSL. Tops are entered above a short line, while any available bases are entered below a short line.

Movement of an area of echoes is indicated by a shaft and barb combination with the shaft indicating the direction and the barbs the speed. A whole barb is 10 knots, a half barb is 5 knots, and a pennant is 50 knots.

Individual cell movement is indicated by an arrow that shows the direction of movement with the speed in knots entered as a number.

ALL
4174. What important information is provided by the Radar Summary Chart that is not shown on other weather charts?

A—Lines and cells of hazardous thunderstorms.
B—Types of precipitation.
C—Areas of cloud cover and icing levels within the clouds.

Radar does show lines of thunderstorms and thunderstorm cells, all of which are hazardous and are not shown on other weather charts. (I60) — AC 00-45E, Chapter 7

Answer (B) is incorrect because other weather charts show types of precipitation. Answer (C) is incorrect because icing conditions are not detected by radar.

ALL
4230. (Refer to Figure 8.) What weather conditions are depicted in the area indicated by arrow A on the Radar Summary Chart?

A—Moderate to strong echoes; echo tops 30,000 feet MSL; line movement toward the northwest.
B—Weak to moderate echoes; average echo bases 30,000 feet MSL; cell movement toward the southeast; rain showers with thunder.
C—Strong to very strong echoes; echo tops 30,000 feet MSL; thunderstorms and rain showers.

Area A indicates an echo intensity level of 3 to 4, strong to very strong, since it is in the second ring. Echo tops in are at 30,000 feet. Thunderstorms and rain showers are indicated by the contraction "TRW." See the figure on the next page. (I60) — AC 00-45E, Chapter 7

Answer (A) and (B) are incorrect because the area indicated is an area of strong to very strong echoes and the tops are 30,000 feet.

ALL
4231. (Refer to Figure 8.) What weather conditions are depicted in the area indicated by arrow D on the Radar Summary Chart?

A—Echo tops 4,100 feet MSL, strong to very strong echoes within the smallest contour, and area movement toward the northeast at 50 knots.
B—Intense to extreme echoes within the smallest contour, echo tops 29,000 feet MSL, and cell movement toward the northeast at 50 knots.
C—Strong to very strong echoes within the smallest contour, echo bases 29,000 feet MSL, and cell in northeast Nebraska moving northeast at 50 knots.

Area D indicates an echo intensity level of 5 to 6, intense to extreme, since it is in the third ring. Echo tops in are at 29,000 feet. Individual cell movement is indicated by an arrow with the speed in knots entered as a number and the arrow shaft pointing in the direction of movement. In area D an individual cell is moving northeast at 50 knots. See the figure on the next page. (I60) — AC 00-45E, Chapter 7

Answers

4174 [A]   4230 [C]   4231 [B]

ALL
4232. (Refer to Figure 8.) What weather conditions are depicted in the area indicated by arrow C on the Radar Summary Chart?

A—Average echo bases 2,800 feet MSL, thundershowers, and intense to extreme echo intensity.
B—Cell movement toward the northwest at 20 knots, intense echoes, and echo bases 28,000 feet MSL.
C—Area movement toward the northeast strong to very strong echoes, and echo tops 28,000 feet MSL.

Area C indicates an echo intensity level of 3 to 4, strong to very strong, since it is in the second ring. Echo tops in are at 28,000 feet. Line or area movement is indicated by a shaft and barb combination with the shaft indicating the direction and the barbs the speed. A whole barb is 10 knots, a half barb is 5 knots, and a pennant is 50 knots. In area C, the area movement is toward the northeast at 20 knots. See the figure below. (I60) — AC 00-45E, Chapter 7

VIP Level	Echo Intensity	Precipitation Intensity	Rainfall Rate in/hr Stratiform	Rainfall Rate in/hr Convective
1	Weak	Light	Less than 0.1	Less than 0.2
2	Moderate	Moderate	0.1 – 0.5	0.2 – 1.1
3	Strong	Heavy	0.5 – 1.0	1.1 – 2.2
4	Very strong	Very heavy	1.0 – 2.0	2.2 – 4.5
5	Intense	Intense	2.0 – 5.0	4.5 – 7.1
6	Extreme	Extreme	More than 5.0	More than 7.1

450
Highest precipitation top in area in hundreds of feet MSL (45,000 Feet MSL)

*The numbers representing the intensity level do not appear on the chart. Beginning from the first contour line, bordering the area, the intensity level is 1–2, second contour is 3–4, and the third contour is 5–6.

Symbols Used on Charts

Symbol	Meaning
R	Rain
RW	Rain shower
Hail	Hail
S	Snow
PL	Ice pellets
SW	Snow shower
L	Drizzle
T	Thunderstorm
ZR, ZL	Freezing precipitation
NE	No echoes observed
NA	Observations unavailbale
OM	Out for Maintenance
STC	STC ON – all precipitation may not be seen
ROBEPS	Radar Operating Below Performance Standards
RHINO	Range Height Indicator Not Operating

Symbol	Meaning
+	Intensity increasing or new echo
–	Intensity decreasing
No Symbol	No change in intensity
↗35	Cell movement to NE at 35 knots
↘	Line or area movement to East at 20 knots
LM	Little movement
MA	Echoes mostly aloft
PA	Echoes partly aloft

Symbol	Meaning
◗	Line of echoes
SLD	8/10 or greater coverage in a line
WS999	Severe thunderstorm watch
WT999	Tornado watch
LEWP	Line echo wave pattern
HOOK	Hook echo
BWER	Bounded weak echo region
PCLL	Persistent cell
FNLN	Fine line

Question 4230, 4231, 4232, 4233, 4234, 4236 and 4237

Answers
4232 [C]

Chapter 2 **Weather Services**

ALL
4233. (Refer to Figure 8.) What weather conditions are depicted in the area indicated by arrow B on the Radar Summary Chart?

A—Weak echoes, heavy rain showers, area movement toward the southeast.
B—Weak to moderate echoes, rain showers increasing in intensity.
C—Strong echoes, moderate rain showers, no cell movement.

Area B indicates an echo intensity of 1 to 2, weak to moderate, since it is in the first ring. The contraction "RW+" indicates rain showers, intensity increasing, or new echo. See the figure on the previous page. (I60) — AC 00-45E, Chapter 7

ALL
4234. (Refer to Figure 8.) What weather conditions are depicted in the area indicated by arrow E on the Radar Summary Chart?

A—Highest echo tops 30,000 feet MSL, weak to moderate echoes, thunderstorms and rain showers, and cell movement toward northwest at 15 knots.
B—Echo bases 29,000 to 30,000 feet MSL, strong echoes, rain showers increasing in intensity, and area movement toward northwest at 15 knots.
C—Thundershowers decreasing in intensity; area movement toward northwest at 15 knots; echo bases 30,000 feet MSL.

Area E indicates an echo intensity of 1 to 2, weak to moderate, since it is in the first ring. Echo tops are at 30,000 feet. Individual cell movement is indicated by an arrow with the speed in knots entered as a number and the arrow shaft pointing in the direction of movement. In area E an individual cell is moving northwest at 15 knots. Thunderstorms and rain showers are indicated by the contraction "TRW." See the figure on the previous page. (I60) — AC 00-45E, Chapter 7

ALL
4236. (Refer to Figure 8.) What weather conditions are depicted in the area indicated by arrow G on the Radar Summary Chart?

A—Echo bases 10,000 feet MSL; cell movement toward northeast at 15 knots; weak to moderate echoes; rain.
B—Area movement toward northeast at 15 knots; rain decreasing in intensity; echo bases 1,000 feet MSL; strong echoes.
C—Strong to very strong echoes; area movement toward northeast at 15 knots; echo tops 10,000 feet MSL; light rain.

Area G indicates an echo intensity of 1 to 2, weak to moderate, since it is in the first ring. Echo bases are at 10,000 feet. Individual cell movement is indicated by an arrow with the speed in knots entered as a number and the arrow shaft pointing in the direction of movement. In area G an individual cell is moving northeast at 15 knots. The contraction "R" indicates rain. See the figure on the previous page. (I60) — AC 00-45E, Chapter 7

ALL
4237. (Refer to Figure 8.) What weather conditions are depicted in the area indicated by arrow F on the Radar Summary Chart?

A— Line of echoes; thunderstorms; highest echo tops 45,000 feet MSL; no line movement indicated.
B—Echo bases vary from 15,000 feet to 46,000 feet MSL; thunderstorms increasing in intensity; line of echoes moving rapidly toward the north.
C—Line of severe thunderstorms moving from south to north; echo bases vary from 4,400 feet to 4,600 feet MSL; extreme echoes.

When echoes are reported as a "LINE," a line will be drawn through them on the chart. When there is at least 8/10 coverage in the line, it is labeled solid "SLD" at both ends of the line. The line in area F is not solid, and it is not moving. Echo tops in area F are at 45,000 feet. Thunderstorms are indicated by the contraction "TRW." See the figure on the previous page. (I60) — AC 00-45E, Chapter 7

Answers

| 4233 | [B] | 4234 | [A] | 4236 | [A] | 4237 | [A] |

Chapter 2 **Weather Services**

ALL
4235. For most effective use of the Radar Summary Chart during preflight planning, a pilot should

A—consult the chart to determine more accurate measurements of freezing levels, cloud cover, and wind conditions between reporting stations.
B—compare it with the charts, reports, and forecasts of a three-dimensional picture of clouds and precipitation.
C—utilize the chart as the only source of information regarding storms and hazardous conditions existing between reporting stations.

The Radar Summary Chart aids in preflight planning by identifying general areas and movement of precipitation and/or thunderstorms. Radar detects only drops or ice particles of precipitation size; it does not detect clouds and fog. Therefore, the absence of echoes does not guarantee clear weather. Furthermore, cloud tops may be higher than precipitation tops detected by radar. The chart must be used in conjunction with other charts, reports and forecasts. (I60) — AC 00-45E, Chapter 7

Answer (A) is incorrect because freezing levels, cloud cover, and wind conditions are not on the Radar Summary Chart. Answer (C) is incorrect because the Radar Summary Chart must be used in conjunction with other charts, reports and forecasts.

Constant Pressure Chart

A Constant Pressure Analysis Chart is an upper air weather map where all the information depicted is at the specified pressure of the chart. Each of the Constant Pressure Analysis Charts (850 MB, 700 MB, 500 MB, 300 MB, 250 MB, and 200 MB) can provide observed temperature, temperature/dewpoint spread, wind, height of the pressure surface, and the height changes over the previous 12-hour period.

ALL
4195. What flight planning information can a pilot derive from constant pressure charts?

A—Clear air turbulence and icing conditions.
B—Levels of widespread cloud coverage.
C—Winds and temperatures aloft.

Plotted at each reporting station (at the level of the specified pressure) on constant pressure charts are the observed temperature, temperature/dewpoint spread, wind, height of the pressure surface, as well as height changes over the previous 12-hour period. (I61) — AC 00-45E, Chapter 8

Answer (A) is incorrect because Low-Level Prog Charts indicate clear air turbulence. Answer (B) is incorrect because Weather Depiction and Surface Analysis Charts indicate areas and levels of widespread cloud coverage.

ALL
4173. What conclusion(s) can be drawn from a 500-millibar Constant Pressure Chart for a planned flight at FL 180?

A—Winds aloft at FL 180 generally flow across the height contours.
B—Observed temperature, wind, and temperature/dewpoint spread along the proposed route can be approximated.
C—Upper highs, lows, troughs, and ridges will be depicted by the use of lines of equal pressure.

Plotted at each reporting station (at the level of the specified pressure) are the observed temperature, temperature/dewpoint spread, wind, and height of the pressure surface, as well as the height changes over the previous 12-hour period. (I61) — AC 00-45E, Chapter 8

Answer (A) is incorrect because winds generally flow parallel to the height contours. Answer (C) is incorrect because heights are identified by solid lines (contours) which depict high height centers and low height centers.

Answers

| 4235 | [B] | 4195 | [C] | 4173 | [B] |

Chapter 2 **Weather Services**

Tropopause Data Chart

The Tropopause Data Chart is a two-panel chart containing a maximum wind prog and a vertical wind shear prog. The chart is prepared for the contiguous 48 states and is available once a day with a valid time of 18Z. It shows the temperature, pressure, and wind at the tropopause.

ALL
4243. (Refer to Figure 20.) What is the maximum wind velocity forecast in the jet stream shown on the high level Significant Weather Prognostic Chart over Canada?

A—80.
B—103.
C—130.

The maximum core speed along the jet stream is depicted by shafts, pennants, and feathers. The maximum wind velocity forecast in the jet stream can be found in Canada, north of Michigan. The two pennants and three lines indicate a forecast maximum speed of 130 knots, at a height of 30,000 feet MSL (FL300). (I64) — AC 00-45E, Chapter 11

ALL
4244. (Refer to Figure 20.) What is the height of the tropopause over Kentucky?

A—FL390.
B—FL300 sloping to FL 400 feet MSL.
C—FL340.

Tropopause heights are depicted in hundreds of feet MSL. The five sided polygon indicates areas of high and low tropopause heights. The height over Kentucky is 34,000 feet MSL, or FL340. (I64) — AC 00-45E, Chapter 11

ALL
4242. (Refer to Figure 7.) The symbol on the U.S. HIGH-LEVEL SIGNIFICANT WEATHER PROG, indicated by arrow G, represents the

A—wind direction at the tropopause (300°).
B—height of the tropopause.
C—height of maximum wind shear (30,000 feet).

Tropopause heights are plotted in hundreds of feet at selected locations. Heights are enclosed by rectangles. (I64) — AC 00-45E, Chapter 11

ALL
4245. (Refer to Figure 7.) The area indicated by arrow H indicates

A—light turbulence below 34,000 feet.
B—isolated embedded cumulonimbus clouds with bases below FL180 and tops at FL340.
C—moderate turbulence at and below 34,000 feet.

Areas of moderate or greater turbulence are enclosed by bold dashed lines. Turbulence intensities are identified by symbols. The vertical extent of turbulence layers is specified by top and base heights in hundreds of feet. Turbulence bases that extend below the layer of the chart are identified with "XXX." For example, "340/XXX" identifies a layer of turbulence from below FL240 to FL340. (I64) — AC 00-45E, Chapter 11

Answers

4243 [C]   4244 [C]   4242 [B]   4245 [C]

Chapter 2 **Weather Services**

Significant Weather Prognostics

The Low-Level Significant Weather Prognostic Chart (surface to 24,000 feet) portrays forecast weather which may influence flight planning, including those areas or activities of most significant turbulence and icing. It is a four-panel chart; the two lower panels are 12- and 24-hour surface progs. The two upper panels are 12- and 24-hour progs of significant weather from the surface to 24,000 feet. The chart is issued four times daily. The chart uses standard weather symbols as shown in Figure 2-6.

The High-Level Significant Weather Prognostic Chart (24,000 feet to 63,000 feet) outlines areas of forecast turbulence and cumulonimbus clouds, shows the expected height of the tropopause, and predicts jet stream location and velocity. The chart depicts clouds and turbulence as shown in Figure 2-7 on the next page.

The height of the tropopause is depicted in hundreds of feet MSL and is enclosed in a rectangular box. Areas of forecast moderate or greater Clear Air Turbulence (CAT) are bounded by heavy dashed lines and are labeled with the appropriate symbol and the vertical extent in hundreds of feet MSL. Cumulonimbus clouds imply moderate or greater turbulence and icing.

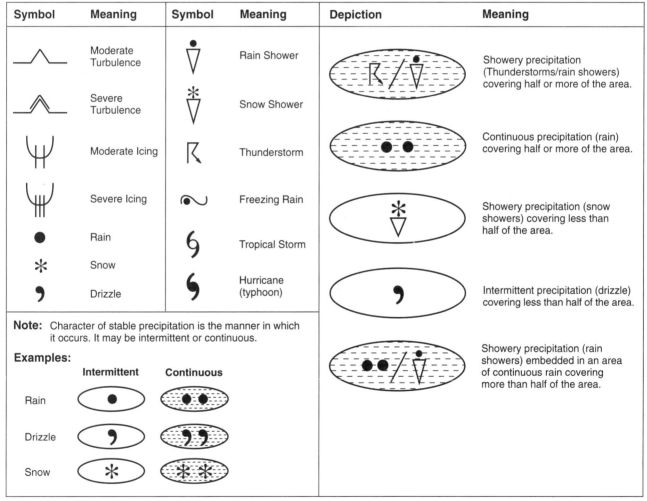

Figure 2-6

2002 Instrument Test Prep ASA **2–23**

ALL
4211. The Low-Level Significant Weather Prognostic Chart depicts weather conditions

A—that are forecast to exist at a valid time shown on the chart.
B—as they existed at the time the chart was prepared.
C—that existed at the time shown on the chart which is about 3 hours before the chart is received.

The Low-Level Prog is a four-panel chart. The two lower panels are 12- and 24-hour surface progs. The two upper panels are 12- and 24-hour progs of significant weather from the surface to 400 millibars (24,000 feet). The charts show conditions as they are forecast to be at the valid time of the chart. (I64) — AC 00-45E, Chapter 11

Answer (B) is incorrect because the Low-Level Prog charts contain 12 and 24-hour forecasts. Answer (C) is incorrect because the Low-Level Prog charts contain forecasts (not previous conditions).

Depiction	Meaning
ISOL EMBD CB 420/XXX	Embedded cumulonimbus, less than 1/8 coverage, bases below 24,000 feet, tops 42,000 feet
OCNL EMBD CB 520/XXX	Embedded cumulonimbus, 1/8 to 4/8 coverage, bases below 24,000 feet, tops 52,000 feet
FRQ CB 330/XXX	Cumulonimbus, 5/8 to 8/8 coverage, bases below 24,000 feet, tops 33,000 feet
TO 330/XXX	Moderate to severe turbulence, below 24,000 feet to 33,000 feet (for turbulence below 24,000 feet see low-level prog)
ABV 630 / 350	Moderate turbulence from 35,000 feet to above upper limit of the prog
(volcano symbol)	Volcanic Eruption Mt. Spurr 61.3 N 152.3 W At 15/1325Z CHECK SIGMETS FOR VOLCANIC ASH

Figure 2-7

ALL
4214. A prognostic chart depicts the conditions

A—existing at the surface during the past 6 hours.
B—which presently exist from the 1,000-millibar through the 700-millibar level.
C—forecast to exist at a specific time in the future.

The Low-Level Significant Weather Prognostic Chart depicts conditions as they are forecast to be at the valid time of the chart (12- and 24-hour forecasts). (I64) — AC 00-45E, Chapter 11

Answers (A) and (B) are incorrect because the Low-Level Prog Charts indicate 12 and 24-hour forecasts (not current or past conditions).

ALL
4212. Which meteorological conditions are depicted by a prognostic chart?

A—Conditions existing at the time of the observation.
B—Interpretation of weather conditions for geographical areas between reporting stations.
C—Conditions forecast to exist at a specific time shown on the chart.

The Low-Level Significant Weather Prognostic Chart depicts conditions as they are forecast to be at the valid time of the chart (12- and 24-hour forecasts). (I64) — AC 00-45E, Chapter 11

Answers (A) and (B) are incorrect because the Low-Level Prog charts contain 12- and 24-hour forecasts (not existing conditions).

Answers

4211 [A]   4214 [C]   4212 [C]

ALL
4213. (Refer to Figure 5.) What is the meaning of the symbol depicted as used on the U.S. Low-Level Significant Weather Prog Chart?

A—Showery precipitation (e.g. rain showers) embedded in an area of continuous rain covering half or more of the area.
B—Continuous precipitation (e.g. rain) covering half or more of the area.
C—Showery precipitation (e.g. thunderstorms/rain showers) covering half or more of the area.

The triangle with a dot over it indicates showery precipitation (i.e., rain showers). The two thick dots indicate continuous rain. The shaded area indicates more than one-half the area is obscured. (I64) — AC 00-45E, Chapter 11

Answer (B) is incorrect because a single dot over the triangle indicates that showery conditions exist (not continuous precipitation) in the area. Answer (C) is incorrect because thunderstorms are indicated by a different symbol resembling an R with an arrow on the leg.

ALL
4216. (Refer to Figure 18, SFC PROG.) A planned low altitude flight from northern Florida to southern Florida at 00Z is likely to encounter

A—intermittent rain or rain showers, moderate turbulence, and freezing temperatures above 8,000 feet.
B—showery precipitation, thunderstorms/rain showers covering half or more of the area.
C—showery precipitation covering less than half the area, no turbulence below 18,000 feet, and freezing temperatures above 12,000 feet.

Use the chart in the lower, left-hand corner. The open triangle with the dot above it indicates rain showers, while the "R" symbol indicates a thunderstorm. The shaded area indicates that the precipitation affects half or more of the area. (I64) — AC 00-45E, Chapter 11

ALL
4217. (Refer to Figure 18, SFC-400 MB.) The 24-Hour Low Level Significant Weather Prog at 12Z indicates that southwestern West Virginia will likely experience

A—ceilings less than 1,000 feet, visibility less than 3 miles.
B—clear sky and visibility greater than 6 miles.
C—ceilings 1,000 to 3,000 feet and visibility 3 to 5 miles.

Use the chart in the upper, right-hand corner. The smooth line enclosing West Virginia indicates a ceiling less than 1,000 feet and/or visibility less than 3 miles. (I64) — AC 00-45E, Chapter 11

ALL
4218. (Refer to Figure 18, SFC-400MB.) The U.S. Low Level Significant Weather Surface Prog Chart at 00Z indicates that northwestern Colorado and eastern Utah can expect

A—moderate or greater turbulence from the surface to FL 240.
B—moderate or greater turbulence above FL 240.
C—no turbulence is indicated.

Use the chart in the upper, left-hand corner. Numbers below and/or above a short line show expected base and/or top of the turbulent layer in hundreds of feet MSL. Absence of a number below the line indicates turbulence from the surface upward. No number above the line indicates turbulence extending above the upper limit of the chart. The 24 above the line in eastern Utah and northwestern Colorado indicates moderate or greater turbulence from the surface to FL240. (I64) — AC 00-45E, Chapter 11

ALL
4219. (Refer to Figure 18, SFC-PROG.) The chart symbols shown in the Gulf of Mexico at 12Z and extending into AL, GA, SC and northern FL indicate a

A—tropical storm.
B—hurricane.
C—tornado originating in the Gulf of Mexico.

Use the chart in the lower, right-hand corner. The symbol in the center of the shaded area indicates a tropical storm. (I64) — AC 00-45E, Chapter 11

Chapter 2 **Weather Services**

ALL
4221. (Refer to Figure 7.) What weather conditions are depicted within the area indicated by arrow E?

A—Occasional cumulonimbus, 1/8 to 4/8 coverage, bases below 24,000 feet MSL, and tops at 40,000 feet MSL.
B—Frequent embedded thunderstorms, less than 1/8 coverage, and tops at FL370.
C—Frequent lightning in thunderstorms at FL370.

Scalloped lines enclose areas of expected cumulonimbus development. "OCNL EMBD CB" indicates 1/8 to 4/8 coverage of embedded cumulonimbus. "400/XXX" indicates tops at 40,000 and bases below 24,000 feet MSL (below the prog's forecast layer). (I64) — AC 00-45E, Chapter 11

ALL
4222. (Refer to Figure 7.) What weather conditions are depicted within the area indicated by arrow D?

A—Forecast isolated thunderstorms, tops at FL 440, more than 1/8 coverage.
B—Existing isolated cumulonimbus clouds, tops above 43,000 feet with less than 1/8 coverage.
C—Forecast isolated embedded cumulonimbus clouds with tops at 43,000 feet MSL, and less than 1/8 coverage.

The scalloped lines enclose areas of expected cumulonimbus development. "ISOL EMBD CB" indicates less than 1/8 coverage of embedded thunderstorms. "430/XXX" indicates tops at 43,000 and bases below 24,000 feet MSL (below the prog's forecast layer). (I64) — AC 00-45E, Chapter 11

ALL
4223. (Refer to Figure 7.) What weather conditions are predicted within the area indicated by arrow C?

A—Light turbulence at FL 370 within the area outlined by dashes.
B—Moderate turbulence at 32,000 feet MSL.
C—Moderate to severe CAT has been reported at FL 320.

Areas of moderate or greater turbulence are enclosed by bold dashed lines. The hat symbol indicates moderate turbulence. The "350/XXX" identifies a layer of turbulence from below FL240 to FL350. (I64) — AC 00-45E, Chapter 11

ALL
4224. (Refer to Figure 7.) What weather conditions are depicted within the area indicated by arrow B?

A—Light to moderate turbulence at and above 37,000 feet MSL.
B—Moderate to severe CAT is forecast to exist at FL 370.
C—Moderate turbulence from below 24,000 feet MSL to 37,000 feet MSL.

Areas of moderate or greater turbulence are enclosed by bold dashed lines. The hat symbol indicates moderate turbulence. The "370/XXX" identifies a layer of turbulence from below FL240 to FL370. (I64) — AC 00-45E, Chapter 11

Answers

4221　[A]　　　4222　[C]　　　4223　[B]　　　4224　[C]

Chapter 2 **Weather Services**

ALL
4225. (Refer to Figure 7.) What information is indicated by arrow A?

A—The height of the tropopause in meters above sea level.
B—The height of the existing layer of CAT.
C—The height of the tropopause in hundreds of feet above MSL.

Tropopause heights are to be depicted in hundreds of feet MSL and are enclosed in small rectangular blocks. (I64) — AC 00-45E, Chapter 11

Answer (A) is incorrect because tropopause heights are given in hundreds of feet MSL. Answer (B) is incorrect because CAT is shown by heavy dashed lines.

ALL
4229. (Refer to Figure 7.) What weather conditions are depicted within the area indicated by arrow F?

A—1/8 to 4/8 coverage, occasional embedded thunderstorms, maximum tops at 51,000 feet MSL.
B—Occasionally embedded cumulonimbus, bases below 24,000 feet with tops to 48,000 feet.
C—2/8 to 6/8 coverage, occasional embedded thunderstorms, tops at FL 540.

Scalloped lines enclose areas of expected cumulonimbus development. "OCNL EMBD CB" indicates 1/8 to 4/8 coverage of embedded cumulonimbus. "480/XXX" indicates tops at 48,000 and bases below 24,000 feet MSL (below the prog's forecast layer). (I64) — AC 00-45E, Chapter 11

Severe Weather Outlook Chart

The Severe Weather Outlook Chart is used primarily for advance planning. It provides an outlook for general and severe thunderstorms, tornadoes, and tornado-watch areas. Single-hatched areas will be annotated to indicate either slight, moderate, or high risk of possible severe thunderstorms. Crosshatched areas indicate a forecast risk of tornadoes.

ALL
4197. (Refer to Figure 9.) The Severe Weather Outlook Chart depicts

A—areas of probable severe thunderstorms by the use of single hatched areas on the chart.
B—areas of forecast, severe or extreme turbulence, and areas of severe icing for the next 24 hours.
C—areas of general thunderstorm activity (excluding severe) by the use of hatching on the chart.

The single-hatched area indicates possible/probable severe thunderstorms. The line with an arrowhead delineates an area of probable general thunderstorm activity. When you face in the direction of the arrow, activity is expected to the right of the line. (I65) — AC 00-45E, Chapter 12

Answer (B) is incorrect because SIGMETs indicate (not forecast) severe and extreme turbulence. Answer (C) is incorrect because when you face in the direction of the arrow, general thunderstorm activity is expected to the right of the line.

ALL
4239. (Refer to Figure 9.) The Severe Weather Outlook Chart, which is used primarily for advance planning, provides what information?

A—An 18-hour categorical outlook with a 48-hour valid time for severe weather watch, thunderstorm lines, and of expected tornado activity.
B—A preliminary 12-hour outlook for severe thunderstorm activity and probable convective turbulence.
C—A 24-hour severe weather outlook for possible thunderstorm activity.

The Severe Weather Outlook Chart is a 48-hour outlook for thunderstorm activity. This chart is presented in two panels. The left panel covers the first 24-hour period beginning at 12Z and depicts areas of possible general thunderstorm activity as well as severe thunderstorms. The right hand panel covers the following day beginning at 12Z and is an outlook for the possibility of severe thunderstorms only. (I65) — AC 00-45E, Chapter 12

Answers

4225 [C] 4229 [B] 4197 [A] 4239 [C]

Chapter 2 Weather Services

ALL
4240. (Refer to Figure 9.) Using the DAY 2 CONVECTIVE OUTLOOK, what type of thunderstorms, if any, may be encountered on a flight from Montana to central California?

A—Moderate risk area, surrounded by a slight risk area, of possible severe turbulence.
B—General.
C—None.

Risk areas come in three varieties and are based on the expected number of severe thunderstorm reports per geographical unit and forecaster confidence. General thunderstorms (non-severe) are outlined, but with no label on the graphic map. (I65) — AC 00-45E, Chapter 12

ALL
4248. (Refer to Figure 9.) What type of thunderstorm activity is expected over Montana on April 4th at 0800Z?

A—None.
B—A slight risk of severe thunderstorms.
C—General.

Use the DAY 1 CONVECTIVE OUTLOOK. No thunderstorms are expected over Montana. (I65) — AC 00-45E, Chapter 12

Answers

4240 [B]   4248 [A]

Chapter 3
Flight Instruments

Airspeeds *3–3*

Altitudes *3–6*

Altimeter Settings *3–9*

Vertical Speed Indicator *3–12*

Attitude Indicator *3–13*

Turn Coordinator *3–16*

Heading Indicator *3–24*

Magnetic Compass *3–25*

Slaved Gyro *3–28*

Instrument Errors *3–29*

Fundamental Skills *3–33*

Attitude Instrument Flying *3–35*

Unusual Attitude Recoveries *3–43*

Chapter 3 Flight Instruments

Chapter 3 **Flight Instruments**

Airspeeds

Airspeed Definitions:

Indicated Airspeed (IAS)—the airspeed that is shown on the airspeed indicator. *See* Figure 3-1.

Calibrated Airspeed (CAS)—the indicated airspeed corrected for position installation error.

Equivalent Airspeed (EAS)—calibrated airspeed corrected for compressibility. EAS will always be lower than CAS.

True Airspeed (TAS)—the equivalent airspeed corrected for temperature and pressure altitude. At speeds below 200 knots, TAS can be found by correcting CAS for temperature and pressure altitude. A rule of thumb for TAS is that it is 2% more than CAS per 1,000 feet of altitude above sea level.

Mach Number—the ratio of aircraft true airspeed to the speed of sound.

Figure 3-1. Airspeed indicator

ALL
4267. (Refer to Figures 27 and 28.) What CAS must be used to maintain the filed TAS at the flight planned altitude if the outside air temperature is -5°C?

A—134 KCAS.
B—139 KCAS.
C—142 KCAS.

To find CAS when TAS is known, the correct OAT and the pressure altitude are required. Since neither are available for the problem, substitute the uncorrected OAT and the planned flight altitude. Using a flight computer and the information from the flight plan in FAA Figure 27:

Altitude 8,000 feet (Block 7 of flight plan)
Temperature -5°C (Given in question)
TAS 155 Knots (Block 4 of flight plan)

Therefore, CAS is 139 KCAS.

(H342) — AC 61-23C, Chapter 8

ALL
4278. (Refer to Figure 32.) What CAS must be used to maintain the filed TAS at the flight planned altitude if the outside air temperature is +8°C?

A—154 KCAS.
B—157 KCAS.
C—163 KCAS.

To find CAS when TAS is known, the correct OAT and the pressure altitude are required. Since neither are available for the problem, substitute the uncorrected OAT and the planned flight altitude. Using a flight computer and the information from the flight plan in FAA Figure 32:

Altitude 8,000 feet (Block 7 of flight plan)
Temperature +8°C (Given in question)
TAS 180 Knots (Block 4 of flight plan)

Therefore, CAS is 157.4 KCAS.

(H342) — AC 61-23C, Chapter 8

Answers

4267 [B]   4278 [B]

Chapter 3 Flight Instruments

ALL
4289. (Refer to Figure 38.) What CAS must be used to maintain the filed TAS at the flight planned altitude if the outside air temperature is +05°C?

A— 129 KCAS.
B— 133 KCAS.
C— 139 KCAS.

To find CAS when TAS is known, the correct OAT and the pressure altitude are required. Since neither are available for the problem, substitute the uncorrected OAT and the planned flight altitude. Using a flight computer and the information from the flight plan in FAA Figure 38:

Altitude 11,000 feet (Block 7 of flight plan)
Temperature +5°C (Given in question)
TAS 156 Knots (Block 4 of flight plan)

Therefore, CAS is 129.4 KCAS.

(H342) — AC 61-23C, Chapter 8

ALL
4301. (Refer to Figure 44.) What CAS must be used to maintain the filed TAS at the flight planned altitude if the outside air temperature is +5°C?

A— 147 KCAS.
B— 150 KCAS.
C— 154 KCAS.

To find CAS when TAS is known, the correct OAT and the pressure altitude are required. Since neither are available for the problem, substitute the uncorrected OAT and the planned flight altitude. Using a flight computer and the information from the flight plan in FAA Figure 44:

Altitude 12,000 feet (Block 7 of flight plan)
Temperature +5°C (Given in question)
TAS 180 Knots (Block 4 of flight plan)

Therefore, CAS is 146.6 KCAS.

(H342) — AC 61-23C, Chapter 8

ALL
4313. (Refer to Figure 50.) What CAS must be used to maintain the filed TAS at the flight planned altitude? (Temperature 0°C.)

A— 136 KCAS.
B— 140 KCAS.
C— 147 KCAS.

To find CAS when TAS is known, the correct OAT and the pressure altitude are required. Since neither are available for the problem, substitute the uncorrected OAT and the planned flight altitude. Using a flight computer and the information from the flight plan in FAA Figure 50:

Altitude 8,000 feet (Block 7 of flight plan)
Temperature +0°C (Given in question)
TAS 158 Knots (Block 4 of flight plan)

Therefore, CAS is 140.1 KCAS.

(H342) — AC 61-23C, Chapter 8

RTC
4323. (Refer to Figure 56.) What CAS should be used to obtain the filed TAS at the flight planned altitude if the outside air temperature is +5°?

A— 97 KCAS.
B— 101 KCAS.
C— 103 KCAS.

To find CAS when TAS is known, the correct OAT and the pressure altitude are required. Since neither are available for the problem, substitute the uncorrected OAT and the planned flight altitude. Using a flight computer and the information from the flight plan in FAA Figure 56:

Altitude 7,000 feet (Block 7 of flight plan)
Temperature +5°C (Given in question)
TAS 110 Knots (Block 4 of flight plan)

Therefore, CAS is 98.4 KCAS.

(H342) — AC 61-23C, Chapter 8

Answers

| 4289 | [A] | 4301 | [A] | 4313 | [B] | 4323 | [A] |

Chapter 3 Flight Instruments

RTC
4334. (Refer to Figure 62.) What CAS should be used to maintain the filed TAS at the flight planned altitude if the outside air temperature is +15°C?

A—91 KCAS.
B—96 KCAS.
C—101 KCAS.

To find CAS when TAS is known, the correct OAT and the pressure altitude are required. Since neither are available for the problem, substitute the uncorrected OAT and the planned flight altitude. Using a flight computer and the information from the flight plan in FAA Figure 62:

Altitude 5,000 feet (Block 7 of flight plan)
Temperature +15°C (Given in question)
TAS 105 Knots (Block 4 of flight plan)

Therefore, CAS is 95.8 KCAS.

(H342) — AC 61-23C, Chapter 8

ALL
4345. (Refer to Figure 69.) What CAS should be used to maintain the filed TAS if the outside air temperature is +05°C?

A—119 KCAS.
B—124 KCAS.
C—126 KCAS.

To find CAS when TAS is known, the correct OAT and the pressure altitude are required. Since neither are available for the problem, substitute the uncorrected OAT and the planned flight altitude. Using a flight computer and the information from the flight plan in FAA Figure 69:

Altitude 5,000 feet (Block 7 of flight plan)
Temperature +5°C (Given in question)
TAS 128 Knots (Block 4 of flight plan)

Therefore, CAS is 118.9 KCAS.

(H342) — AC 61-23C, Chapter 8

ALL
4359. (Refer to Figure 74.) What CAS should be used to maintain the filed TAS at the flight planned altitude if the outside air temperature is +5°C?

A—129 KCAS.
B—133 KCAS.
C—139 KCAS.

To find CAS when TAS is known, the correct OAT and the pressure altitude are required. Since neither are available for the problem, substitute the uncorrected OAT and the planned flight altitude. Using a flight computer and the information from the flight plan in FAA Figure 74:

Altitude 11,000 feet (Block 7 of flight plan)
Temperature +5°C (Given in question)
TAS 160 Knots (Block 4 of flight plan)

Therefore, CAS is 132.8 KCAS.

(H342) — AC 61-23C, Chapter 8

ALL
4864. What information does a Mach meter present?

A—The ratio of aircraft true airspeed to the speed of sound.
B—The ratio of aircraft indicated airspeed to the speed of sound.
C—The ratio of aircraft equivalent airspeed, corrected for installation error, to the speed of sound.

The Mach meter indicates the ratio of aircraft true airspeed to the speed of sound at flight altitude. (H810) — FAA-H-8083-15, Chapter 3

Answers (B) and (C) are incorrect because the Mach meter uses true airspeed (not indicated or equivalent airspeed).

Answers

| 4334 | [B] | 4345 | [A] | 4359 | [B] | 4864 | [A] |

Chapter 3 **Flight Instruments**

Altitudes

Indicated Altitude—the altitude above mean sea level indicated on an altimeter that is set to the current local altimeter setting. Better vertical separation of aircraft in a particular area results when all pilots use the local altimeter setting.

Pressure Altitude—the altitude indicated on an altimeter when it is set to the standard sea level pressure of 29.92 inches of mercury. Above 18,000 feet MSL, flight levels, which are pressure altitudes, are flown.

Density Altitude—the result of a given pressure altitude that is corrected for a non-standard temperature.

True Altitude—the exact height above sea level. The altimeter setting yields true altitude at field elevation.

Pressure decreases with altitude most rapidly in cold air and least rapidly in warm air. However, the standard pressure lapse rate is a decrease of 1" Hg for each 1,000 feet of increase in altitude. *See* Figures 3-2 and 3-3.

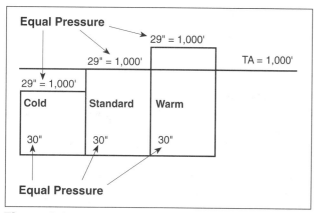

Figure 3-2

ALL
4089. Under what condition is pressure altitude and density altitude the same value?

A—At standard temperature.
B—When the altimeter setting is 29.92" Hg.
C—When indicated, and pressure altitudes are the same value on the altimeter.

Density altitude is pressure altitude corrected for non-standard temperature. Under standard atmospheric conditions, each level of air in the atmosphere has a specific density, and under standard conditions, pressure altitude and density altitude identify the same level. (I22) — AC 00-6A, page 19

Answer (B) is incorrect because pressure altitude must be corrected for nonstandard temperature. Answer (C) is incorrect because pressure and indicated altitudes are not corrected for nonstandard temperature, and density altitude is.

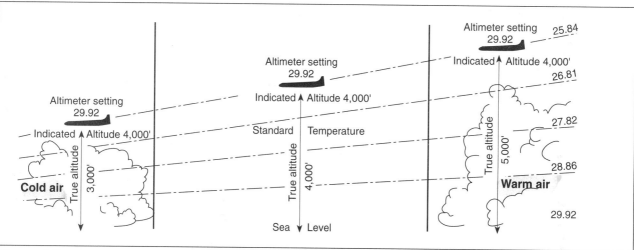

Figure 3-3

Answers
4089 [A]

ALL
4090. Under which condition will pressure altitude be equal to true altitude?

A—When the atmospheric pressure is 29.92" Hg.
B—When standard atmospheric conditions exist.
C—When indicated altitude is equal to the pressure altitude.

Pressure altitude will be equal to true altitude when standard atmospheric conditions exist for the same altitude. (I22) — AC 00-6A, page 19

Answer (A) is incorrect because it does not consider the possibility of nonstandard temperature. Answer (C) is incorrect because indicated altitude only considers air pressure, not both air pressure and temperature.

ALL
4091. Which condition would cause the altimeter to indicate a lower altitude than actually flown (true altitude)?

A—Air temperature lower than standard.
B—Atmospheric pressure lower than standard.
C—Air temperature warmer than standard.

Effects of nonstandard conditions can result in a difference of as much as 2,000 feet between true and indicated altitude. On a warm day, the pressure levels are raised, and the altimeter will indicate a lower altitude than the one actually flown (true altitude). (I22) — AC 00-6A, page 17

Answer (A) is incorrect because air temperature colder than standard has lower pressure levels, and will cause the altimeter to indicate higher than true altitude. Answer (B) is incorrect because the altimeter will adjust for nonstandard pressure settings.

ALL
4109. Under what condition will true altitude be lower than indicated altitude with an altimeter setting of 29.92" Hg?

A—In warmer than standard air temperature.
B—In colder than standard air temperature.
C—When density altitude is higher than indicated altitude.

Effects of nonstandard conditions can result in a difference of as much as 2,000 feet between true and indicated altitude. On a cold day, the pressure levels are lowered, and the altimeter will indicate a higher altitude than the one actually flown (true altitude). (I22) — AC 00-6A, page 17

Answer (A) is incorrect because in warmer than standard air, the true altitude will be higher than the indicated altitude. Answer (C) is incorrect because a high density altitude indicates warmer than standard air, which would mean that the true altitude will be higher than the indicated altitude.

ALL
4111. Altimeter setting is the value to which the scale of the pressure altimeter is set so the altimeter indicates

A—true altitude at field elevation.
B—pressure altitude at field elevation.
C—pressure altitude at sea level.

Since the altitude scale is adjustable, you can set the altimeter to read true altitude at some specified height. Takeoff and landing are the most critical phases of flight; therefore, airport elevation is the most desirable altitude for a true reading of the altimeter. Altimeter setting is the value to which the scale of the pressure altimeter is set so the altimeter indicates true altitude at the field elevation. (H808) — FAA-H-8083-15, Chapter 3

Answers (B) and (C) are incorrect because altimeter setting is the value to which the scale of the pressure altimeter is set so the altimeter indicates true altitude (not pressure altitude) at field elevation.

ALL
4477. How can you obtain the pressure altitude on flights below 18,000 feet?

A—Set your altimeter to 29.92" Hg.
B—Use your computer to change the indicated altitude to pressure altitude.
C—Contact an FSS and ask for the pressure altitude.

Pressure altitude can always be determined from the altimeter whether in flight or on the ground. Set the altimeter at the standard altimeter setting of 29.92 inches, and the altimeter will indicate pressure altitude. (H808) — FAA-H-8083-15, Chapter 3

ALL
4478. How can you determine the pressure altitude on an airport without a tower or FSS?

A—Set the altimeter to 29.92" Hg and read the altitude indicated.
B—Set the altimeter to the current altimeter setting of a station within 100 miles and correct this indicated altitude with local temperature.
C—Use your computer and correct the field elevation for temperature.

Pressure altitude can always be determined from the altimeter whether in flight or on the ground. Set the altimeter at the standard altimeter setting of 29.92 inches, and the altimeter will indicate pressure altitude. (H808) — FAA-H-8083-15, Chapter 3

Answers

| 4090 [B] | 4091 [C] | 4109 [B] | 4111 [A] | 4477 [A] | 4478 [A] |

Chapter 3 **Flight Instruments**

ALL
4479. Which altitude is indicated when the altimeter is set to 29.92" Hg?

A—Density.
B—Pressure.
C—Standard.

With the altimeter set at the standard altimeter setting of 29.92 inches, the altimeter will indicate pressure altitude. (H808) — FAA-H-8083-15, Chapter 3

ALL
4911. At an altitude of 6,500 feet MSL, the current altimeter setting is 30.42" Hg. The pressure altitude would be approximately

A—7,500 feet.
B—6,000 feet.
C—6,500 feet.

Pressure altitude is the altitude indicated when the altimeter setting window is adjusted to 29.92" Hg. (the theoretical Standard Datum Plane). To find pressure altitude, follow these steps:

1. *Find the difference between the given pressure and standard pressure:*

   30.42 – 29.92 = 0.50" Hg.

2. *Find the change in indicated altitude. The Standard Pressure Lapse Rate is about 1 inch per 1,000 feet:*

   0.50 x 1000 = 500 feet difference

3. *Find pressure altitude. The Standard Datum Plane pressure is lower than the current station setting and therefore, pressure altitude will also be lower:*

   *6,500 feet MSL – 500 feet =
   6,000 feet Pressure Altitude*

(H808) — FAA-H-8083-15, Chapter 3

ALL
4912. The pressure altitude at a given location is indicated on the altimeter after the altimeter is set to

A—the field elevation.
B—29.92" Hg.
C—the current altimeter setting.

Pressure Altitude is the altitude read on the altimeter when the instrument is adjusted to indicate height above the Standard Datum Plane. The Standard Datum Plane is a theoretical level where the weight of the atmosphere is 29.92" Hg. (H808) — FAA-H-8083-15, Chapter 3

Answer (A) is incorrect because the altimeter is set to standard pressure, not field elevation. Answer (C) is incorrect because the current altimeter setting would give indicated, not pressure, altitude.

ALL
4913. If the outside air temperature increases during a flight at constant power and at a constant indicated altitude, the true airspeed will

A—decrease and true altitude will increase.
B—increase and true altitude will decrease.
C—increase and true altitude will increase.

As temperature increases, pressure levels increase. An increase in pressure causes true airspeed and true altitude to increase. (H808) — FAA-H-8083-15, Chapter 3

Answer (A) is incorrect because true airspeed will increase (not decrease). Answer (B) is incorrect because true altitude will increase (not decrease).

ALL
4922. Altimeter setting is the value to which the scale of the pressure altimeter is set so the altimeter indicates

A—pressure altitude at sea level.
B—true altitude at field elevation.
C—pressure altitude at field elevation.

The altimeter setting broadcast by each Flight Service Station is a computed correction for nonstandard surface pressure only, for a specific location and elevation. Consequently, altimeter indications based upon a local altimeter setting indicate true altitude at field elevation. (I04) — FAA-H-8083-15

Answers (A) and (C) are incorrect because the altimeter setting causes the altimeter to indicate true (not pressure) altitude at field (not sea) level.

ALL
4923. Pressure altitude is the altitude read on your altimeter when the instrument is adjusted to indicate height above

A—sea level.
B—the standard datum plane.
C—ground level.

Pressure altitude is the altitude read on your altimeter when the instrument is adjusted to indicate height above the Standard Datum Plane. The Standard Datum Plane is a theoretical level where the weight of the atmosphere is 29.92" of mercury as measured by a barometer. As atmospheric pressure changes, the Standard Datum Plane may be below, at, or above sea level. Pressure altitude is important as a basis for determining aircraft performance as well as for assigning flight levels to aircraft operating at high altitude. (H808) — FAA-H-8083-15, Chapter 3

Answers

| 4479 | [B] | 4911 | [B] | 4912 | [B] | 4913 | [C] | 4922 | [B] | 4923 | [B] |

Chapter 3 **Flight Instruments**

Altimeter Settings

Before departure, the pilot should obtain the altimeter setting given for the local airport. En route, the pilot should set the altimeter to the closest station within 100 NM. The pilot will be periodically advised by Air Traffic Control (ATC) of the proper altimeter setting.

At or above 18,000 feet, all aircraft use 29.92" Hg. The change to or from 29.92" Hg should be made when passing through 18,000 feet. If this change is not made, the altimeter will read erroneously.

Aside from incorrectly setting an altimeter, one of the most common errors is to misread it. The small hand indicates thousand of feet. The long, thin hand indicates hundreds of feet. The very thin hand with the triangular tip indicates tens of thousand of feet. *See* Figure 3-4.

To preflight the altimeter, the current reported altimeter setting should be set in the Kollsman window. Note any variation between the known field elevation and the altimeter indication. If the variation exceeds ±75 feet, the accuracy of the altimeter is questionable and the problem should be referred to an appropriately-rated repair station for evaluation and possible correction.

ALL
4093. When an altimeter is changed from 30.11" Hg to 29.96" Hg, in which direction will the indicated altitude change and by what value?

A—Altimeter will indicate 15 feet lower.
B—Altimeter will indicate 150 feet lower.
C—Altimeter will indicate 150 feet higher.

The altimeter indication changes 1,000 feet for every one inch change in pressure. When the pressure setting is changed to a lower number, the altitude indicated on the altimeter will decrease. The altimeter setting in this question has changed 0.15 (30.11 − 29.96 = 0.15), which will cause a 150-foot change in the altimeter reading (.15 x 1,000). (I22) — AC 00-6A, page 12

Answers (A) and (C) are incorrect because a .15" pressure change would cause the altimeter to indicate 150 (not 15) feet lower (not higher).

ALL
4445. En route at FL290, the altimeter is set correctly, but not reset to the local altimeter setting of 30.57" Hg during descent. If the field elevation is 650 feet and the altimeter is functioning properly, what is the approximate indication upon landing?

A—715 feet.
B—1,300 feet.
C—Sea level.

En route at FL290, the pilot would have the altimeter set to 29.92" Hg, since the altitude is at or above FL180. If the pilot failed to reset it to the local altimeter setting of 30.57 during descent, the altimeter will not indicate field elevation upon landing:

Since 30.57
      − 29.92
      ───────
        0.65

and since 1 inch per 1,000 feet equals 0.1 inch per 100 feet and 0.01 inch per 10 feet, the altimeter will, in this case, indicate an error of 650 feet. Because the flight went from a low-pressure area (29.92) to a higher-pressure area (30.57), the altimeter error is subtracted from the field elevation of 650 feet.

650 − 650 = 0 feet, or the altimeter will indicate sea level

(J26) — FAA-H-8083-15

Figure 3-4. Altimeter

Answers
4093 [B]   4445 [C]

2002 Instrument Test Prep ASA **3–9**

Chapter 3 **Flight Instruments**

ALL
4481. En route at FL290, your altimeter is set correctly, but you fail to reset it to the local altimeter setting of 30.26" Hg during descent. If the field elevation is 134 feet and your altimeter is functioning properly, what will it indicate after landing?

A—100 feet MSL.
B—474 feet MSL.
C—206 feet below MSL.

En route at FL290, the pilot would have the altimeter set to 29.92 inches, since the altitude is above FL180. Since the pilot failed to reset it to the local altimeter setting of 30.26 during descent, the altimeter will not indicate field elevation upon landing. Since:

$$\begin{array}{r} 30.26 \\ -29.92 \\ \hline 0.34 \end{array}$$

and one inch per 1,000 feet equals 0.1 inch per 100 feet and 0.01 inch per 10 feet, the altimeter will, in this case, indicate an error of 340 feet.

Because the flight went from a low-pressure area (29.92) to a higher-pressure area (30.26), the altimeter error is subtracted from the field elevation of 134 feet.

$$\begin{array}{rl} 134 & \text{airport elevation} \\ -340 & \text{error} \\ \hline -206 & \text{feet below sea level} \end{array}$$

(J26) — FAA-H-8083-15

ALL
4110. Which of the following defines the type of altitude used when maintaining FL210?

A—Indicated.
B—Pressure.
C—Calibrated.

At or above 18,000 feet MSL, the altimeter is set to 29.92" Hg. You can always determine pressure altitude from your altimeter whether in flight or on the ground. Simply set the altimeter to 29.92 inches, and the altimeter indicates pressure altitude. (I22) — AC 00-6A, page 19

Answer (A) is incorrect because indicated altitude is what is read from the altimeter at any setting. Answer (C) is incorrect because all altimeters are calibrated.

ALL
4444. What is the procedure for setting the altimeter when assigned an IFR altitude of 18,000 feet or higher on a direct flight off airways?

A—Set the altimeter to 29.92" Hg before takeoff.
B—Set the altimeter to the current altimeter setting until reaching the assigned altitude, then set to 29.92" Hg.
C—Set the altimeter to the current reported setting for climb-out and 29.92" Hg upon reaching 18,000 feet.

The cruising altitude or flight level of aircraft shall be maintained by reference to an altimeter which shall be set, when operating at or above 18,000 feet MSL, to 29.92" Hg (standard setting). (J26) — AIM ¶7-2-2

Answer (A) is incorrect because the local altimeter settings or indicated altitude should be used up to FL180. Answer (B) is incorrect because the altimeter should be set to 29.92 upon reaching FL180 (not the assigned altitude).

ALL
4446. While you are flying at FL250, you hear ATC give an altimeter setting of 28.92" Hg in your area. At what pressure altitude are you flying?

A—24,000 feet.
B—25,000 feet.
C—26,000 feet.

Pressure altitude is the altitude indicated when the altimeter setting window is adjusted to 29.92. Because the pilot is flying at FL250 the altimeter will be set at 29.92" Hg (since the pilot is above FL180). Consequently the pressure altitude will be 25,000 feet. (J26) — FAA-H-8083-15

Answers (A) and (C) are incorrect because FL250 implies a pressure altitude of 25,000 feet, by definition.

ALL
4480. If you are departing from an airport where you cannot obtain an altimeter setting, you should set your altimeter

A—on 29.92" Hg.
B—on the current airport barometric pressure, if known.
C—to the airport elevation.

In the case of an aircraft not equipped with a radio, set the altimeter to the elevation of the departure airport or use an appropriate altimeter setting available prior to departure. (J26) — 14 CFR §91.121

Answers

| 4481 | [C] | 4110 | [B] | 4444 | [C] | 4446 | [B] | 4480 | [C] |

ALL
4482. How does a pilot normally obtain the current altimeter setting during an IFR flight in Class E airspace below 18,000 feet?

A—The pilot should contact ARTCC at least every 100 NM and request the altimeter setting.
B—FSS's along the route broadcast the weather information at 15 minutes past the hour.
C—ATC periodically advises the pilot of the proper altimeter setting.

For flight below 18,000 feet MSL, the altimeter should be set to the current reported altimeter setting of a station along the route, and within 100 NM of the aircraft, or if there is no station within this area, the current reported altimeter setting of an appropriate available station. When an aircraft is en route on an instrument flight plan, ATC will furnish this information to the pilot at least once while the aircraft is in its area of jurisdiction. (J26) — AIM ¶7-2-2(a)

Answer (A) is incorrect because ATC provides the altimeter setting without being asked. Answer (B) is incorrect because the FSS is not responsible for broadcasting altimeter settings.

ALL
4910. The local altimeter setting should be used by all pilots in a particular area, primarily to provide for

A—the cancellation of altimeter error due to nonstandard temperatures aloft.
B—better vertical separation of aircraft.
C—more accurate terrain clearance in mountainous areas.

The altimeter setting system provides the means that must be used to correct the altimeter for pressure variations. The system is necessary to ensure safe terrain clearance for instrument approaches and landings, and to maintain vertical separation between aircraft during instrument weather conditions. (H312) — AC 61-23C, Chapter 3

Answer (A) is incorrect because a local altimeter setting does not cancel error due to nonstandard temperatures aloft. Answer (C) is incorrect because a local altimeter setting will not provide more accurate terrain clearance since it does not cancel error due to nonstandard temperatures aloft.

ALL
4402. How should you preflight check the altimeter prior to an IFR flight?

A—Set the altimeter to 29.92" Hg. With current temperature and the altimeter indication, determine the true altitude to compare with the field elevation.
B—Set the altimeter first with 29.92" Hg and then the current altimeter setting. The change in altitude should correspond to the change in setting.
C—Set the altimeter to the current altimeter setting. The indication should be within 75 feet of the actual elevation for acceptable accuracy.

With the altimeter set to the current reported altimeter setting, note any variation between the known field elevation and the altimeter indication. If the variation exceeds ±75 feet, the accuracy of the altimeter is questionable and the problem should be referred to an appropriately-rated repair station for evaluation and possible correction. (J26) — AIM ¶7-2-1

Answer (A) is incorrect because setting the altimeter to 29.92 and adjusting for temperature give only the density (not true) altitude. Answer (B) is incorrect because only the indicated altitude (not change in altitude) is checked.

ALL
4880. How should you preflight check the altimeter prior to an IFR flight?

A—Set the altimeter to the current temperature. With current temperature and the altimeter indication, determine the calibrated altitude to compare with the field elevation.
B—Set the altimeter first with 29.92" Hg and then the current altimeter setting. The change in altitude should correspond to the change in setting.
C—Set the altimeter to the current altimeter setting. The indication should be within 75 feet of the actual elevation for acceptable accuracy.

With the altimeter set to the current reported altimeter setting, note any variation between the known field elevation and the altimeter indication. If the variation exceeds ±75 feet, the accuracy of the altimeter is questionable and the problem should be referred to an appropriately-rated repair station for evaluation and possible correction. (H812) — FAA-H-8083-15, Chapter 3

Answer (A) is incorrect because setting the altimeter to 29.92 and adjusting for temperature would indicate density (not true) altitude. Answer (B) is incorrect because only the indicated altitude (not change in altitude) is checked.

Answers

| 4482 | [C] | 4910 | [B] | 4402 | [C] | 4880 | [C] |

Chapter 3 Flight Instruments

ALL
4483. (Refer to Figure 83.) Which altimeter depicts 12,000 feet?

A—2.
B—3.
C—4.

The small hand indicates thousands of feet. The long, thin hand indicates hundreds of feet. The very thin hand with the triangular tip indicates tens of thousands of feet. An altimeter indicating 12,000 feet has the small hand on the 2, the long, thin hand on the 0, and the very thin hand with the triangular tip past the 1. (H808) — FAA-H-8083-15, Chapter 3

ALL
4484. (Refer to Figure 84.) Which altimeter depicts 8,000 feet?

A—1.
B—2.
C—3.

The small hand indicates thousands of feet. The long, thin hand indicates hundreds of feet. The very thin hand with the triangular tip indicates tens of thousands of feet. An altimeter indicating 8,000 feet has the small hand on the 8, the long, thin hand on the 0, and the very thin hand with the triangular tip before the 1. (H808) — FAA-H-8083-15, Chapter 3

Vertical Speed Indicator

The Vertical Speed Indicator (VSI) is designed to indicate the rate of climb or descent. Although not required by regulation to be on board, it is commonly found on instrument airplanes. A VSI pre-takeoff check is made to check for a "zero" reading. If it is indicating a climb or descent, note the error and apply it to readings in flight. *See* Figure 3-5.

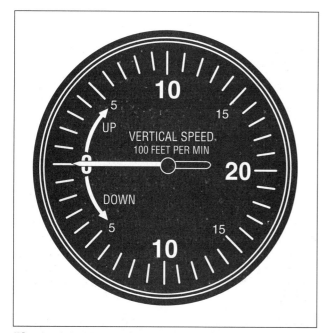

Figure 3-5. Vertical Speed Indicator (VSI)

ALL
4056. You check the flight instruments while taxiing and find that the vertical speed indicator (VSI) indicates a descent of 100 feet per minute. In this case, you

A—must return to the parking area and have the instrument corrected by an authorized instrument repairman.
B—may take off and use 100 feet descent as the zero indication.
C—may not take off until the instrument is corrected by either the pilot or a mechanic.

The needle of the vertical velocity indicator should indicate zero when the aircraft is on the ground or maintaining a constant pressure level in flight. If this adjustment cannot be made, you must allow for the error when interpreting the indications in flight. Since the VSI is not a required instrument, there are no regulation requirements concerning calibration. (H814) — FAA-H-8083-15, Chapter 5

Answers (A) and (C) are incorrect because the VSI may be used by noting any discrepancies and accommodating for them in flight. A VSI is not a required for instrument flight, so even if it was completely inaccurate, the flight could still proceed.

Answers
4483 [C]   4484 [B]   4056 [B]

Attitude Indicator

Gyroscopes (gyros) exhibit two important principles: rigidity in space, and precession. There are three instruments controlled by gyroscopes: attitude indicator, turn coordinator, and the heading indicator.

The attitude indicator provides an immediate, direct and corresponding indication of any change of aircraft pitch and bank attitude in relation to the natural horizon. It is the pilot's primary instrument during transitions of pitch or bank attitudes. *See* Figure 3-6.

The attitude indicator exhibits several inherent errors:

- Following a 180° turn, it indicates a slight climb and bank in the opposite direction (turn error). Following a 360° turn, this error cancels itself out and is not apparent.
- During acceleration, the horizon bar moves down, indicating a climb (acceleration error).
- During deceleration, the horizon bar moves up, indicating a descent (deceleration error).

An attitude indicator pre-takeoff check verifies that the horizon bar stabilizes within five minutes, and does not dip more than 5° during taxiing turns.

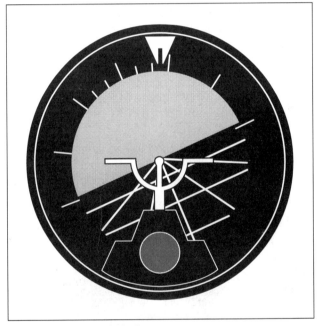

Figure 3-6. Attitude indicator

RTC
4861. The primary reason the pitch attitude must be increased, to maintain a constant altitude during a co-ordinated turn, is because the

A— use of pedals has increased the drag.
B— vertical component of lift has decreased as the result of the bank.
C— thrust is acting in a different direction, causing a reduction in airspeed and loss of lift.

The primary reason the pitch attitude must be increased, to maintain a constant altitude during a coordinated turn, is because the vertical component of lift has decreased (and horizontal component of lift increased) as a result of the bank. (H825) — FAA-H-8083-15, Chapter 6

ALL
4902. One characteristic that a properly functioning gyro depends upon for operation is the

A— ability to resist precession 90° to any applied force.
B— resistance to deflection of the spinning wheel or disc.
C— deflecting force developed from the angular velocity of the spinning wheel.

Any rotating body exhibits gyroscopic properties according to Newton's laws of motion. The first law states: a body at rest will remain at rest; or if in motion in a straight line, it will continue in motion in a straight line unless acted upon by an outside force. The second law states: the deflection of a moving body is proportional to the deflective force applied and is inversely proportional to its weight and speed. A gyro is a wheel or disc designed to utilize these principles. (H810) — FAA-H-8083-15, Chapter 3

Answer (A) is incorrect because a gyro uses (not resists) precession. Answer (C) is incorrect because deflecting forces are applied (not developed) to gyros.

Answers
4861 [B]   4902 [B]

ALL
4901. If a 180° steep turn is made to the right and the aircraft is rolled out to straight-and-level flight by visual references, the attitude indicator

A—should immediately show straight-and-level flight.
B—will show a slight skid and climb to the right.
C—may show a slight climb and turn.

Attitude indicators are free from most errors, but there is a possibility of a small bank angle and pitch error after a 180° turn. (H810) — FAA-H-8083-15, Chapter 3

ALL
4820. As a rule of thumb, altitude corrections of less than 100 feet should be corrected by using a

A—full bar width on the attitude indicator.
B—half bar width on the attitude indicator.
C—two bar width on the attitude indicator.

As a rule-of-thumb, for errors less than 100 feet, use a half-bar-width correction. (H814) — FAA-H-8083-15, Chapter 5

ALL
4835. Which condition during taxi is an indication that an attitude indicator is unreliable?

A—The horizon bar tilts more than 5° while making taxi turns.
B—The horizon bar vibrates during warmup.
C—The horizon bar does not align itself with the miniature airplane after warmup.

If the attitude indicator's horizon bar erects to the horizontal position and remains at the correct position for the attitude of the airplane, or if it begins to vibrate after this attitude is reached and then slowly stops vibrating altogether, the instrument is operating properly. If the horizon bar fails to remain in the horizontal position during straight taxiing, or tips in excess of 5° during taxi turns, the instrument is unreliable. (H812) — FAA-H-8083-15, Chapter 3

Answer (B) is incorrect because during warm-up the horizon bar will normally vibrate as the gyros gather speed. Answer (C) is incorrect because the horizon bar will align itself with the center of the dial thus indicating level flight. At this point, the miniature airplane can be adjusted to align with the horizon bar from the pilot's viewpoint.

ALL
4842. What pretakeoff check should be made of the attitude indicator in preparation for an IFR flight?

A—The horizon bar does not vibrate during warmup.
B—The miniature airplane should erect and become stable within 5 minutes.
C—The horizon bar should erect and become stable within 5 minutes.

If the horizon bar erects to the horizontal position and remains at the correct position for the attitude of the airplane, or if it begins to vibrate after this attitude is reached and then slowly stops vibrating altogether, the instrument is operating properly. If the horizon bar fails to remain in the horizontal position during straight taxiing, or tips in excess of 5° during taxi turns, the instrument is unreliable. (H812) — FAA-H-8083-15, Chapter 3

Answer (A) is incorrect because during warm-up, the horizon will normally vibrate as the gyros begin to gather speed. Answer (B) is incorrect because the miniature airplane can be set manually.

ALL
4857. During normal operation of a vacuum-driven attitude indicator, what attitude indication should you see when rolling out from a 180° skidding turn to straight-and-level coordinated flight?

A—A straight-and-level coordinated flight indication.
B—A nose-high indication relative to level flight.
C—The miniature aircraft shows a turn in the direction opposite the skid.

After return of the aircraft to straight-and-level, coordinated flight, the miniature aircraft shows a turn in the direction opposite the skid. (H810) — FAA-H-8083-15, Chapter 3

Answer (A) is incorrect because the attitude indicator shows a small turn in the opposite direction and does not reflect coordinated flight. Answer (B) is incorrect because a nose-high indication would be found when rolling out of a coordinated turn.

Answers

| 4901 | [C] | 4820 | [B] | 4835 | [A] | 4842 | [C] | 4857 | [C] |

ALL
4860. During normal coordinated turns, what error due to precession should you observe when rolling out to straight-and-level flight from a 180° steep turn to the right?

A—A straight-and-level coordinated flight indication.
B—The miniature aircraft would show a slight turn indication to the left.
C—The miniature aircraft would show a slight descent and wings-level attitude.

If a 180° steep turn is made to the right and the aircraft is rolled-out to straight-and-level flight by visual references, the miniature aircraft will show a slight climb and turn to the left. (H810) — FAA-H-8083-15, Chapter 3

Answer (A) is incorrect because the attitude indicator will show a slight climb to the left and does not reflect coordinated flight. Answer (C) is incorrect because it will show a slight climb and a turning error.

ALL
4900. Errors in both pitch and bank indication on an attitude indicator are usually at a maximum as the aircraft rolls out of a

A—180° turn.
B—270° turn.
C—360° turn.

Errors in both pitch and bank indications occur during normal coordinated turns. These errors are caused by the movement of the pendulous vanes by centrifugal force, resulting in the precession of the gyro towards the inside of the turn. The error is greatest in a 180° steep turn. (H810) — FAA-H-8083-15, Chapter 3

ALL
4918. When an aircraft is accelerated, some attitude indicators will precess and incorrectly indicate a

A—climb.
B—descent.
C—right turn.

Acceleration and deceleration induce precession errors, depending upon the amount and extent of the force applied. During acceleration, the horizon bar moves down, indicating a climb. (H810) — FAA-H-8083-15, Chapter 3

ALL
4919. When an aircraft is decelerated, some attitude indicators will precess and incorrectly indicate a

A—left turn.
B—climb.
C—descent.

Acceleration and deceleration induce precession errors, depending upon the amount and extent of the force applied. During deceleration, the horizon bar moves up, indicating a descent. (H810) — FAA-H-8083-15, Chapter 3

ALL
4928. While cruising at 160 knots, you wish to establish a climb at 130 knots. When entering the climb (full panel), it is proper to make the initial pitch change by increasing back elevator pressure until the

A—attitude indicator, airspeed, and vertical speed indicate a climb.
B—vertical speed indication reaches the predetermined rate of climb.
C—attitude indicator shows the approximate pitch attitude appropriate for the 130-knot climb.

To enter a constant airspeed climb from cruising airspeed, raise the miniature aircraft on the attitude indicator to the approximate nose-high indication appropriate to the predetermined climb speed. The attitude will vary according to the type of airplane you are flying. (H815) — FAA-H-8083-15, Chapter 5

Answers (A) and (B) are incorrect because initial pitch changes should be made using the attitude indicator.

ALL
4929. While cruising at 190 knots, you wish to establish a climb at 160 knots. When entering the climb (full panel), it would be proper to make the initial pitch change by increasing back elevator pressure until the

A—attitude indicator shows the approximate pitch attitude appropriate for the 160-knot climb.
B—attitude indicator, airspeed, and vertical speed indicate a climb.
C—airspeed indication reaches 160 knots.

To enter a constant airspeed climb from cruising airspeed, raise the miniature aircraft on the attitude indicator to the approximate nose-high indication appropriate to the predetermined climb speed. The attitude will vary according to the type of airplane you are flying. (H815) — FAA-H-8083-15, Chapter 5

Answers (B) and (C) are incorrect because initial pitch changes should be made using the attitude indicator.

Answers

| 4860 | [B] | 4900 | [A] | 4918 | [A] | 4919 | [C] | 4928 | [C] | 4929 | [A] |

RTC
4841. Which initial pitch attitude change on the attitude indicator should be made to correct altitude while at normal cruise in a helicopter?

A—Two bar width.
B—One and one-half bar width.
C—One bar width.

When making initial pitch attitude corrections to maintain altitude, the changes of attitude should be small and smoothly applied. The initial movement of the horizon bar should not exceed one bar high or low. If further change is required, an additional correction of one-half bar will normally correct any deviation from the desired altitude. This correction (one and one-half bars) is normally the maximum. (H822) — FAA-H-8083-15, Chapter 6

RTC
4846. During the initial acceleration on an instrument takeoff in a helicopter, what flight attitude should be established on the attitude indicator?

A—Level flight attitude.
B—Two bar widths low.
C—One bar width high.

During the initial acceleration, the pitch attitude of the helicopter, as read on the attitude indicator, should be one to two bar widths low. (H828) — FAA-H-8083-15, Chapter 6

RTC
4852. During a stabilized autorotation, approximately what flight attitude should be established on the attitude indicator?

A—Two bar widths below the artificial horizon.
B—A pitch attitude that will give an established rate of descent of not more than 500 feet per minute.
C—Level flight attitude.

The pitch attitude of the helicopter should be approximately level as shown by the attitude indicator. (H827) — FAA-H-8083-15, Chapter 6

Turn Coordinator

The turn coordinator is a gyroscopically-operated instrument that is designed to show roll rate, rate of turn, and quality of turn. It acts as a backup system in case of a failure of the vacuum powered attitude indicator. *See* Figure 3-7.

Before starting the engine, the turn needle should be centered and the race full of fluid. During a taxiing turn, the needle will indicate a turn in the proper direction and the ball will show a skid.

An airplane turns because of the horizontal component of lift in a banked attitude. The greater the horizontal lift at any airspeed, the greater the rate of turn. The angle of attack must be increased to maintain altitude during a turn because the vertical component of lift decreases as a result of the bank. When airspeed is decreased in a turn, either a decrease in the bank angle or an increase in the angle of attack is required to maintain level flight. During a constant bank level turn, an increase in airspeed would result in a decrease in rate of turn and an increase in turn radius.

In a coordinated turn, horizontal lift and centrifugal force are equal. In a skid, the rate of turn is too great for the angle of bank, and excessive centrifugal force causes the ball to move to the outside of the turn. To correct to coordinated flight, the pilot should increase the bank or decrease the rate of turn, or a combination of both. In a slip, the rate of turn is too slow for the angle of bank, and the lack of centrifugal force causes the ball to move to the inside of the turn. To return to coordinated flight, the pilot needs to decrease the bank or increase the rate of turn, or a combination of both.

A standard rate turn (3°/second) takes 2 minutes to complete a 360° turn. A half-standard rate turn (1.5°/second) takes 4 minutes to complete a 360° turn.

Answers
4841 [C]   4846 [B]   4852 [C]

Figure 3-7. Turn coordinator

ALL
4882. Prior to starting an engine, you should check the turn-and-slip indicator to determine if the

A—needle indication properly corresponds to the angle of the wings or rotors with the horizon.
B—needle is approximately centered and the tube is full of fluid.
C—ball will move freely from one end of the tube to the other when the aircraft is rocked.

The following are items that should be checked before starting the engine(s):

1. *Check turn-and-slip indicator and magnetic compass for fluid level (should be full).*

2. *If instruments are electrical, turn on and listen for any unusual or irregular mechanical noise.*

3. *Check instruments for poor condition, mounting, marking, broken or loose knobs. Also check the power-off indications of the instrument pointers and warning flags.*

(L59) — AC 91-46(5)(a)(2)

Answer (A) is incorrect because the needle is related to rate and direction of turn (not wings or rotors). Answer (C) is incorrect because the ball is checked during taxi and it is not necessary to rock the aircraft.

ALL
4831. What indication should be observed on a turn coordinator during a left turn while taxiing?

A—The miniature aircraft will show a turn to the left and the ball remains centered.
B—The miniature aircraft will show a turn to the left and the ball moves to the right.
C—Both the miniature aircraft and the ball will remain centered.

When an aircraft makes a taxiing left turn, the turn coordinator will show the same indications as a level, no-bank left turn in flight. Those indications are: The miniature airplane will show a left turn (the direction of turn and rate) while the ball will show an uncoordinated turn (a skid), by moving to the right. (H812) — FAA-H-8083-15, Chapter 3

Answer (A) is incorrect because the ball would move to the outside of the turn due to centrifugal force. Answer (C) is incorrect because the miniature aircraft will show a turn to the left and the ball will move to the right.

ALL
4883. What indications should you observe on the turn-and-slip indicator during taxi?

A—The ball moves freely opposite the turn, and the needle deflects in the direction of the turn.
B—The needle deflects in the direction of the turn, but the ball remains centered.
C—The ball deflects opposite the turn, but the needle remains centered.

You should observe the following on the turn-and-slip indicator during taxi:

1. *Check and ensure the turn needle indicates proper direction of turn.*

2. *Check the ball for freedom of movement in the glass tube. Centrifugal force causes the ball to move to the outside of the turn.*

(L59) — AC 91-46, page 89

Answer (B) is incorrect because the ball will move opposite to the direction of the turn. Answer (C) is incorrect because the needle moves in the direction of the turn.

Answers

4882   [B]   4831   [B]   4883   [A]

Chapter 3 **Flight Instruments**

ALL
4839. What does the miniature aircraft of the turn coordinator directly display?

A—Rate of roll and rate of turn.
B—Angle of bank and rate of turn.
C—Angle of bank.

The miniature aircraft of the turn coordinator displays rate of roll and rate of turn. (H810) — FAA-H-8083-15, Chapter 3

Answers (B) and (C) are incorrect because the miniature aircraft of the turn coordinator does not directly display the bank angle of the aircraft.

ALL
4847. What indications are displayed by the miniature aircraft of a turn coordinator?

A—Rate of roll and rate of turn.
B—Direct indication of bank angle and pitch attitude.
C—Indirect indication of bank angle and pitch attitude.

The miniature aircraft of the turn coordinator displays only rate-of-roll and rate-of-turn. It does not directly display the bank angle of the aircraft. (H810) — FAA-H-8083-15, Chapter 3

Answer (B) is incorrect because the turn coordinator is an indirect indication of bank angle and is not related to pitch attitude. Answer (C) is incorrect because the turn coordinator does not reflect pitch attitude.

ALL
4856. What indication is presented by the miniature aircraft of the turn coordinator?

A—Indirect indication of the bank attitude.
B—Direct indication of the bank attitude and the quality of the turn.
C—Quality of the turn.

The miniature aircraft of the turn coordinator displays only rate-of-roll and rate-of-turn. It only indirectly displays the bank angle of the aircraft. (H816) — FAA-H-8083-15, Chapter 5

Answer (B) is incorrect because the turn coordinator does not provide a direct indication of bank. Answer (C) is incorrect because the ball in the turn coordinator displays the quality of the turn.

ALL
4884. Which instrument indicates the quality of a turn?

A—Attitude indicator.
B—Heading indicator or magnetic compass.
C—Ball of the turn coordinator.

The ball of the turn coordinator is a balance indicator, and is used as a visual aid to determine coordinated use of the aileron and rudder control. During a turn it indicates the relationship between the angle of bank and rate-of-turn. It indicates the "quality" of the turn, or whether the aircraft has the correct angle of bank for the rate-of-turn. When the ball is centered, the turn is coordinated, and neither slipping nor skidding. (H807) — FAA-H-8083-15, Chapter 2

Answer (A) is incorrect because the attitude indicator provides both pitch and bank information. Answer (B) is incorrect because the heading indicator and/or magnetic compass indicates the current or change in direction.

ALL
4870-1. What force causes an airplane to turn?

A—Rudder pressure or force around the vertical axis.
B—Vertical lift component.
C—Horizontal lift component.

An aircraft requires a sideward force to make it turn. In a normal turn, this force is supplied by banking the aircraft so that lift is exerted inward as well as upward. The horizontal lift component is the sideward force that causes the aircraft to turn. (H807) — FAA-H-8083-15, Chapter 2

Answer (A) is incorrect because rudder pressure is used to coordinate the airplane when it is banked. Answer (B) is incorrect because the vertical component of lift will act against weight, which affects altitude.

RTC
4870-2. What force causes a helicopter to turn?

A—Rudder pressure or force around the vertical axis.
B—Vertical lift component.
C—Horizontal lift component.

When the helicopter is banked, the rotor disc is tilted sideward resulting in lift being separated into two components. Lift acting vertically, in opposition to weight, is the vertical component of lift. Lift acting horizontally, in opposition to inertia (centrifugal force), is the horizontal component of lift (centripetal force). (H703) — FAA-H-8083-21, Chapter 3

Answers
4839 [A] 4847 [A] 4856 [A] 4884 [C] 4870-1 [C] 4870-2 [C]

ALL
4843. The rate of turn at any airspeed is dependent upon

A—the horizontal lift component.
B—the vertical lift component.
C—centrifugal force.

The rate-of-turn at any given airspeed depends on the amount of sideward force causing the turn; that is, the horizontal lift component, which varies directly in proportion to bank in a correctly executed turn. (H807) — FAA-H-8083-15, Chapter 2

Answer (B) is incorrect because the vertical lift component determines altitude and change in altitude. Answer (C) is incorrect because centrifugal force will act against the horizontal lift component.

ALL
4833-1. When airspeed is decreased in a turn, what must be done to maintain level flight?

A—Decrease the angle of bank and/or increase the angle of attack.
B—Increase the angle of bank and/or decrease the angle of attack.
C—Increase the angle of attack.

The angle of bank necessary for a given rate-of-turn is proportional to the true airspeed. Since the turns are executed at standard rate, the angle of bank must be varied in direct proportion to the airspeed change in order to maintain a constant rate-of-turn. During a reduction of airspeed, you must decrease the angle of bank and increase the pitch attitude to maintain altitude and a standard rate turn. (H816) — FAA-H-8083-15, Chapter 5

Answer (B) is incorrect because the increased vertical lift required must come from a decrease in angle of bank and/or increase in angle of attack. Answer (C) is incorrect because it is possible to decrease the angle of bank as well as increase the angle of attack.

ALL
4833-2. When airspeed is decreased in a turn, what must be done to maintain level flight?

A—Increase the pitch attitude and/or increase the angle of bank.
B—Increase the angle of bank and/or decrease the pitch attitude.
C—Decrease the angle of bank and/or increase the pitch attitude.

During a reduction of airspeed, you must decrease the angle of bank and increase the pitch attitude to maintain altitude and a standard-rate turn. (H825) — FAA-H-8083-15, Chapter 6

ALL
4844. During a skidding turn to the right, what is the relationship between the component of lift, centrifugal force, and load factor?

A—Centrifugal force is less than horizontal lift and the load factor is increased.
B—Centrifugal force is greater than horizontal lift and the load factor is increased.
C—Centrifugal force and horizontal lift are equal and the load factor is decreased.

A skidding turn results from excess centrifugal force over the horizontal lift component, pulling the aircraft toward the outside of the turn. As centrifugal force increases, the load factor also increases. See the figure on the next page. (H807) — FAA-H-8083-15, Chapter 2

Answer (A) is incorrect because a slipping turn will occur if centrifugal force is less than horizontal lift. Answer (C) is incorrect because centrifugal force and horizontal lift are equal in a coordinated turn and load factor will increase.

Answers

4843 [A]   4833-1 [A]   4833-2 [C]   4844 [B]

Chapter 3 **Flight Instruments**

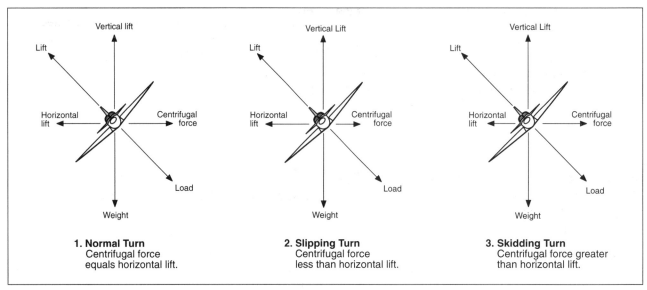

Question 4844

ALL
4868. What is the relationship between centrifugal force and the horizontal lift component in a coordinated turn?

A—Horizontal lift exceeds centrifugal force.
B—Horizontal lift and centrifugal force are equal.
C—Centrifugal force exceeds horizontal lift.

In a coordinated turn, horizontal lift equals centrifugal force. This is indicated by the ball on the turn-and-slip indicator being centered. See the figure below. (H807) — FAA-H-8083-15, Chapter 2

Answer (A) is incorrect because if horizontal lift exceeds centrifugal force, a slipping turn will occur. Answer (C) is incorrect because if centrifugal force exceeds horizontal lift, a skidding turn will occur.

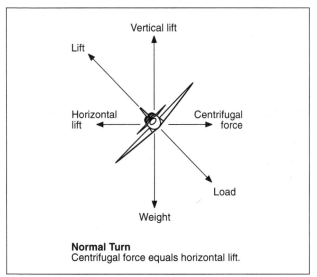

Question 4868

ALL
4878-1. When airspeed is increased in a turn, what must be done to maintain a constant altitude?

A—Decrease the angle of bank.
B—Increase the angle of bank and/or decrease the angle of attack.
C—Decrease the angle of attack.

If the airspeed is increased in a turn, the angle of attack must be decreased and/or the angle of bank increased in order to maintain level flight. As airspeed is increased in a constant-rate level turn, both the radius of turn and centrifugal force increase. This increase in centrifugal force must be balanced by an increase in the horizontal lift component, which can be accomplished only by increasing the angle of bank. Thus, to maintain a turn at a constant rate, the angle of bank must be varied with changes in airspeed. (H816) — FAA-H-8083-15, Chapter 5

Answer (A) is incorrect because the angle of bank must be increased. Answer (C) is incorrect because the angle of bank could also be increased.

ALL
4878-2. During standard-rate turns, which instrument is considered "primary" for bank?

A—Heading indicator.
B—Turn and slip indicator or turn coordinator.
C—Attitude indicator.

Answers

4868 [B]   4878-1 [B]   4878-2 [B]

On the roll-in, use the attitude indicator to establish the approximate angle of bank, then check the turn coordinator's miniature aircraft for a standard-rate turn indication. Maintain the bank for this rate of turn, using the turn coordinator's miniature aircraft as the primary bank reference and the attitude indicator as the supporting bank instrument. (H816) — FAA-H-8083-15, Chapter 5

RTC
4878-3. When airspeed is increased in a turn, what must be done to maintain a constant altitude?

A—Decrease the angle of attack.
B—Increase the angle of bank and/or decrease the pitch attitude.
C—Decrease the angle of bank.

During an increase in airspeed, you must increase the angle of bank and decrease the pitch attitude to maintain altitude and a standard-rate turn. (H825) — FAA-H-8083-15, Chapter 6

ALL
4895. If a half-standard rate turn is maintained, how long would it take to turn 360°?

A—1 minute.
B—2 minutes.
C—4 minutes.

A standard rate turn (3°/second) takes 2 minutes to complete a 360° turn. A half-standard rate turn (1.5°/second) takes 4 minutes complete a 360° turn. (H810) — FAA-H-8083-15, Chapter 3

ALL
4896. If a standard rate turn is maintained, how long would it take to turn 180°?

A—1 minute.
B—2 minutes.
C—3 minutes.

A standard rate turn is one during which the heading changes 3° per second. Therefore, at a rate of 3° of turn each second, 180° of turn would require 180/3 or 60 seconds (1 minute). (H810) — FAA-H-8083-15, Chapter 3

ALL
4897. If a half-standard rate turn is maintained, how much time would be required to turn clockwise from a heading of 090° to a heading of 180°?

A—30 seconds.
B—1 minute.
C—1 minute 30 seconds.

A half-standard rate turn is one in which the heading changes 1.5° per second. A heading change of 090° to 180° is a total turn of 90°. At a rate-of-turn of 1.5° each second would require 90/1.5 or 60 seconds (1 minute). (H810) — FAA-H-8083-15, Chapter 3

ALL
4898. During a constant-bank level turn, what effect would an increase in airspeed have on the rate and radius of turn?

A—Rate of turn would increase, and radius of turn would increase.
B—Rate of turn would decrease, and radius of turn would decrease.
C—Rate of turn would decrease, and radius of turn would increase.

Both the rate of turn and the radius of turn vary with changes in airspeed. If airspeed is increased, then the rate of turn is decreased, and the radius of the turn is increased. (H807) — FAA-H-8083-15, Chapter 2

Answer (A) is incorrect because rate of turn decreases. Answer (B) is incorrect because the radius of turn increases.

ALL
4903. If a standard rate turn is maintained, how much time would be required to turn to the right from a heading of 090° to a heading of 270°?

A—1 minute.
B—2 minutes.
C—3 minutes.

A standard rate turn is one during which the heading changes 3° per second. A heading change from 090° to 270° is a total turn of 180°. At a rate-of-turn of 3° each second, this turn would require 180/3 or 60 seconds (1 minute). (H810) — FAA-H-8083-15, Chapter 3

Answers

4878-3 [B]   4895 [C]   4896 [A]   4897 [B]   4898 [C]   4903 [A]

Chapter 3 Flight Instruments

ALL
4904. If a standard rate turn is maintained, how much time would be required to turn to the left from a heading of 090° to a heading of 300°?

A—30 seconds.
B—40 seconds.
C—50 seconds.

A standard rate turn is one during which the heading changes 3° per second. A turn to the left from 090° to 300° is a total turn of 150°. At a turn rate of 3° each second, this turn would require 150/3 or 50 seconds. (H807) — FAA-H-8083-15, Chapter 2

ALL
4905. If a half-standard rate turn is maintained, how long would it take to turn 135°?

A—1 minute.
B—1 minute 20 seconds.
C—1 minute 30 seconds.

A standard rate turn is one during which the heading changes 3° per second. In this case, the aircraft is turning one-half that rate, or 1.5° per second. Therefore, a turn of 135° would require 135/1.5 or 90 seconds (1 minute 30 seconds). (H807) — FAA-H-8083-15, Chapter 2

ALL
4914. Rate of turn can be increased and radius of turn decreased by

A—decreasing airspeed and shallowing the bank.
B—decreasing airspeed and increasing the bank.
C—increasing airspeed and increasing the bank.

The horizontal component of lift varies directly in proportion to bank in a correctly executed turn. Thus, the rate of turn increases (and radius of turn decreases) as the angle of bank increases, and the airspeed decreases. (H807) — FAA-H-8083-15, Chapter 2

Answer (A) is incorrect because shallowing the bank would cause a decrease in the rate of turn. Answer (C) is incorrect because increasing the airspeed would cause a decreasing rate of turn.

ALL
4915. The primary reason the angle of attack must be increased, to maintain a constant altitude during a coordinated turn, is because the

A—thrust is acting in a different direction, causing a reduction in airspeed and loss of lift.
B—vertical component of lift has decreased as the result of the bank.
C—use of ailerons has increased the drag.

The division of lift into horizontal and vertical components during a turn reduces the amount of lift supporting the weight of the aircraft. Consequently, the reduced vertical component results in the loss of altitude unless the total lift is increased by:

1. *Increasing the angle of attack of the wing,*
2. *Increasing the airspeed, or*
3. *Increasing the angle of attack and airspeed in combination.*

(H807) — FAA-H-8083-15, Chapter 2

Answer (A) is incorrect because thrust does not act in a different direction during a turn. Answer (C) is incorrect because ailerons will only increase drag when entering or recovering from a turn.

ALL
4921. The displacement of a turn coordinator during a coordinated turn will

A—indicate the angle of bank.
B—remain constant for a given bank regardless of airspeed.
C—increase as angle of bank increases.

The miniature aircraft of the turn coordinator displays only rate-of-roll and rate-of-turn. It does not directly display the bank angle of the aircraft. The displacement of a turn coordinator increases as angle of bank increases. (H816) — FAA-H-8083-15, Chapter 5

Answer (A) is incorrect because the angle of bank is indirectly shown. Answer (B) is incorrect because the rate-of-turn varies with airspeed for any given bank angle, therefore the displacement of the turn coordinator will change with a change in airspeed.

Answers

4904 [C]   4905 [C]   4914 [B]   4915 [B]   4921 [C]

ALL
4931. (Refer to Figure 144.) What changes in control displacement should be made so that "2" would result in a coordinated standard rate turn?

A—Increase left rudder and increase rate of turn.
B—Increase left rudder and decrease rate of turn.
C—Decrease left rudder and decrease angle of bank.

Illustration 2 shows a slipping, half-standard rate turn. To perform a coordinated standard rate turn, we must increase our rate-of-turn and correct the slip by applying additional left rudder. (H814) — FAA-H-8083-15, Chapter 5

Answer (B) is incorrect because the rate of turn needs to be increased to obtain a standard rate turn. Answer (C) is incorrect because left rudder must be increased to correct for the current slipping condition.

ALL
4932. (Refer to Figure 144.) Which illustration indicates a coordinated turn?

A—3.
B—1.
C—2.

Illustration 3 indicates a coordinated turn, since the ball is centered. (H814) — FAA-H-8083-15, Chapter 5

Answer (B) is incorrect because illustration 1 shows a skidding turn. Answer (C) is incorrect because illustration 2 shows a slipping turn.

ALL
4933. (Refer to Figure 144.) Which illustration indicates a skidding turn?

A—2.
B—1.
C—3.

Illustration 1 indicates a skidding turn, since the ball is to the outside of the direction of the turn. In a skid, the rate of turn is too great for the angle of bank, and excessive centrifugal force causes the ball to move to the outside of the turn. (H814) — FAA-H-8083-15, Chapter 5

Answer (A) is incorrect because illustration 2 shows a slipping turn. Answer (C) is incorrect because illustration 3 shows a coordinated turn.

ALL
4934. (Refer to Figure 144.) What changes in control displacement should be made so that "1" would result in a coordinated standard rate turn?

A—Increase right rudder and decrease rate of turn.
B—Increase right rudder and increase rate of turn.
C—Decrease right rudder and increase angle of bank.

Illustration 1 shows a skidding, half-standard rate turn. To perform a coordinated standard rate turn, we must increase our rate of turn and correct the skid by applying additional right rudder. (H814) — FAA-H-8083-15, Chapter 5

Answer (A) is incorrect because the rate of turn must be increased for a standard rate turn. Answer (C) is incorrect because the right rudder must be increased to correct the current skidding condition.

ALL
4935. (Refer to Figure 144.) Which illustration indicates a slipping turn?

A—1.
B—3.
C—2.

Illustration 2 indicates a slipping turn, since the ball is to the inside of the direction of the turn. In a slip, the rate of turn is too slow for the angle of bank, and the lack of centrifugal force causes the ball to move to the inside of the turn. (H814) — FAA-H-8083-15, Chapter 5

Answer (A) is incorrect because illustration 1 shows a skidding turn. Answer (B) is incorrect because illustration 3 shows a coordinated turn.

Answers

| 4931 | [A] | 4932 | [A] | 4933 | [B] | 4934 | [B] | 4935 | [C] |

Chapter 3 **Flight Instruments**

Heading Indicator

The heading indicator is a gyroscopically-operated instrument. Its purpose is to indicate the aircraft's heading without the errors that are inherent in the magnetic compass. *See* Figure 3-8. Due to precessional error, however, the heading indicator should be regularly compared to the magnetic compass in flight.

The heading indicator pre-takeoff check is made by setting it and checking for proper alignment after making taxiing turns.

Figure 3-8. Heading indicator

ALL
4885. What pretakeoff check should be made of a vacuum-driven heading indicator in preparation for an IFR flight?

A—After 5 minutes, set the indicator to the magnetic heading of the aircraft and check for proper alignment after taxi turns.
B—After 5 minutes, check that the heading indicator card aligns itself with the magnetic heading of the aircraft.
C—Determine that the heading indicator does not precess more than 2° in 5 minutes of ground operation.

Allow 5 minutes after starting engines for the gyro rotor of the vacuum-operated heading indicator to attain normal operating speed. Before taxiing, or while taxiing straight, set the heading indicator to correspond with the magnetic compass heading. Be sure the instrument is fully uncaged if it has a caging feature. Before takeoff, recheck the heading indicator. If your magnetic compass and deviation card are accurate, the heading indicator should show the known taxiway or runway direction when the airplane is aligned with them (within 5°). (H812) — FAA-H-8083-15, Chapter 3

Answer (B) is incorrect because a non-slaved heading indicator must be set manually to the correct magnetic heading. Answer (C) is incorrect because a precession error of no greater than 3° (not 2°) in 15 minutes is acceptable.

Answers
4885   [A]

Magnetic Compass

The magnetic compass is the only self-contained directional instrument in the aircraft. It is influenced by magnetic dip which causes northerly turning error and acceleration/deceleration error.

When northerly turning error occurs, the compass will lag behind the actual aircraft heading while turning through headings in the northern half of the compass rose, and lead the aircraft's actual heading in the southern half. The error is most pronounced when turning through north or south, and is approximately equal in degrees to the latitude. *See* Figure 3-9.

The acceleration/deceleration error is most pronounced on headings of east and west. When accelerating, the compass indicates a turn toward the north, and when decelerating it indicates a turn toward the south. The acronym ANDS is a good memory aid:

Accelerate

North

Decelerate

South

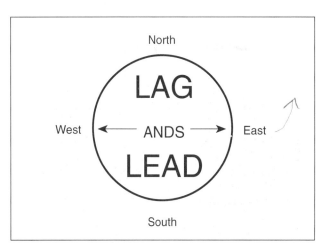

Figure 3-9

A magnetic compass pre-takeoff check verifies that:

1. The liquid chamber is full of fluid.
2. The compass is free turning.
3. No bubbles appear in the fluid.
4. The compass agrees with a known heading with radios and electrical systems on.

ALL
4834. On the taxi check, the magnetic compass should

A—swing opposite to the direction of turn when turning from north.
B—exhibit the same number of degrees of dip as the latitude.
C—swing freely and indicate known headings.

Check the magnetic compass card for freedom of movement and be sure that the bowl is full of fluid. Determine compass accuracy by comparing the indicated heading against a known heading, while the airplane is stopped or taxiing straight. (H812) — FAA-H-8083-15, Chapter 3

Answer (A) is incorrect because turning errors are only associated with an airplane that is airborne and in a bank. Answer (B) is incorrect because compass turning errors are the result of magnetic dip in flight.

ALL
4877. What should be the indication on the magnetic compass as you roll into a standard rate turn to the left from an east heading in the Northern Hemisphere?

A—The compass will initially indicate a turn to the right.
B—The compass will remain on east for a short time, then gradually catch up to the magnetic heading of the aircraft.
C—The compass will indicate the approximate correct magnetic heading if the roll into the turn is smooth.

When you are on an east or west heading, the compass indicates correctly as you start a turn in either direction. (H809) — FAA-H-8083-15, Chapter 3

Answer (A) is incorrect because the compass will initially indicate a turn in the opposite direction only when on a northerly heading. Answer (B) is incorrect because there will not be a lag time when turning from an east heading.

Answers

4834 [C]   4877 [C]

Chapter 3 **Flight Instruments**

ALL
4886. What should be the indication on the magnetic compass as you roll into a standard rate turn to the right from an easterly heading in the Northern Hemisphere?

A—The compass will initially indicate a turn to the left.
B—The compass will remain on east for a short time, then gradually catch up to the magnetic heading of the aircraft.
C—The compass will indicate the approximate correct magnetic heading if the roll into the turn is smooth.

When on an east or west heading, no error is apparent while entering a turn to north or south. (H314) — AC 61-23C, Chapter 3

Answer (A) is incorrect because the compass will initially indicate a turn in the opposite direction of a bank while on a northerly heading. Answer (B) is incorrect because there is no lag when turning from an east heading.

ALL
4887. What should be the indication on the magnetic compass as you roll into a standard rate turn to the right from a south heading in the Northern Hemisphere?

A—The compass will indicate a turn to the right, but at a faster rate than is actually occurring.
B—The compass will initially indicate a turn to the left.
C—The compass will remain on south for a short time, then gradually catch up to the magnetic heading of the aircraft.

If on a southerly heading and a turn is made toward the east or west, the initial indication of the compass needle will indicate a greater amount of turn than is actually made. This lead diminishes as the turn progresses toward east or west where there is no turn error. (H314) — AC 61-23C, Chapter 3

Answer (B) is incorrect because the compass will initially indicate a turn to the opposite direction if the turn is made from a north (not south) heading. Answer (C) is incorrect because the turn leads from a southerly heading (not lags).

ALL
4888. On what headings will the magnetic compass read most accurately during a level 360° turn, with a bank of approximately 15°?

A—135° through 225°.
B—90° and 270°.
C—180° and 0°.

A 15° bank is an approximate standard rate turn for most general aviation aircraft. Either lag or lead diminishes as the turn progresses toward east or west, where there is no turn error. (H314) — AC 61-23C, Chapter 3

Answer (A) is incorrect because the compass leads the heading through a southerly heading. Answer (C) is incorrect because the compass has the greatest turning errors at south and north headings.

ALL
4889. What causes the northerly turning error in a magnetic compass?

A—Coriolis force at the mid-latitudes.
B—Centrifugal force acting on the compass card.
C—The magnetic dip characteristic.

Northerly turning error is the most pronounced of the dip errors. Due to the mounting of the magnetic compass, its center of gravity is below the pivot point on the pedestal and the card is well balanced in the fluid. When the aircraft is banked, the card is also banked as a result of centrifugal force. While the card is in the banked attitude, the vertical component of the Earth's magnetic field causes the north-seeking ends of the compass to dip to the low side of the turn, giving an erroneous turn indication. (H809) — FAA-H-8083-15, Chapter 3

Answer (A) is incorrect because Coriolis force affects winds, not compass indications. Answer (B) is incorrect because centrifugal force does not cause turning errors in magnetic compasses.

ALL
4890. What should be the indication on the magnetic compass when you roll into a standard rate turn to the left from a south heading in the Northern Hemisphere?

A—The compass will indicate a turn to the left, but at a faster rate than is actually occurring.
B—The compass will initially indicate a turn to the right.
C—The compass will remain on south for a short time, then gradually catch up to the magnetic heading of the aircraft.

If on a southerly heading and a turn is made toward the east or west, the initial indication of the compass needle will indicate a greater amount of turn than is actually made. This lead diminishes as the turn progresses toward east or west where there is no turn error. (H314) — AC 61-23C, Chapter 3

Answers (B) and (C) are incorrect because the magnetic compass will lead the heading during turns from the south or through southerly headings.

Answers
4886 [C]   4887 [A]   4888 [B]   4889 [C]   4890 [A]

ALL
4891. What should be the indication on the magnetic compass as you roll into a standard rate turn to the right from a westerly heading in the Northern Hemisphere?

A—The compass will initially show a turn in the opposite direction, then turn to a northerly indication but lagging behind the actual heading of the aircraft.
B—The compass will remain on a westerly heading for a short time, then gradually catch up to the actual heading of the aircraft.
C—The compass will indicate the approximate correct magnetic heading if the roll into the turn is smooth.

When on an east or west heading, no error is apparent while entering a turn to north or south. (H314) — AC 61-23C, Chapter 3

Answer (A) is incorrect because the magnetic compass will only indicate a turn in the opposite direction when turning from a north heading. Answer (B) is incorrect because the magnetic compass will only lag when turning from a north heading.

ALL
4892. What should be the indication on the magnetic compass as you roll into a standard rate turn to the right from a northerly heading in the Northern Hemisphere?

A—The compass will indicate a turn to the right, but at a faster rate than is actually occurring.
B—The compass will initially indicate a turn to the left.
C—The compass will remain on north for a short time, then gradually catch up to the magnetic heading of the aircraft.

If on a northerly heading and a turn is made toward east or west, the initial indication of the compass lags or indicates a turn in the opposite direction. This lag diminishes as the turn progresses toward east or west where there is no turn error. (H314) — AC 61-23C, Chapter 3

Answer (A) is incorrect because the compass will lead the turn only if turning from a southerly heading. Answer (C) is incorrect because the compass will only indicate the approximate correct heading if turning from the east or west.

ALL
4893. What should be the indication on the magnetic compass as you roll into a standard rate turn to the left from a west heading in the Northern Hemisphere?

A—The compass will initially indicate a turn to the right.
B—The compass will remain on west for a short time, then gradually catch up to the magnetic heading of the aircraft.
C—The compass will indicate the approximate correct magnetic heading if the roll into the turn is smooth.

When on an east or west heading, no error is apparent while entering a turn to north or south. (H314) — AC 61-23C, Chapter 3

Answer (A) is incorrect because the compass will only indicate a turn in the opposite direction from a northerly heading. Answer (B) is incorrect because the compass will only lag the turn from a southerly heading.

ALL
4894. What should be the indication on the magnetic compass as you roll into a standard rate turn to the left from a north heading in the Northern Hemisphere?

A—The compass will indicate a turn to the left, but at a faster rate than is actually occurring.
B—The compass will initially indicate a turn to the right.
C—The compass will remain on north for a short time, then gradually catch up to the magnetic heading of the aircraft.

If on a northerly heading and a turn is made toward east or west, the initial indication of the compass lags or indicates a turn in the opposite direction. This lag diminishes as the turn progresses toward east or west where there is no turn error. (H314) — AC 61-23C, Chapter 3

Answer (A) is incorrect because the compass will lead the turn only if turning from a southerly heading. Answer (C) is incorrect because the compass will initially indicate a turn in the opposite direction if turning from a northerly heading.

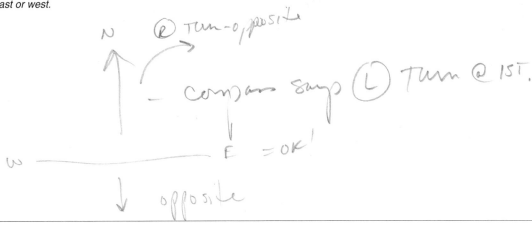

Answers
4891 [C]   4892 [B]   4893 [C]   4894 [B]

Chapter 3 **Flight Instruments**

Slaved Gyro

Many aircraft have a slaved gyro system where a remotely-mounted magnetic compass is used to automatically correct the directional gyro for precession errors. The slaving meter indicates any difference between the displayed heading and the magnetic heading. Right deflection indicates a clockwise error of the compass card. Left deflection indicates a counterclockwise error of the compass card. Whenever the aircraft is in a turn and the card rotates, it is normal for this meter to show a full deflection to one side or the other.

During level flight, it is normal for the needle of the meter to continuously move from side to side and to be fully deflected during a turn. If the needle stays fully deflected left or right, during level flight, the free gyro mode can be used to center it.

When the slave and free gyro button is depressed, the system is in the slaved gyro mode. When the button is in the outer position (not engaged), the system is in the free gyro mode.

When the system is in the free gyro mode, depressing the clockwise manual heading drive button will rotate the compass card to the right to eliminate left compass card error (or a "clockwise adjustment").

When the system is in the free gyro mode, depressing the counterclockwise manual heading drive button will rotate the compass card to the left to eliminate right compass card error (or a "counterclockwise adjustment").

A preflight check of any electrically-driven attitude instrument consists of turning on electrical power before engine start and listening for any unusual or irregular noise.

ALL
4827. (Refer to Figure 143.) The heading on a remote indicating compass is 120° and the magnetic compass indicates 110°. What action is required to correctly align the heading indicator with the magnetic compass?

A—Select the free gyro mode and depress the counterclockwise heading drive button.
B—Select the slaved gyro mode and depress the clockwise heading drive button.
C—Select the free gyro mode and depress the clockwise heading drive button.

The slaving meter indicates the difference between the displayed heading and the magnetic heading. A right deflection indicates a clockwise error of the compass card; a left deflection indicates a counterclockwise error. In this case, the compass card is indicating 120°, when it should be indicating 110°. It has a left deflection, or a counterclockwise error. If the heading on the compass is too far right, then the compass has rotated too far left. The compass card may be aligned with the magnetic compass by selecting the free-gyro mode and depressing the clockwise heading drive button, which will cause the compass card to rotate in a clockwise direction. (H809) — FAA-H-8083-15, Chapter 3

ALL
4828. (Refer to Figure 143.) When the system is in the free gyro mode, depressing the clockwise manual heading drive button will rotate the remote indicating compass card to the

A—right to eliminate left compass card error.
B—right to eliminate right compass card error.
C—left to eliminate left compass card error.

When the system is in the free-gyro mode, depressing the clockwise manual heading drive button will rotate the compass card to the right to eliminate left compass card error (clockwise adjustment). (H809) — FAA-H-8083-15, Chapter 3

Answers
4827 [C]   4828 [A]

ALL
4829. (Refer to Figure 143.) The heading on a remote indicating compass is 5° to the left of that desired. What action is required to move the desired heading under the heading reference?

A—Select the free gyro mode and depress the clockwise heading drive button.
B—Select the slaved gyro mode and depress the clockwise heading drive button.
C—Select the free gyro mode and depress the counterclockwise heading drive button.

The slaving meter indicates the difference between the displayed heading and the magnetic heading. A right deflection indicates a clockwise error of the compass card; a left deflection indicates a counterclockwise error. In this case, the compass has rotated too far to the right, which is a right compass error. The compass card may be aligned with the magnetic compass by selecting the free-gyro mode and depressing the counterclockwise heading drive button, which will cause the compass card to rotate in a counterclockwise direction. (H809) — FAA-H-8083-15, Chapter 3

ALL
4881. Which practical test should be made on the electric gyro instruments prior to starting an engine?

A—Check that the electrical connections are secure on the back of the instruments.
B—Check that the attitude of the miniature aircraft is wings level before turning on electrical power.
C—Turn on the electrical power and listen for any unusual or irregular mechanical noise.

The following are items that should be checked before starting the engine(s):

1. *If instruments are electrical, turn on and listen for any unusual or irregular mechanical noise.*
2. *Check instruments for poor condition, mounting, marking, broken or loose knobs. Also check the power-off indications of the instrument pointers and warning flags.*

(L59) — AC 91-46, 5(a)(4)

Answer (A) is incorrect because the pilot should not be physically moving electrical connections behind the instrument panel. Answer (B) is incorrect because the miniature aircraft will probably not be wings level before the gyro begins spinning.

Instrument Errors

The pitot-static system provides the source of air pressure for the operation of the altimeter, airspeed indicator, and vertical speed indicator. See Figure 3-10. Pitot-static system failures will present indications as shown in Figure 3-11 on the next page.

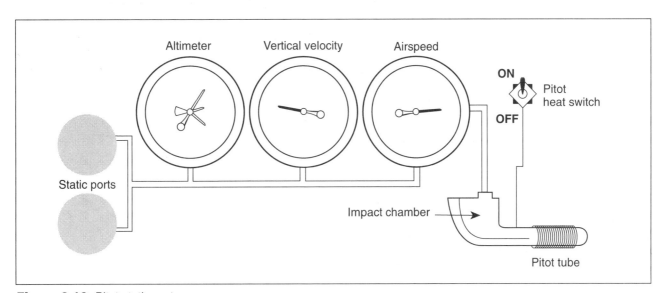

Figure 3-10. Pitot-static system

Answers
4829 [C]   4881 [C]

Chapter 3 **Flight Instruments**

Situation	Airspeed	Altimeter	VSI
1. Blocked pitot.	zero	works	works
2. Blocked pitot and drain hole. Open static.	**high** in climb **low** in descent	works	works
3. Blocked static — open pitot.	**low** in climb **high** in descent	frozen	frozen
4. Using alternate cockpit static air.	reads high	reads high	momentarily shows a climb
5. Broken VSI glass.	reads high	reads high	reverses

Figure 3-11. Pitot-static system failures

ALL
4821. If both the ram air input and drain hole of the pitot system are blocked, what airspeed indication can be expected?

A—No variation of indicated airspeed in level flight even if large power changes are made.
B—Decrease of indicated airspeed during a climb.
C—Constant indicated airspeed during a descent.

If both the ram air input and drainhole are completely blocked by ice, the pressure is trapped in the system and the airspeed indicator may react as an altimeter:

1. *During level flight, airspeed indication will not change even when actual airspeed is varied by large power changes.*
2. *During climb, airspeed indication will increase.*
3. *During descent, airspeed indication will decrease.*

(L57) — AC 91-43

ALL
4830. If both the ram air input and the drain hole of the pitot system are blocked, what reaction should you observe on the airspeed indicator when power is applied and a climb is initiated out of severe icing conditions?

A—The indicated airspeed would show a continuous deceleration while climbing.
B—The airspeed would drop to, and remain at, zero.
C—No change until an actual climb rate is established, then indicated airspeed will increase.

If both the ram air input and the drain hole are blocked, the pressure is trapped in the system and the airspeed indicator may react as an altimeter:

1. *During level flight, the airspeed indication will not change even when actual airspeed is varied by large power changes;*
2. *During climb, the airspeed indication will increase; and*
3. *During descent, the airspeed indication will decrease.*

(L57) — AC 91-43(3)(b)

Answer (A) is incorrect because the airspeed indicator will show an increase in airspeed with an increase in altitude. Answer (B) is incorrect because this would only occur if the ram air input was blocked, not the drain hole.

ALL
4854. What indication should a pilot observe if an airspeed indicator ram air input and drain hole are blocked?

A—The airspeed indicator will react as an altimeter.
B—The airspeed indicator will show a decrease with an increase in altitude.
C—No airspeed indicator change will occur during climbs or descents.

If both the ram air input and the drain hole are blocked, the pressure is trapped in the system and the airspeed indicator may react as an altimeter:

1. *During level flight, the airspeed indication will not change even when actual airspeed is varied by large power changes;*
2. *During climb, the airspeed indication will increase; and*
3. *During descent, the airspeed indication will decrease.*

(L57) — AC 91-43(3)(b)

Answer (B) is incorrect because indicated airspeed will increase given an increase in altitude. Answer (C) is incorrect because the differential pressure between the pitot tube and static air source changes; therefore, so does the indicated airspeed.

Answers

4821 [A]   4830 [C]   4854 [A]

ALL
4879. What would be the indication on the VSI during entry into a 500 FPM actual descent from level flight if the static ports were iced over?

A—The indication would be in reverse of the actual rate of descent (500 FPM climb).
B—The initial indication would be a climb, then descent at a rate in excess of 500 FPM.
C—The VSI pointer would remain at zero regardless of the actual rate of descent.

The vertical speed indicator operates by indicating a change in pressure between the instantaneous static pressure in the diaphragm and the trapped static pressure inside the case. If the static ports are iced over prior to entering a climb or descent, the VSI pointer will remain at zero because the diaphragm static pressure and the trapped static pressure inside the case remain equal to each other. (H808) — FAA-H-8083-15, Chapter 3

Answers (A) and (B) are incorrect because without a change in pressure, the VSI would remain at zero, indicating level flight.

ALL
4908. If, while in level flight, it becomes necessary to use an alternate source of static pressure vented inside the airplane, which of the following should the pilot expect?

A—The altimeter and airspeed indicator to become inoperative.
B—The gyroscopic instruments to become inoperative.
C—The vertical speed to momentarily show a climb.

If the alternate static source is vented inside the airplane, where static pressure is usually lower than outside static pressure, selection of the alternate source may result in the following instrument indications:

1. *The altimeter reads higher than normal;*
2. *Indicated airspeed greater than normal; and*
3. *The vertical-velocity indicator momentarily shows a climb.*

(H859) — FAA-H-8083-15, Chapter 11

ALL
4909. During flight, if the pitot tube becomes clogged with ice, which of the following instruments would be affected?

A—The airspeed indicator only.
B—The airspeed indicator and the altimeter.
C—The airspeed indicator, altimeter, and Vertical Speed Indicator.

Clogging of the pitot opening by ice or dirt (or failure to remove the pitot cover) affects the airspeed indicator only. (H808) — FAA-H-8083-15, Chapter 3

Answers (B) and (C) are incorrect because the pitot tube is only connected to the airspeed indicator.

ALL
4930. If while in level flight, it becomes necessary to use an alternate source of static pressure vented inside the airplane, which of the following variations in instrument indications should the pilot expect?

A—The altimeter will read lower than normal, airspeed lower than normal, and the VSI will momentarily show a descent.
B—The altimeter will read higher than normal, airspeed greater than normal, and the VSI will momentarily show a climb.
C—The altimeter will read lower than normal, airspeed greater than normal, and the VSI will momentarily show a climb and then a descent.

If the alternate static source is vented inside the airplane, where static pressure is usually lower than outside static pressure, selection of the alternate source may result in the following instrument indications:

1. *The altimeter reads higher than normal;*
2. *Indicated airspeed greater than normal; and*
3. *The vertical-velocity indicator momentarily shows a climb.*

(H312) — AC 61-23C, Chapter 3

Answers
4879 [C]   4908 [C]   4909 [A]   4930 [B]

Chapter 3 **Flight Instruments**

ALL
4937. (Refer to Figure 146.) Identify the system that has failed and determine a corrective action to return the airplane to straight-and-level flight.

A—Static/pitot system is blocked; lower the nose and level the wings to level-flight attitude by use of attitude indicator.
B—Vacuum system has failed; reduce power, roll left to level wings, and pitch up to reduce airspeed.
C—Electrical system has failed; reduce power, roll left to level wings, and raise the nose to reduce airspeed.

In this situation, the vacuum instruments both show a climbing right turn. They are supported by the turn coordinator which shows a right turn, and the altimeter and VSI which indicate a climb. These are contradicted by the airspeed indicator which is rising (airspeed would not be increasing in this situation). The airspeed indicator then, is the one instrument which conflicts with the remaining unrelated systems and indicates that the pitot tube and drain holes are blocked, and the airspeed indicator is acting as an altimeter. The correct recovery procedure for a nose-high attitude is to lower the nose and level the wings to level flight attitude by use of the attitude indicator. (H818) — FAA-H-8083-15, Chapter 5

ALL
4941. (Refer to Figure 150.) What is the flight attitude? One instrument has malfunctioned.

A—Climbing turn to the right.
B—Climbing turn to the left.
C—Descending turn to the right.

The question states that one instrument (not system) has malfunctioned. The attitude indicator disagrees with both the turn coordinator and the DG in its bank indication. It also disagrees with the altimeter, VSI, and airspeed indicator in its pitch indication. Elimination of the attitude indicator shows the airplane to be in a climbing turn to the right. (H818) — FAA-H-8083-15, Chapter 5

ALL
4939. (Refer to Figure 148.) What is the flight attitude? One system which transmits information to the instruments has malfunctioned.

A—Climbing turn to left.
B—Climbing turn to right.
C—Level turn to left.

The attitude indicator and directional gyro show a right turn. The airspeed indicator, altimeter, and VSI show a climb. The turn coordinator (showing a lack of turn) is not in agreement with the other instruments. (H818) — FAA-H-8083-15, Chapter 5

ALL
4940. (Refer to Figure 149.) What is the flight attitude? One system which transmits information to the instruments has malfunctioned.

A—Level turn to the right.
B—Level turn to the left.
C—Straight-and-level flight.

The attitude indicator and the heading indicator (both vacuum driven) are presenting conflicting indications. The attitude indicator shows a bank to the right, while the heading indicator shows a left turn; therefore, there must be a vacuum system malfunction. All other instruments are in agreement with straight-and-level flight. (H818) — FAA-H-8083-15, Chapter 5

RTC
4942. (Refer to Figure 151.) What is the flight attitude? One instrument has malfunctioned.

A—Climbing turn to the right.
B—Level turn to the right.
C—Level turn to the left.

The question states that one instrument (not system) has malfunctioned. Although the airspeed indicator shows a decrease, the other pitch instruments (attitude indicator, altimeter, and VSI) all agree and show the aircraft in level flight. The airspeed indicator has malfunctioned. The three bank instruments are in agreement, showing a level turn to the right. (H826) — FAA-H-8083-15, Chapter 6

Answers

4937 [A]   4941 [A]   4939 [B]   4940 [C]   4942 [B]

Chapter 3 **Flight Instruments**

Fundamental Skills

During attitude instrument training, the pilot must develop three fundamental skills involved in all instrument flight maneuvers:

1. Instrument cross-check.
2. Instrument interpretation.
3. Aircraft control.

When executing an ILS approach, the pilot must keep the aircraft on the electronic glide slope. This requires the ability to establish the proper rate of descent for the ground speed. As ground speed increases, the rate of descent required to maintain the glide slope must be increased; as ground speed decreases, the rate of descent required to maintain the glide slope also decreases. By first cross-checking the instruments and then interpreting them, the pilot is able to precisely control the aircraft to a successful landing.

ALL
4840. What is the correct sequence in which to use the three skills used in instrument flying?

A—Aircraft control, cross-check, and instrument interpretation.
B—Instrument interpretation, cross-check, and aircraft control.
C—Cross-check, instrument interpretation, and aircraft control.

During attitude instrument training, a pilot must develop three fundamental skills involved in all instrument flight maneuvers:

1. *instrument cross-check*
2. *instrument interpretation*
3. *aircraft control.*

(H813) — FAA-H-8083-15, Chapter 4

Answer (A) is incorrect because aircraft control is the third skill used in instrument flying. Answer (B) is incorrect because instrument interpretation is the second skill and cross-check is the first skill used in instrument flying.

ALL
4855. What are the three fundamental skills involved in attitude instrument flying?

A—Instrument interpretation, trim application, and aircraft control.
B—Cross-check, instrument interpretation, and aircraft control.
C—Cross-check, emphasis, and aircraft control.

The three fundamental skills involved in all instrument flight maneuvers are:

1. *instrument cross-check*
2. *instrument interpretation*
3. *aircraft control.*

(H813) — FAA-H-8083-15, Chapter 4

Answer (A) is incorrect because application of trim is just one aspect of aircraft control. Answer (C) is incorrect because emphasis is a common error in instrument cross-checking.

ALL
4859. What is the third fundamental skill in attitude instrument flying?

A—Instrument cross-check.
B—Power control.
C—Aircraft control.

The three fundamental skills involved in all instrument flight maneuvers are:

1. *instrument cross-check*
2. *instrument interpretation*
3. *aircraft control.*

(H813) — FAA-H-8083-15, Chapter 4

Answer (A) is incorrect because instrument cross-check is the first skill in instrument flying. Answer (B) is incorrect because the use of power control is only one aspect of aircraft control.

Answers

4840 [C]   4855 [B]   4859 [C]

Chapter 3 **Flight Instruments**

ALL
4862. What is the first fundamental skill in attitude instrument flying?

A—Aircraft control.
B—Instrument cross-check.
C—Instrument interpretation.

The three fundamental skills involved in all instrument flight maneuvers are:

1. *instrument cross-check*
2. *instrument interpretation*
3. *aircraft control.*

(H813) — FAA-H-8083-15, Chapter 4

Answer (A) is incorrect because aircraft control is the third fundamental skill in instrument flight. Answer (C) is incorrect because instrument interpretation is the second fundamental skill in instrument flight.

ALL
4721. What effect will a change in wind direction have upon maintaining a 3° glide slope at a constant true airspeed?

A—When ground speed decreases, rate of descent must increase.
B—When ground speed increases, rate of descent must increase.
C—Rate of descent must be constant to remain on the glide slope.

The vertical speed indicator (VSI) shows rate of change of altitude. The faster an aircraft descends, the greater the rate of descent shown on the VSI. The slower an aircraft descends, the slower the rate of descent shown on the VSI. The vertical speed required is dependent on the ground speed when flying on a fixed glide slope. Therefore, when ground speed increases, the rate of descent must increase. (I10) — AC 00-54A

ALL
4745. The rate of descent required to stay on the ILS glide slope

A—must be increased if the ground speed is decreased.
B—will remain constant if the indicated airspeed remains constant.
C—must be decreased if the ground speed is decreased.

The vertical speed required is dependent on the ground speed when flying on a fixed glide slope. In order to stay on the glide slope, the vertical speed must be decreased if the ground speed is decreased. (K04) — AC 00-54A, ¶6(a)

Answers (A) and (B) are incorrect because the rate of descent must be decreased with decreasing ground speed or increased with increasing ground speed.

ALL
4748. To remain on the ILS glidepath, the rate of descent must be

A—decreased if the airspeed is increased.
B—decreased if the ground speed is increased.
C—increased if the ground speed is increased.

The vertical speed required is dependent on the ground speed when flying on a fixed glide slope. In order to stay on the glide slope, the vertical speed must be increased if the ground speed is increased. (K04) — AC 00-54A, ¶6(a)

Answer (A) is incorrect because the rate of descent is dependent on ground speed (not airspeed). Answer (B) is incorrect because if the rate of descent is decreased when the ground speed increases, the aircraft will rise above the glidepath.

ALL
4752. The rate of descent on the glide slope is dependent upon

A—true airspeed.
B—calibrated airspeed.
C—ground speed.

The vertical speed required is dependent on the ground speed when flying on a fixed glide slope. For example, in order to stay on the glide slope, the vertical speed must be increased if the ground speed is increased. (K04) — AC 00-54A, ¶6(a)

Answers (A) and (B) are incorrect because the vertical speed is dependent on ground speed (not true or calibrated airspeed).

Answers

4862 [B]   4721 [B]   4745 [C]   4748 [C]   4752 [C]

ALL
4756. The glide slope and localizer are centered, but the airspeed is too fast. Which should be adjusted initially?

A—Pitch and power.
B—Power only.
C—Pitch only.

The heaviest demand on pilot technique occurs during descent from the outer marker to the middle marker, when you maintain the localizer course, adjust pitch attitude to maintain the proper rate of descent, and adjust power to maintain proper airspeed. (H815) — FAA-H-8083-15, Chapter 5

Answer (A) is incorrect because the pitch and power adjustments are closely coordinated, but the power should be adjusted first. Answer (C) is incorrect because a pitch adjustment only will cause you to fly above the glide path.

ALL
4772. During a precision radar or ILS approach, the rate of descent required to remain on the glide slope will

A—remain the same regardless of ground speed.
B—increase as the ground speed increases.
C—decrease as the ground speed increases.

The vertical speed required is dependent on the ground speed when flying on a fixed glide slope. For example, in order to stay on the glide slope, the vertical speed must be increased if the ground speed is increased. (K04) — AC 00-54A

Answer (A) is incorrect because vertical speed is dependent on ground speed. Answer (C) is incorrect because the vertical speed will increase (not decrease) if the ground speed increases.

Attitude Instrument Flying

The attitude of an aircraft is controlled by movement around its lateral (pitch), longitudinal (roll), and vertical (yaw) axes. In instrument flying, attitude requirements are determined by correctly interpreting the flight instruments. Instruments are grouped as to how they relate to control, function and aircraft performance. Attitude control is discussed in terms of pitch, bank, and power control. The three pitot-static instruments, the three gyroscopic instruments, and the tachometer or manifold pressure gauge are grouped into the following categories:

Pitch Instruments:
- Attitude indicator
- Altimeter
- Airspeed indicator
- Vertical speed indicator

Bank Instruments:
- Attitude indicator
- Heading indicator
- Turn coordinator

Power Instruments:
- Manifold pressure gauge
- Tachometer
- Airspeed indicator

The instruments which provide the best information for controlling the aircraft in any given maneuver are referred to as "primary" instruments. These are usually the instruments that should be held at a constant indication. The remaining instruments should help maintain the primary instruments at the desired indications.

Figure 3-12 (on the next page), which is equally applicable to both airplanes and helicopters, shows examples of primary and supporting instruments in selected maneuvers.

When climbing and descending, it is necessary to begin level-off in enough time to avoid overshooting the desired altitude. The amount of lead to level-off from a climb varies with the rate of climb and pilot technique. If the aircraft is climbing at 1,000 feet per minute, it will continue to climb at a descending rate throughout the transition to level flight. An effective practice is to lead the altitude by 10% of the vertical speed (500 fpm would have a 50 foot lead; 1,000 fpm would have a 100 foot lead).

Answers
4756 [B]   4772 [B]

Chapter 3 **Flight Instruments**

Maneuver	Pitch		Bank		Power	
	Primary	Supporting	Primary	Supporting	Primary	Supporting
Straight & Level	Altimeter	VSI, Attitude Indicator	Directional Gyro	Turn & Slip, Compass, Attitude Indicator	Airspeed	
Level Standard Rate Turn	Altimeter	VSI, Attitude Indicator	Turn & Slip		Airspeed	
Climb (Constant Airspeed)	Airspeed	Attitude Indicator	Directional Gyro	Turn & Slip, Attitude Indicator	Engine Gauge	
Descent (Constant Rate)	VSI	Attitude Indicator	Directional Gyro	Turn & Slip, Attitude Indicator	Airspeed	Attitude Indicator

Figure 3-12

The amount of lead to level-off from a descent also depends upon the rate of descent and control technique. To level-off from a descent at descent airspeed, lead the desired altitude by approximately 10%. For level-off at an airspeed higher than descending airspeed, lead the level-off by approximately 25%.

When making initial pitch attitude corrections to maintain altitude during straight-and-level flight, the changes of attitude should be small and smoothly applied. As a rule-of-thumb for airplanes, use a half-bar-width correction for errors of less than 100 feet and a full-bar-width correction for errors in excess of 100 feet.

ALL
4832. The gyroscopic heading indicator is inoperative. What is the primary bank instrument in unaccelerated straight-and-level flight?

A—Magnetic compass.
B—Attitude indicator.
C—Miniature aircraft of turn coordinator.

The heading indicator provides the most pertinent banking information in unaccelerated straight-and-level flight, since banking means turning, and it is primary for bank. With gyroscopic heading indicator failure, the magnetic compass becomes the only heading indicator. (H814) — FAA-H-8083-15, Chapter 5

Answer (B) is incorrect because the attitude indicator does not provide heading information needed to maintain straight flight. Answer (C) is incorrect because the miniature aircraft is the primary bank indicator in established standard rate turns (not in straight flight).

ALL
4836. What instruments are considered supporting bank instruments during a straight, stabilized climb at a constant rate?

A—Attitude indicator and turn coordinator.
B—Heading indicator and attitude indicator.
C—Heading indicator and turn coordinator.

The attitude indicator is primary only during transitions from one attitude to another. Once stabilized, the attitude indicator becomes a support instrument. The turn coordinator is a supporting instrument in straight flight. (H815) — FAA-H-8083-15, Chapter 5

Answers (B) and (C) are incorrect because the heading indicator is used as the primary bank instrument in a straight climb.

ALL
4837. What instruments are primary for pitch, bank, and power, respectively, when transitioning into a constant airspeed climb from straight-and-level flight?

A—Attitude indicator, heading indicator, and manifold pressure gauge or tachometer.
B—Attitude indicator for both pitch and bank; airspeed indicator for power.
C—Vertical speed, attitude indicator, and manifold pressure or tachometer.

Primary pitch information during the transition is attitude indicator for pitch, heading indicator for bank, and tachometer or manifold pressure for power. When the airspeed has stabilized, the airspeed indicator becomes primary for pitch. (H815) — FAA-H-8083-15, Chapter 5

Answer (B) is incorrect because the heading indicator is a primary instrument for bank, and the manifold pressure gauge is primary for power. Answer (C) is incorrect because the attitude indicator is the primary instrument for pitch, and the heading indicator is primary for bank.

Answers

4832 [A]   4836 [A]   4837 [A]

ALL
4838. What is the primary bank instrument once a standard rate turn is established?

A—Attitude indicator.
B—Turn coordinator.
C—Heading indicator.

The turn coordinator displays the movement of the aircraft on the roll axis that is proportional to the roll rate. Therefore, the turn coordinator is the primary bank instrument once a standard rate turn is established. (H816) — FAA-H-8083-15, Chapter 5

Answer (A) is incorrect because the attitude indicator will be used as a primary bank instrument in establishing a standard rate turn, but not for maintaining the turn once it is established. Answer (C) is incorrect because the heading indicator is a primary bank instrument for straight flight.

ALL
4845-1. As power is increased to enter a 500 feet per minute rate of climb in straight flight, which instruments are primary for pitch, bank, and power respectively?

A—Attitude indicator, heading indicator, and manifold pressure gauge or tachometer.
B—VSI, attitude indicator, and airspeed indicator.
C—Airspeed indicator, attitude indicator, and manifold pressure gauge or tachometer.

As the power is increased to the approximate setting for the desired rate, simultaneously raise the miniature aircraft to the climbing attitude for the desired airspeed and rate of climb. As the power is increased, the AI is primary for pitch control until the vertical speed approaches the desired value. The heading indicator is primary for bank and the tachometer, or manifold pressure gauge, is primary for power. As the vertical speed needle stabilizes, it becomes primary for pitch. (H815) — FAA-H-8083-15, Chapter 5

Answer (B) is incorrect because the vertical speed indicator becomes the primary instrument for pitch once a constant rate climb has been established. Answer (C) is incorrect because the heading indicator is primary for bank in straight flight.

RTC
4845-2. As power is increased to enter a 500 feet per minute rate of climb in straight flight, which instruments are primary for pitch, bank, and power respectively?

A—Airspeed indicator, attitude indicator, and manifold pressure gauge or tachometer.
B—VSI, attitude indicator, and airspeed indicator.
C—Airspeed indicator, heading indicator, and manifold pressure gauge or tachometer.

To enter a constant-rate climb, increase power to the approximate setting for the desired rate. As power is applied, the airspeed indicator is primary for pitch until the vertical speed approaches the desired rate. (H823) — FAA-H-8083-15, Chapter 6

ALL
4848. What is the primary pitch instrument during a stabilized climbing left turn at cruise climb airspeed?

A—Attitude indicator.
B—VSI.
C—Airspeed indicator.

For a constant airspeed climb (at cruising airspeed), once the airplane stabilizes at a constant airspeed and attitude, the airspeed indicator becomes the primary pitch instrument. (H815) — FAA-H-8083-15, Chapter 5

Answers (A) and (B) are incorrect because the attitude indicator and vertical speed indicator are each used as supporting pitch instruments in a stabilized climb.

ALL
4850-1. What is the primary pitch instrument when establishing a constant altitude standard rate turn?

A—Altimeter.
B—VSI.
C—Airspeed indicator.

The altimeter is the primary pitch instrument when establishing a level standard rate turn. (H816) — FAA-H-8083-15, Chapter 5

Answer (B) is incorrect because the VSI is used as a supporting pitch instrument for establishing a level standard rate turn. Answer (C) is incorrect because the airspeed indicator is used as the primary power instrument when establishing a constant altitude standard rate turn.

ALL
4850-2. As a rule of thumb, altitude corrections of less than 100 feet should be corrected by using

A—two bar widths on the attitude indicator.
B—less than a full bar width on the attitude indicator.
C—less than half bar width on the attitude indicator.

As a rule of thumb, for errors of less than 100 feet, use a half-bar-width correction. (H822) — FAA-H-8083-15, Chapter 6

Answers
4838 [B]   4845-1 [A]   4845-2 [C]   4848 [C]   4850-1 [A]   4850-2 [C]

Chapter 3 Flight Instruments

ALL

4851. What is the initial primary bank instrument when establishing a level standard rate turn?

A—Turn coordinator.
B—Heading indicator.
C—Attitude indicator.

The attitude indicator is the initial primary bank instrument when establishing a level standard rate turn. (H816) — FAA-H-8083-15, Chapter 5

Answer (A) is incorrect because after the turn has been established, the turn coordinator becomes the primary bank instrument. Answer (B) is incorrect because the heading indicator is used as the primary bank instrument for straight flight.

ALL

4853. What instrument(s) is(are) supporting bank instrument when entering a constant airspeed climb from straight-and-level flight?

A—Heading indicator.
B—Attitude indicator and turn coordinator.
C—Turn coordinator and heading indicator.

The attitude indicator and turn coordinator are the supporting bank instruments when entering a constant airspeed climb from straight-and-level flight. (H815) — FAA-H-8083-15, Chapter 5

Answers (A) and (C) are incorrect because the heading indicator is used as the primary bank instrument for straight flight.

ALL

4858. What is the primary bank instrument while transitioning from straight-and-level flight to a standard rate turn to the left?

A—Attitude indicator.
B—Heading indicator.
C—Turn coordinator (miniature aircraft).

The attitude indicator is the initial primary bank instrument when establishing a level standard rate turn. (H816) — FAA-H-8083-15, Chapter 5

Answer (B) is incorrect because the heading indicator is used as the primary bank instrument for straight flight. Answer (C) is incorrect because the turn coordinator is the primary bank instrument only after the turn has been established.

ALL

4863. As power is reduced to change airspeed from high to low cruise in level flight, which instruments are primary for pitch, bank, and power, respectively?

A—Attitude indicator, heading indicator, and manifold pressure gauge or tachometer.
B—Altimeter, attitude indicator, and airspeed indicator.
C—Altimeter, heading indicator, and manifold pressure gauge or tachometer.

As the power is reduced to change airspeed from high to low cruise in level flight the primary instruments are:

Pitch — Altimeter
Bank — Heading indicator
Power — Manifold pressure gauge (momentarily)

(H814) — FAA-H-8083-15, Chapter 4

Answer (A) is incorrect because the altimeter is primary for pitch (not attitude indicator). Answer (B) is incorrect because the heading indicator is primary for bank (not attitude indicator), and manifold pressure gauge is momentarily the primary for power (not the airspeed indicator).

ALL

4865. Which instrument provides the most pertinent information (primary) for bank control in straight-and-level flight?

A—Turn-and-slip indicator.
B—Attitude indicator.
C—Heading indicator.

The primary instruments in straight-and-level flight are:

Primary Pitch—Altimeter
Primary Bank—Heading Indicator
Primary Power—Airspeed Indicator

See the figure on the next page. (H814) — FAA-H-8083-15, Chapter 5

Answer (A) is incorrect because the turn-and-slip indicator is a supporting bank instrument in straight-and-level flight. Answer (B) is incorrect because the attitude indicator is a supporting bank and pitch instrument in straight-and-level flight.

Answers

4851 [C]   4853 [B]   4858 [A]   4863 [C]   4865 [C]

Question 4865

ALL
4866. Which instruments are considered primary and supporting for bank, respectively, when establishing a level standard rate turn?

A—Turn coordinator and attitude indicator.
B—Attitude indicator and turn coordinator.
C—Turn coordinator and heading indicator.

On the roll-in to a level standard rate turn, use the attitude indicator to establish the approximate angle of bank (primary), then check the miniature aircraft of the turn coordinator for a standard rate turn indication (secondary). (H816) — FAA-H-8083-15, Chapter 5

Answer (A) is incorrect because the turn coordinator is the primary bank instrument, and the attitude instrument the supporting bank instrument, only after the standard rate turn has been established. Answer (C) is incorrect because the turn coordinator is the supporting bank instrument, and the heading indicator is neither a primary or supporting instrument, when establishing a standard rate turn.

ALL
4869. Which instruments, in addition to the attitude indicator, are pitch instruments?

A—Altimeter and airspeed only.
B—Altimeter and VSI only.
C—Altimeter, airspeed indicator, and vertical speed indicator.

In addition to the attitude indicator, the altimeter, airspeed indicator, and vertical speed indicator provide pitch information. (H813) — FAA-H-8083-15, Chapter 4

Answer (A) is incorrect because the vertical speed indicator and airspeed indicator also provide pitch information. Answer (B) is incorrect because the airspeed indicator also provides pitch information.

Answers

4866 [B]   4869 [C]

Chapter 3 Flight Instruments

ALL
4871. Which instrument provides the most pertinent information (primary) for pitch control in straight-and-level flight?

A—Attitude indicator.
B—Airspeed indicator.
C—Altimeter.

The primary instruments in straight-and-level flight are:

Primary Pitch — Altimeter
Primary Bank — Heading Indicator
Primary Power — Airspeed Indicator

(H814) — FAA-H-8083-15, Chapter 5

Answer (A) is incorrect because the attitude indicator is the supporting pitch instrument in straight-and-level flight. Answer (B) is incorrect because the airspeed indicator is the primary power instrument in straight-and-level flight.

ALL
4872. Which instruments are considered to be supporting instruments for pitch during change of airspeed in a level turn?

A—Airspeed indicator and VSI.
B—Altimeter and attitude indicator.
C—Attitude indicator and VSI.

The attitude indicator and the vertical speed indicator are supporting instruments for pitch during change of airspeed in a level turn. See the figure that follows. (H816) — FAA-H-8083-15, Chapter 5

Answer (A) is incorrect because the airspeed indicator is a supporting power instrument during a change of airspeed in a level turn. It then becomes the primary power instrument as the desired airspeed is obtained. Answer (B) is incorrect because the altimeter is the primary (not secondary) instrument for pitch in level flight.

Question 4872

Answers
4871 [C]   4872 [C]

ALL
4874. Which instrument is considered primary for power as the airspeed reaches the desired value during change of airspeed in a level turn?

A—Airspeed indicator.
B—Attitude indicator.
C—Altimeter.

The manifold pressure gauge (or tachometer) is primary for power control while the airspeed is changing in a level turn. As the airspeed approaches the new indication, the airspeed indicator becomes primary for power control. (H816) — FAA-H-8083-15, Chapter 5

Answer (B) is incorrect because the attitude indicator is a supporting pitch and bank instrument in this scenario. Answer (C) is incorrect because the altimeter is a primary pitch instrument in this scenario.

ALL
4876. Which instruments should be used to make a pitch correction when you have deviated from your assigned altitude?

A—Altimeter and VSI.
B—Manifold pressure gauge and VSI.
C—Attitude indicator, altimeter, and VSI.

The pitch instruments are the attitude indicator, the altimeter, the vertical-speed indicator and the airspeed indicator. When a pitch error is detected, corrective action should be taken promptly, but with light control pressures. (H813) — FAA-H-8083-15, Chapter 4

Answer (A) is incorrect because an attitude indicator would also be used to make a pitch correction. Answer (B) is incorrect because the manifold pressure gauge is used as a power instrument (not pitch).

ALL
4899. Conditions that determine the pitch attitude required to maintain level flight are

A—airspeed, air density, wing design, and angle of attack.
B—flightpath, wind velocity, and angle of attack.
C—relative wind, pressure altitude, and vertical lift component.

Factors that affect the attitude in maintaining level flight include airspeed, air density, wing design, and angle of attack. (H807) — FAA-H-8083-15, Chapter 2

Answer (B) is incorrect because flight path and wind velocity do not determine pitch attitude. Answer (C) is incorrect because relative wind, pressure altitude, and vertical lift component do not determine the pitch attitude.

ALL
4906. Approximately what percent of the indicated vertical speed should be used to determine the number of feet to lead the level-off from a climb to a specific altitude?

A—10 percent.
B—20 percent.
C—25 percent.

The amount of lead varies with rate of climb and pilot technique. An effective practice is to lead the altitude by 10 percent of the vertical speed shown (500 fpm/50-foot lead—1,000 fpm/100-foot lead). (H815) — FAA-H-8083-15, Chapter 5

ALL
4907. To level off from a descent to a specific altitude, the pilot should lead the level-off by approximately

A—10 percent of the vertical speed.
B—30 percent of the vertical speed.
C—50 percent of the vertical speed.

The level-off from a descent at descent airspeed must be started before you reach the desired altitude. The amount of lead depends upon the rate of descent and control technique. With too little lead, you will tend to overshoot the selected altitude, unless your technique is rapid. An effective practice is to lead the altitude by 10 percent of the vertical speed shown. Assuming a 500-fpm rate of descent, lead the desired altitude by approximately 50 feet. (H815) — FAA-H-8083-15, Chapter 5

ALL
4920. For maintaining level flight at constant thrust, which instrument would be the least appropriate for determining the need for a pitch change?

A—Altimeter.
B—VSI.
C—Attitude indicator.

The attitude indicator would be the least appropriate instrument for determining the need for a pitch change in level flight at constant thrust. Until level flight, as indicated by the attitude indicator, is identified and established by reference to the altimeter and VSI, there is no way of knowing if it is truly level flight. With constant thrust, any change in the altimeter or VSI indications shows a need for a pitch change. (H814) — FAA-H-8083-15, Chapter 5

Answers

| 4874 [A] | 4876 [C] | 4899 [A] | 4906 [A] | 4907 [A] | 4920 [C] |

Chapter 3 Flight Instruments

ALL
4924. To enter a constant-airspeed descent from level-cruising flight, and maintain cruising airspeed, the pilot should

A—first adjust the pitch attitude to a descent using the attitude indicator as a reference, then adjust the power to maintain the cruising airspeed.
B—first reduce power, then adjust the pitch using the attitude indicator as a reference to establish a specific rate on the VSI.
C—simultaneously reduce power and adjust the pitch using the attitude indicator as a reference to maintain the cruising airspeed.

The following method for entering descents is effective either with or without an attitude indicator:

1. *Reduce airspeed to your selected descent airspeed while maintaining straight-and-level flight, then make a further reduction in power (to a predetermined setting); and*
2. *As the power is adjusted, simultaneously lower the nose to maintain constant airspeed, and trim off control pressures.*

(H815) — FAA-H-8083-15, Chapter 5

Answer (A) is incorrect because adjusting the pitch attitude first will result in an increased airspeed. Answer (B) is incorrect because adjusting the power first will result in a decreased airspeed, and the airspeed indicator (not VSI) is used to maintain a constant airspeed.

ALL
4925. To level off at an airspeed higher than the descent speed, the addition of power should be made, assuming a 500 FPM rate of descent, at approximately

A—50 to 100 feet above the desired altitude.
B—100 to 150 feet above the desired altitude.
C—150 to 200 feet above the desired altitude.

The level-off from a descent must be started before you reach the desired altitude. The amount of lead depends upon the rate of descent and control technique. Assuming a 500-fpm rate of descent, lead the altitude by 100–150 feet for level-off at an airspeed higher than descending speed. At the lead point, add power to the appropriate level flight cruise setting. (H815) — FAA-H-8083-15, Chapter 5

ALL
4926. To level off from a descent maintaining the descending airspeed, the pilot should lead the desired altitude by approximately

A—20 feet.
B—50 feet.
C—60 feet.

To level-off from a descent at descent airspeed, lead the desired altitude by approximately 50 feet, simultaneously adjusting the pitch attitude to level flight and adding power to a setting that will hold the airspeed constant. (H815) — FAA-H-8083-15, Chapter 5

RTC
4849. What is the primary pitch instrument during a stabilized autorotation?

A—Altimeter.
B—Airspeed indicator.
C—VSI.

The airspeed indicator is the primary pitch instrument and should be adjusted to the recommended autorotation speed. (H827) — FAA-H-8083-15, Chapter 6

Answers
4924 [C]   4925 [B]   4926 [B]   4849 [B]

Unusual Attitude Recoveries

When recovering from an unusual attitude without the aid of the attitude indicator, approximate level pitch attitude is reached when the airspeed indicator and altimeter stop moving and the vertical speed indicator reverses its trend.

The following procedures are accomplished to recover from a nose-low attitude:

1. Reduce power.
2. Level the wings.
3. Raise the nose to the horizon.

The following procedures are accomplished to recover from a nose-high attitude:

1. Add power.
2. Apply forward elevator pressure.
3. Level the wings.

ALL
4867. While recovering from an unusual flight attitude without the aid of the attitude indicator, approximate level pitch attitude is reached when the

A—airspeed and altimeter stop their movement and the VSI reverses its trend.
B—airspeed arrives at cruising speed, the altimeter reverses its trend, and the vertical speed stops its movement.
C—altimeter and vertical speed reverse their trend and the airspeed stops its movement.

When the rate of movement of altimeter and airspeed indicator needles decreases, and the vertical speed indicator reverses its trend, the aircraft is approaching level pitch attitude. (H818) — FAA-H-8083-15, Chapter 5

Answer (B) is incorrect because the vertical speed indicator will lag and only show a decrease in vertical movement when it has stopped. Answer (C) is incorrect because the altimeter must stop (not just reverse its trend) in order to indicate a level pitch attitude.

AIR
4873-1. If an airplane is in an unusual flight attitude and the attitude indicator has exceeded its limits, which instruments should be relied on to determine pitch attitude before starting recovery?

A—Turn indicator and VSI.
B—Airspeed and altimeter.
C—VSI and airspeed to detect approaching V_{S1} or V_{MO}.

As soon as the unusual attitude is detected, the recovery should be initiated primarily by reference to the airspeed indicator (for pitch attitude), altimeter (for pitch attitude), vertical-speed indicator, and turn coordinator. (H818) — FAA-H-8083-15, Chapter 5

Answer (A) is incorrect because the turn coordinator does not provide information about pitch attitude. Answer (C) is incorrect because the VSI is not as reliable as the altimeter in determining a climb or descent.

RTC
4873-2. If a helicopter is in an unusual flight attitude and the attitude indicator has exceeded its limits, which instruments should be relied on to determine pitch attitude before starting recovery?

A—Turn indicator and VSI.
B—Airspeed, VSI and altimeter.
C—VSI and airspeed to detect approaching VSI or VMO.

To recover from an unusual attitude, correct bank-and-pitch attitude, and adjust power as necessary. Pitch attitude should be corrected by reference to the altimeter, airspeed indicator, vertical speed indicator, and attitude indicator. (H826) — FAA-H-8083-15, Chapter 6

AIR
4875-1. Which is the correct sequence for recovery from a spiraling, nose-low, increasing airspeed, unusual flight attitude?

A—Increase pitch attitude, reduce power, and level wings.
B—Reduce power, correct the bank attitude, and raise the nose to a level attitude.
C—Reduce power, raise the nose to level attitude, and correct the bank attitude.

Reduce power to prevent excessive airspeed and loss of altitude, correct the bank attitude with coordinated aileron and rudder pressure to straight flight by refer-

Continued

Answers
4867 [A]   4873-1 [B]   4873-2 [B]   4875-1 [B]

Chapter 3 Flight Instruments

ring to the turn coordinator. Raise the nose to level-flight attitude by smooth back-elevator pressure. (H818) — FAA-H-8083-15, Chapter 5

Answer (A) is incorrect because power should be decreased first and then wings leveled. Answer (C) is incorrect because the wings should be level before the nose is pulled up, in order to minimize the effects of an excessive load factor.

RTC
4875-2. Which is the correct sequence for recovery from a spiraling, nose low, increasing airspeed, unusual flight attitude?

A— Increase pitch attitude, reduce power, and level wings.
B— Correct the bank attitude, raise the nose to a level attitude and reduce power.
C— Reduce power, raise the nose to level attitude, and correct the bank attitude.

To recover from an usual attitude, correct bank-and-pitch attitude, and adjust power as necessary. (H826) — FAA-H-8083-15, Chapter 6

ALL
4927. During recoveries from unusual attitudes, level flight is attained the instant

A— the horizon bar on the attitude indicator is exactly overlapped with the miniature airplane.
B— a zero rate of climb is indicated on the VSI.
C— the altimeter and airspeed needles stop prior to reversing their direction of movement.

A level-pitch attitude is indicated by the reversal and stabilization of the airspeed indicator and altimeter needles. (H818) — FAA-H-8083-15, Chapter 5

Answer (A) is incorrect because the precessing tendency of the attitude indicator makes it unreliable after recovery from an unusual attitude. Answer (B) is incorrect because the vertical speed indicator has a short lag time which does not allow it to show level flight at the instant it is achieved.

ALL
4936. (Refer to Figure 145.) What is the correct sequence for recovery from the unusual attitude indicated?

A— Reduce power, increase back elevator pressure, and level the wings.
B— Reduce power, level the wings, bring pitch attitude to level flight.
C— Level the wings, raise the nose of the aircraft to level flight attitude, and obtain desired airspeed.

The conditions associated with this unusual attitude are: airspeed is high, nose is below horizon, vertical speed shows excessive rate of descent, compass indicates right turn, and turn coordinator shows uncoordinated greater-than-standard rate turn to the right. To correct for this nose-low condition the correct sequence is: reduce power to prevent excessive airspeed and loss of altitude, correct the bank attitude with coordinated aileron and rudder pressure to straight flight by referring to the turn coordinator, and raise the nose to level flight attitude by smooth back elevator pressure. (H818) — FAA-H-8083-15, Chapter 5

Answer (A) is incorrect because the wings must be leveled before increasing the back elevator pressure to decrease the load factor. Answer (C) is incorrect because the power must be reduced before leveling the wings to prevent additional airspeed.

ALL
4938. (Refer to Figure 147.) Which is the correct sequence for recovery from the unusual attitude indicated?

A— Level wings, add power, lower nose, descend to original attitude, and heading.
B— Add power, lower nose, level wings, return to original attitude and heading.
C— Stop turn by raising right wing and add power at the same time, lower the nose, and return to original attitude and heading.

The conditions associated with this unusual attitude are: airspeed is decreasing, altitude is increasing, nose is above the horizon, vertical speed indicator shows an excessive rate of climb, compass indicates a right turn, and turn coordinator shows uncoordinated greater-than-standard rate turn to the right. To correct for this nose-high condition the correct sequence is: increase power, apply forward elevator pressure to lower the nose and prevent a stall, and correct the bank by applying coordinated aileron and rudder pressure to level the miniature aircraft and center the ball of the turn coordinator. (H818) — FAA-H-8083-15, Chapter 5

Answers (A) and (C) are incorrect because power should be increased and the nose lowered before leveling the wings.

Answers

4875-2 [B]	4927 [C]	4936 [B]	4938 [B]

Chapter 4
Navigation

NAVAID Classes *4–3*

DME *4–4*

VOR *4–7*

HSI *4–18*

ADF *4–24*

RMI *4–29*

RNAV and LORAN *4–32*

ILS *4–33*

MLS and GPS *4–38*

Chapter 4 **Navigation**

NAVAID Classes

VOR and VORTAC facilities are classed according to their operational use. There are three classes:

1. T (Terminal).
2. L (Low altitude).
3. H (High altitude).

The class defines the service volume of the NAVAID (which is the reception limit to which an unrestricted NAVAID may be used for random or unpublished route navigation).

The service volume of VOR/VORTACs can be found in the Airport/Facility Directory and in the AIM. The NAVAID class and any restrictions to use (unusable radials) will also be found in the Airport/Facility Directory. *See* Figure 4-1.

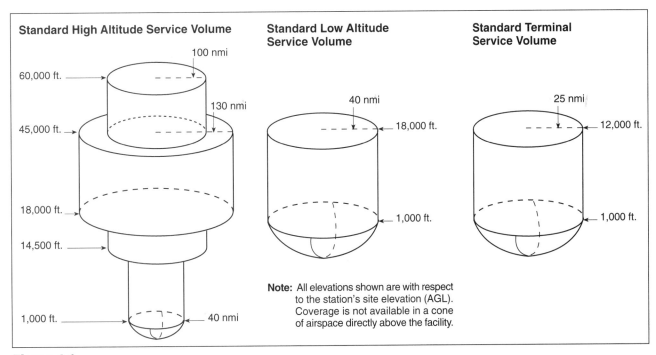

Figure 4-1

ALL
4271. (Refer to Figure 30.) Which restriction to the use of the OED VORTAC would be applicable to the (GNATS1.MOURN) departure?

A—R-333 beyond 30 NM below 6,500 feet.
B—R-210 beyond 35 NM below 8,500 feet.
C—R-251 within 15 NM below 6,100 feet.

In FAA Figure 30, the A/FD lists restrictions to the use of radio aids under the name of that radio aid, found under Radio Aids to Navigation. In this case, 333° falls within the unusable ranges between 280° and 345°. (J40) — AIM ¶1-1-8

Answers
4271 [A]

Chapter 4 **Navigation**

ALL
4400. For operations off established airways at 17,000 feet MSL in the contiguous U.S., (H) Class VORTAC facilities used to define a direct route of flight should be no farther apart than

A—75 NM.
B—100 NM.
C—200 NM.

Normal usable radius distance for high-altitude VOR/VORTAC NAVAIDS (H) within the contiguous United States between 14,500 feet and 17,999 feet is 100 NM. A direct-route flight should have a distance of no more than 200 NM between two H-class NAVAIDS. (J01) — AIM ¶1-1-8

ALL
4273. (Refer to Figures 27 and 30.) To which maximum service volume distance from the OED VORTAC should you expect to receive adequate signal coverage for navigation at the flight planned altitude?

A—100 NM.
B—80 NM.
C—40 NM.

Looking at FAA Figure 30, the OED (Medford) VORTAC class is written in parenthesis in front of the facility in the A/FD. In this case it is (H), or high. By referring to the radio class designations in FAA Legend 27, note that at 8,000 feet (flight planned altitude shown in FAA Figure 27), the High altitude VORTAC normal usable range is 40 NM. (J01) — AIM ¶1-1-8

DME

Distance Measuring Equipment (DME) operates on the principle of a timed UHF signal that is transmitted from the aircraft to the ground station and back to the aircraft. This time is translated into a distance. The DME readout is presented in nautical miles (NM) and is slant range distance, not actual horizontal distance. When passing over a station, the range indicator will decrease until it indicates the height above the station. After crossing the station, it will start to increase. (If an aircraft passes over a station at 6,000 feet, the DME readout will be 1 NM).

The greatest error in DME indications occur when an aircraft is very close to the station at a high altitude. The accuracy of the DME unit is valid only for 1 or more nautical miles from the ground facility for each 1,000 feet of altitude.

The identifier heard on the VORTAC or VOR-DME frequency is actually the identifiers of two separate radios on a time-shared basis. The DME portion identifies itself once every thirty seconds. The remaining identifiers are the VOR portion. Assuming both components are operating normally, there will be an uninterrupted series of identifiers.

ALL
4397. Which distance is displayed by the DME indicator?

A—Slant range distance in NM.
B—Slant range distance in SM.
C—Line-of-sight direct distance from aircraft to VORTAC in SM.

The DME indicator displays slant range distance in nautical miles. (J01) — AIM ¶1-1-7

ALL
4399. Where does the DME indicator have the greatest error between ground distance to the VORTAC and displayed distance?

A—High altitudes far from the VORTAC.
B—High altitudes close to the VORTAC.
C—Low altitudes far from the VORTAC.

The greatest slant-range error occurs when flying directly over the DME facility (VORTAC, VOR/DME), when the displayed distance is the height above the facility in nautical miles. Slant-range error is negligible if the aircraft is 1 mile or more from the ground facility for each 1,000 feet of altitude above the elevation of the facility. (J01) — AIM ¶1-1-7

Answers

| 4400 | [C] | 4273 | [C] | 4397 | [A] | 4399 | [B] |

Chapter 4 **Navigation**

ALL
4674. (Refer to Figure 128.) How should a pilot determine when the DME at Price/Carbon County Airport is inoperative?

A—The airborne DME will always indicate "0" mileage.
B—The airborne DME will "search," but will not "lock on."
C—The airborne DME may appear normal, but there will be no code tone.

A DME tone is heard once for each three or four VOR/LOC tones. A single code each 30 seconds indicates DME is operative. No ident indicates the NAVAID is unreliable even though signals are received. (J01) — AIM ¶1-1-7(f)

ALL
4413. Which DME indication should you receive when you are directly over a VORTAC site at approximately 6,000 feet AGL?

A—0.
B—1.
C—1.3.

Distance information received from DME equipment is actual distance and not horizontal distance. It is measured in nautical miles, and since 6,000 feet equals 1 NM, the DME will show 1.0 NM above the VORTAC. (J01) — AIM ¶1-1-7(b)

Answer (A) is incorrect because the DME would only indicate zero at ground level next to the VORTAC. Answer (C) is incorrect because 1.3 would indicate an altitude of approximately 8,000 feet AGL (6,000 x 1.3).

ALL
4487. As a rule of thumb, to minimize DME slant range error, how far from the facility should you be to consider the reading as accurate?

A—Two miles or more for each 1,000 feet of altitude above the facility.
B—One or more miles for each 1,000 feet of altitude above the facility.
C—No specific distance is specified since the reception is line-of-sight.

Slant-range error is negligible if the aircraft is 1 mile or more from the ground facility for each 1,000 feet of altitude above the elevation of the facility. (I07) — FAA-H-8083-15

ALL
4472. As a rule of thumb, to minimize DME slant range error, how far from the facility should you be to consider the reading as accurate?

A—Two miles or more for each 1,000 feet of altitude above the facility.
B—One or more miles for each 1,000 feet of altitude above the facility.
C—No specific distance is specified since the reception is line-of-sight.

Slant-range error is negligible if the aircraft is 1 mile or more from the ground facility for each 1,000 feet of altitude above the elevation of the facility. (H832) — FAA-H-8083-15, Chapter 7

ALL
4663. When a VOR/DME is collocated under frequency pairings and the VOR portion is inoperative, the DME identifier will repeat at an interval of

A—20 second intervals at 1020 Hz.
B—30 second intervals at 1350 Hz.
C—60 second intervals at 1350 Hz.

VOR/DME, VORTAC, ILS/DME and LOC/DME facilities are identified by synchronized identifications which are transmitted on a time share basis when either the VOR or the DME is inoperative. It is important to recognize which identifier is retained for the operative facility. A single coded identification with a repetition interval of approximately 30 seconds indicates that the DME is operative. DME is identified by a coded tone modulated at 1350 Hz. (J01) — AIM ¶1-1-7(f)

Answers

4674 [C] 4413 [B] 4487 [B] 4472 [B] 4663 [B]

Chapter 4 **Navigation**

ALL
4412. What is the meaning of a single coded identification received only once approximately every 30 seconds from a VORTAC?

A—The VOR and DME components are operative.
B—VOR and DME components are both operative, but voice identification is out of service.
C—The DME component is operative and the VOR component is inoperative.

VOR/DME, VORTAC, ILS/DME, and LOC/DME facilities are identified by synchronized identifications which are transmitted on a time-share basis. When either the VOR or the DME is inoperative, it is important to recognize which identifier is retained for the operative facility. A single-coded identification with a repetition interval of approximately 30 seconds indicates that the DME is operative. (J01) — AIM ¶1-1-7(f)

Answer (A) is incorrect because a constant series of identity codes indicates that the VOR and DME are both working properly. Answer (B) is incorrect because voice identification operates independently of the identity codes.

ALL
4320. (Refer to Figure 55.) As a guide in making range corrections, how many degrees of relative bearing change should be used for each one-half mile deviation from the desired arc?

A—2° to 3°.
B—5° maximum.
C—10° to 20°.

As a guide in making range corrections, change the relative bearing 10 to 20° for each 1/2-mile deviation from the desired arc. (H832) — FAA-H-8083-15, Chapter 7

ALL
4664. When installed with the ILS and specified in the approach procedures, DME may be used

A—in lieu of the OM.
B—in lieu of visibility requirements.
C—to determine distance from TDZ.

When installed with the ILS and specified in the approach procedure, DME may be used in lieu of the OM. (J01) — AIM ¶1-1-10(e)

ALL
4669. How does a pilot determine if DME is available on an ILS/LOC?

A—IAP indicate DME\TACAN channel in LOC frequency box.
B—LOC\DME are indicated on en route low altitude frequency box.
C—LOC\DME frequencies available in the Airman's Information Manual.

DME/TACAN channels are associated with LOC frequencies and are on the plan view of the approach chart. (J42) — AIM ¶1-1-7

Answers (B) and (C) are incorrect because the frequencies are on the instrument approach charts (not low altitude enroute charts or the AIM).

Answers
4412 [C]   4320 [C]   4664 [A]   4669 [A]

VOR

VOR accuracy may be checked by means of a VOR Test Facility (VOT), ground or airborne checkpoints, or by checking dual VORs against each other. A VOT location and frequency can be found in the Airport/Facility Directory (A/FD) and on the Air-to-Ground Communications Panel of the Low Altitude Enroute Chart. To use the VOT, tune to the appropriate frequency and center the CDI. The omni-bearing selector should read 0° with a FROM indication, or 180° with a TO indication. The allowable error is ±4°. VOR receiver checkpoints are listed in the A/FD. With the appropriate frequency tuned and the OBS set to the published certified radial, the CDI should center with a FROM indication when the aircraft is over the designated check point. Allowable accuracy is ±4° for a ground check, and ±6° for an airborne check. If the aircraft is equipped with dual VORs, they may be checked against each other. The maximum permissible variation when tuned to the same VOR is 4°.

The pilot must log the results of the VOR accuracy test in the aircraft logbook or other record. The log must include the date, place, bearing error, if any, and a signature.

All VOR stations transmit an identifier. It is a three-letter Morse code signal interrupted only by a voice identifier on some stations, or to allow the controlling flight service station to speak on the frequency. Absence of a VOR identifier indicates maintenance is being performed on the station and the signal may not be reliable.

All VOR receivers have at least the essential components shown in Figure 4-2.

The pilot may select the desired course or radial by turning the Omni Bearing Selector (OBS). The Course Deviation Indicator (CDI) centers when the aircraft is on the selected radial or its reciprocal. A full-scale deflection of the CDI from the center represents a deviation of approximately 10° to 12°. The TO/FROM Indicator (ambiguity indicator) shows whether the selected course will take the aircraft TO or FROM the station. A TO indication shows that the OBS selection is on the other side of the VOR station. A FROM indication shows that the OBS selection and the aircraft are on the same side of the VOR station. When an aircraft flies over a VOR, the TO/FROM indicator will reverse, indicating station passage.

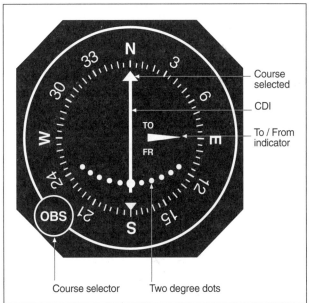

Figure 4-2. VOR indicators

The position of the aircraft can always be determined by rotating the OBS until the CDI centers with a FROM indication. The course displayed indicates the radial FROM the station. The VOR indicator displays information as though the aircraft were going in the direction of the course selected. However, actual heading does not influence the display. See Figure 4-3 on the next page.

VOR radials, all of which originate at the VOR antenna, diverge as they radiate outward. For example, while the 011° radial and the 012° radial both start at the same point, 1 NM from the antenna, they are 100 feet apart. When they are 2 NM from the antenna, they are 200 feet apart. So at 60 NM, the radials would be 1 NM (6,000 feet) apart. See Figure 4-4 on the next page.

Chapter 4 **Navigation**

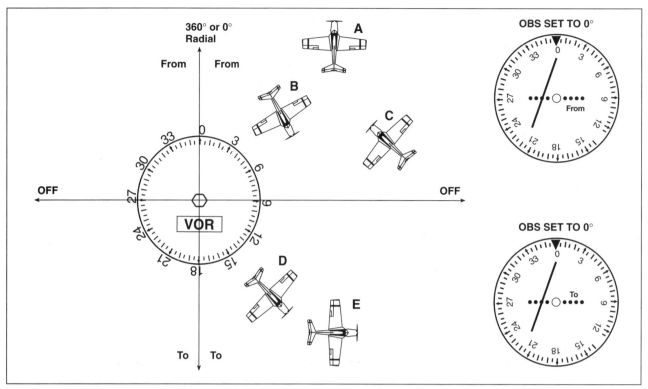

Figure 4-3. VOR display

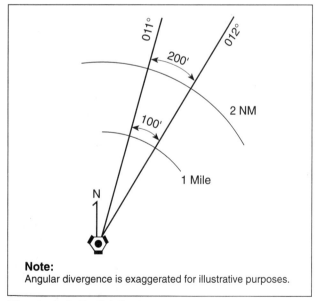

Note:
Angular divergence is exaggerated for illustrative purposes.

Figure 4-4. Radial divergence

Chapter 4 **Navigation**

The VOR indicator uses a series of dots to indicate any deviation from the selected course, with each dot equal to approximately 2° of deviation. Thus, a one-dot deviation at a distance of 30 NM from the station would indicate that the aircraft was 1 NM from the selected radial (200 feet x 30 = 6,000 feet).

To orient where the aircraft is in relation to the VOR, first determine which radial is selected (look at the OBS setting). Next, determine whether the aircraft is flying to or away from the station (look at the TO/FROM indicator), to find which hemisphere the aircraft is in. Last, determine how far off course the aircraft is from the selected course (look at the CDI needle deflection) to find which quadrant the aircraft is in. Remember that aircraft heading does not affect orientation to the VOR.

ALL
4044. Which data must be recorded in the aircraft log or other appropriate log by a pilot making a VOR operational check for IFR operations?

A—VOR name or identification, date of check, amount of bearing error, and signature.
B—Place of operational check, amount of bearing error, date of check, and signature.
C—Date of check, VOR name or identification, place of operational check, and amount of bearing error.

Each person making the VOR operational check shall enter the date, place, bearing error, and sign the aircraft log or other record. (B10) — 14 CFR §91.171(d)

Answers (A) and (C) are incorrect because the VOR name or identification is not required.

ALL
4046. What record shall be made in the aircraft log or other permanent record by the pilot making the VOR operational check?

A—The date, place, bearing error, and signature.
B—The date, frequency of VOR or VOT, number of flight hours since last check, and signature.
C—The date, place, bearing error, aircraft total time, and signature.

Each person making the VOR operational check shall enter the date, place, bearing error, and sign the aircraft log or other record. (B10) — 14 CFR §91.171(d)

Answer (B) is incorrect because it is not necessary to log the frequency of VOR or VOT, or the number of flight hours since last check. Answer (C) is incorrect because it is not necessary to log the aircraft total time.

ALL
4054. When making an airborne VOR check, what is the maximum allowable tolerance between the two indicators of a dual VOR system (units independent of each other except the antenna)?

A—4° between the two indicated radials of a VOR.
B—Plus or minus 4° when set to identical radials of a VOR.
C—6° between the two indicated radials of a VOR.

If dual system VOR (units independent of each other except for the antenna) is installed in the aircraft, the pilot shall tune both systems to the same VOR ground facility and note the indicated radials TO that station. The maximum permissible variation between the two indicated radials is 4° on the ground or in the air. (J01) — 14 CFR §91.171(c)

Answer (B) is incorrect because the VORs are not set to identical radials of a VOR, but rather the CDI is centered and the bearing noted. Answer (C) is incorrect because the maximum allowable tolerance between the two VORs is 4 degrees (not 6).

ALL
4362. (Refer to Figure 76.) Which indication would be an acceptable accuracy check of both VOR receivers when the aircraft is located on the VOR receiver checkpoint at the Helena Regional Airport?

A—A.
B—B.
C—C.

If a test signal is not available at the airport of intended departure, use a point on an airport surface designated as a VOR system checkpoint by the Administrator, the maximum permissible bearing error for a ground check is ±4°. The tail of the RMI will show which radial that the aircraft is on. The ground check point for Helena is located on the 237° radial from the HLN VOR, therefore, both RMI tails should be indicating 237° ±4°. (J01) — AIM ¶1-1-4

Answers
4044 [B]　　　4046 [A]　　　4054 [A]　　　4362 [C]

Chapter 4 Navigation

ALL

4383. While airborne, what is the maximum permissible variation between the two indicated bearings when checking one VOR system against the other?

A—Plus or minus 4° when set to identical radials of a VOR.
B—4° between the two indicated bearings to a VOR.
C—Plus or minus 6° when set to identical radials of a VOR.

If dual system VOR (units independent of each other except for the antenna) is installed in the aircraft, the pilot may check one system against the other. Both systems shall be tuned to the same VOR ground facility and note the indicated bearings to that station. The maximum permissible variation between the two indicated bearings is 4° on the ground or in the air. (J01) — AIM ¶1-1-4

Answer (A) is incorrect because the CDI needles must be centered (not set to identical radials). Answer (C) is incorrect because the maximum permissible variation is 4° (not 6°).

ALL

4382. (Refer to Figure 81.) When checking a dual VOR system by use of a VOT, which illustration indicates the VOR's are satisfactory?

A—1.
B—2.
C—4.

The maximum permissible indicated bearing error is ±4°, when using a VOT to check the equipment. A valid VOT signal should be indicated as the 360° radial FROM the station. Since RMI arrows point TO the station, an RMI should indicate 180° ±4° on any OBS setting when using a VOT facility, which is the case in RMI 1. (J01) — AIM ¶1-1-4

Answer (B) is incorrect because RMI 2 indicates 0° and 180° TO (not 360° FROM). Answer (C) is incorrect because RMI 4 indicates 001° and 0° TO (not 360° FROM).

ALL

4377. How should the pilot make a VOR receiver check when the aircraft is located on the designated checkpoint on the airport surface?

A—Set the OBS on 180° plus or minus 4°; the CDI should center with a FROM indication.
B—Set the OBS on the designated radial. The CDI must center within plus or minus 4° of that radial with a FROM indication.
C—With the aircraft headed directly toward the VOR and the OBS set to 000°, the CDI should center within plus or minus 4° of that radial with a TO indication.

A VOR receiver ground checkpoint requires setting the OBS on a designated radial (found in the A/FD) from a nearby VOR. It requires ±4° accuracy with a FROM indication. (J01) — AIM ¶1-1-4

Answer (A) is incorrect because the designated radial (not 180°) is used. Answer (C) is incorrect because the aircraft heading does not affect the VOR indications.

ALL

4385. (Refer to Figure 82.) Which is an acceptable range of accuracy when performing an operational check of dual VOR's using one system against the other?

A—1.
B—2.
C—4.

Using an RMI to perform a dual VOR check, the bearing pointers must be within 4° of each other, which is the case in RMI 4. (J01) — AIM ¶1-1-4

Answer (A) is incorrect because RMI 1 shows needles with 180° difference. Answer (B) is incorrect because RMI 2 shows needles with 10° difference.

ALL

4386. Where can the VOT frequency for a particular airport be found?

A—On the IAP Chart and in the Airport/Facility Directory.
B—Only in the Airport/Facility Directory.
C—In the Airport/Facility Directory and on the A/G Voice Communication Panel of the En Route Low Altitude Chart.

Locations of airborne checkpoints, ground checkpoints and VOTs are published in the Airport/Facility Directory and on the A/G Voice Communication Panel of the Enroute Low Altitude Chart. (J01) — AIM ¶1-1-4

Answers

| 4383 | [B] | 4382 | [A] | 4377 | [B] | 4385 | [C] | 4386 | [C] |

ALL

4387. Which indications are acceptable tolerances when checking both VOR receivers by use of the VOT?

A—360° TO and 003° TO, respectively.
B—001° FROM and 005° FROM, respectively.
C—176° TO and 003° FROM, respectively.

To test for VOR receiver accuracy center the CDI needle. The omnibearing selector should read 0° with the TO/FROM flag showing FROM and 180° with the flag showing TO. The allowable bearing error is ±4°. Therefore, a 176° TO indication for one of the VOR receivers and a 003° FROM for the other VOR is an acceptable combination. (J01) — AIM ¶1-1-4

ALL

4388. In which publication can the VOR receiver ground checkpoint(s) for a particular airport be found?

A—Aeronautical Information Manual.
B—En Route Low Altitude Chart.
C—Airport/Facility Directory.

A list of VOR receiver check points and VOR Test Facilities (VOT) can be found in the back of the Airport/Facility Directory. (J01) — AIM ¶1-1-4

ALL

4389. Which is the maximum tolerance for the VOR indication when the CDI is centered and the aircraft is directly over the airborne checkpoint?

A—Plus or minus 6° of the designated radial.
B—Plus 6° or minus 4° of the designated radial.
C—Plus or minus 4° of the designated radial.

The indicated VOR bearing when over the airborne (or over the ground point) checkpoint cannot be more than ±6° from the designated radial. The flag will have a FROM indication. (J01) — AIM ¶1-1-4

ALL

4391. When making an airborne VOR check, what is the maximum allowable tolerance between the two indicators of a dual VOR system (units independent of each other except the antenna)?

A—4° between the two indicated radials of a VOR.
B—Plus or minus 4° when set to identical radials of a VOR.
C—6° between the two indicated radials of a VOR.

If dual system VOR (units independent of each other except for the antenna) is installed in the aircraft, the pilot may check one system against the other. Both systems shall be tuned to the same VOR ground facility and note the indicated bearings to that station. The maximum permissible variation between the two indicated bearings is 4° on the ground or in the air. (J01) — AIM ¶1-1-4

ALL

4372. What is the maximum tolerance allowed for an operational VOR equipment check when using a VOT?

A—Plus or minus 4°.
B—Plus or minus 6°.
C—Plus or minus 8°.

The maximum tolerance allowed for an operational VOR equipment check when using a VOT is ±4°. (J01) — AIM ¶1-1-4

ALL

4384. How should the pilot make a VOR receiver check when the airplane is located on the designated checkpoint on the airport surface?

A—With the aircraft headed directly toward the VOR and the OBS set to 000°, the CDI should center within plus or minus 4° of that radial with a TO indication.
B—Set the OBS on the designated radial. The CDI must center within plus or minus 4° of that radial with a FROM indication.
C—Set the OBS on 180° plus or minus 4°; the CDI should center with a FROM indication.

If a test signal is not available at the airport of intended departure, use a point on the airport surface designated as a VOR system checkpoint. The maximum permissible bearing error is ±4° with a FROM indication. (J01) — AIM ¶1-1-4

Answer (A) is incorrect because the heading of the aircraft does not affect the VOR indications. Answer (C) is incorrect because the designated radial (not 180°) should be used.

Answers

4387 [C]   4388 [C]   4389 [A]   4391 [A]   4372 [A]   4384 [B]

Chapter 4 Navigation

ALL
4376. When using VOT to make a VOR receiver check, the CDI should be centered and the OBS should indicate that the aircraft is on the

A—090 radial.
B—180 radial.
C—360 radial.

To use a VOT service, tune in the VOT frequency with the VOR receiver. With the Course Deviation Indicator (CDI) centered, the omni-bearing selector (OBS) should read 0° with a FROM flag, or 180° with a TO flag. The question asks for radial, so you are "0° or 360° FROM." (J01) — AIM ¶1-1-4

ALL
4378. When the CDI needle is centered during an airborne VOR check, the omni-bearing selector and the OBS indicator should read

A—within 4° of the selected radial.
B—within 6° of the selected radial.
C—0° TO, only if you are due south of the VOR.

If no check signal or point is available while in flight, select a VOR radial that lies along the centerline of an established VOR airway and maneuver the aircraft directly over a prominent ground point. Note the VOR bearing indicated by the receiver. The maximum permissible variation between the published radial and the indicated bearing is ±6°. (J01) — AIM ¶1-1-4

ALL
4326. (Refer to Figure 58.) Which indications on the VOR receivers and DME at the Easterwood Field VOR receiver checkpoint would meet the regulatory requirement for this flight?

	VOR No. 1	TO/FROM	VOR No. 2	TO/FROM	DME
A—	097°	FROM	101°	FROM	3.3
B—	097°	TO	096°	TO	3.2
C—	277°	FROM	280°	FROM	3.3

The bottom of FAA Figure 58 gives the VOR receiver checkpoints. Looking at College Station (Easterwood Field), the checkpoint is on the ground (G), the azimuth is on R-097, and the distance from the facility is 3.2 NM. While on R-097, the CDI should be centered with an OBS setting of 097 FROM or 277 TO the station, with an acceptable error of ±4. Therefore the acceptable VOR indications are 097 FROM and 101 FROM the station. (J01) — AIM ¶1-1-4

ALL
4337. (Refer to Figure 64.) The course deviation indicator (CDI) are centered. Which indications on the No. 1 and No. 2 VOR receivers over the Lafayette Regional Airport would meet the requirements for the VOR receiver check?

	VOR No. 1	TO/FROM	VOR No. 2	TO/FROM
A—	162°	TO	346°	FROM
B—	160°	FROM	162°	FROM
C—	341°	FROM	330°	FROM

The top-half of the A/FD excerpt for LFT contains the VOR receiver checkpoints. The Lafayette checkpoint is airborne at 1,000 feet over the rotating beacon and the azimuth from the VORTAC is on the R-340. Once established on R-340, the CDI needle should be centered with an OBS setting of 160° TO or 340° FROM, with an acceptable error being ±6°. (J34) — AIM ¶1-1-4

Answer (B) is incorrect because the radial from the station is 340° which would result in a TO indication with the OBS set to 160°. Answer (C) is incorrect because the No. 2 VOR OBS exceeds the acceptable error of ±6°.

ALL
4410. What indication should a pilot receive when a VOR station is undergoing maintenance and may be considered unreliable?

A—No coded identification, but possible navigation indications.
B—Coded identification, but no navigation indications.
C—A voice recording on the VOR frequency announcing that the VOR is out of service for maintenance.

The only positive method of identifying a VOR is by its Morse Code identification or by the recorded automatic voice identification which is always indicated by use of "VOR" following the range's name. During periods of maintenance, the coded facility identification is removed. If a navigational signal is being received without the corresponding Morse Code identification, the VOR is out for maintenance. (J01) — AIM ¶1-1-3(c)

Answer (B) is incorrect because the coded identification is unavailable when the station is undergoing maintenance. Answer (C) is incorrect because an out-of-service VOR is not identified by a voice recording.

Answers

| 4376 | [C] | 4378 | [B] | 4326 | [A] | 4337 | [A] | 4410 | [A] |

ALL
4411. A particular VOR station is undergoing routine maintenance. This is evidenced by

A—removal of the navigational feature.
B—broadcasting a maintenance alert signal on the voice channel.
C—removal of the identification feature.

The only positive method of identifying a VOR is by its Morse Code identification or by the recorded automatic voice identification which is always indicated by use of "VOR" following the range's name. During periods of maintenance, the coded facility identification is removed. If a navigational signal is being received without the corresponding Morse Code identification, the VOR is out for maintenance. (J01) — AIM ¶1-1-3(c)

Answer (A) is incorrect because the navigational signal may still be transmitted, although it will be inaccurate. Answer (B) is incorrect because an out-of-service VOR is not identified by a voice recording.

ALL
4548. What angular deviation from a VOR course centerline is represented by a full-scale deflection of the CDI?

A—4°.
B—5°.
C—10°.

Full-needle deflection from the center position to either side of the dial indicates the aircraft is 10° or more off course, assuming normal needle sensitivity. (H831) — FAA-H-8083-15, Chapter 7

ALL
4549. When using VOR for navigation, which of the following should be considered as station passage?

A—The first movement of the CDI as the airplane enters the zone of confusion.
B—The moment the TO-FROM indicator becomes blank.
C—The first positive, complete reversal of the TO-FROM indicator.

Approach to the station is indicated by flickering of the TO/FROM indicator and CDI as the aircraft flies into the "cone of confusion" (no-signal area). Station passage is shown by the first positive, complete reversal of the TO/FROM indicator. (H831) — FAA-H-8083-15, Chapter 7

Answer (A) is incorrect because the first movements of the CDI as the airplane enters the zone of confusion indicates the airplane is approaching the station. Answer (B) is incorrect because the moment the TO/FROM indicator becomes blank indicates the airplane is in the cone of confusion over the VOR.

ALL
4550. Which of the following should be considered as station passage when using VOR?

A—The first flickering of the TO-FROM indicator and CDI as the station is approached.
B—The first full-scale deflection of the CDI.
C—The first complete reversal of the TO-FROM indicator.

Station passage is shown by the first positive, complete reversal of the TO/FROM indicator. (H831) — FAA-H-8083-15, Chapter 7

Answer (A) is incorrect because the first flickering of the TO/FROM indicator means the aircraft is approach the station. Answer (B) is incorrect because the first full-scale deflection of the CDI needle indicates the aircraft is in the cone of confusion.

ALL
4551. When checking the sensitivity of a VOR receiver, the number of degrees in course change as the OBS is rotated to move the CDI from center to the last dot on either side should be between

A—5° and 6°.
B—8° and 10°.
C—10° and 12°.

Course sensitivity may be checked by noting the number of degrees of change in the course selected as you rotate the OBS to move the CDI from center to the last dot on either side. This should be between 10° and 12°. (H831) — FAA-H-8083-15, Chapter 7

ALL
4552. A VOR receiver with normal five-dot course sensitivity shows a three-dot deflection at 30 NM from the station. The aircraft would be displaced approximately how far from the course centerline?

A—2 NM.
B—3 NM.
C—5 NM.

Aircraft displacement from course is approximately 200 feet per dot per nautical mile. For example, at 30 NM from the station, one dot deflection indicates approximately 1 NM displacement of the aircraft from the course centerline. Therefore, a three-dot deflection would mean the aircraft is approximately 3 NM from the course centerline. (H576) — FAA-H-8083-3, Chapter 11

Answers (A) and (C) are incorrect because 3 dots per 200 feet per 30 NM = 3 NM.

Answers

| 4411 | [C] | 4548 | [C] | 4549 | [C] | 4550 | [C] | 4551 | [C] | 4552 | [B] |

Chapter 4 Navigation

ALL
4553. An aircraft which is located 30 miles from a VOR station and shows a 1/2 scale deflection on the CDI would be how far from the selected course centerline?

A— 1 1/2 miles.
B— 2 1/2 miles.
C— 3 1/2 miles.

Aircraft displacement from course is approximately 200 feet per dot per nautical mile. For example, at 30 NM from the station, a 1-dot deflection indicates approximately 1 NM displacement from the course centerline. Assuming a receiver with normal course sensitivity and full-scale deflection being five dots, a half-scale deflection would be two and one-half dots. Therefore, at 30 NM, the aircraft would be approximately 2.5 NM from course centerline. (I08) — FAA-H-8083-15

ALL
4554. What angular deviation from a VOR course centerline is represented by a 1/2 scale deflection of the CDI?

A— 2°.
B— 4°.
C— 5°.

Full-needle deflection from the center position to either side of the dial indicates the aircraft is 10° or more off course, assuming normal needle sensitivity. Therefore, a half-scale deflection would indicate an angular deviation of one-half of 10°, or 5°. (H831) — FAA-H-8083-15, Chapter 7

ALL
4556. After passing a VORTAC, the CDI shows 1/2 scale deflection to the right. What is indicated if the deflection remains constant for a period of time?

A— The airplane is getting closer to the radial.
B— The OBS is erroneously set on the reciprocal heading.
C— The airplane is flying away from the radial.

Aircraft displacement from course is approximately 200 feet per dot per nautical mile. Therefore, in this instance, the aircraft is two and one-half dots, or 500 feet, from the radial per nautical mile from the station (at 10 miles, 5,000 feet; at 20 miles, 10,000 feet, and so on). The constant deflection indicates the airplane is flying away from the radial. (H831) — FAA-H-8083-15, Chapter 7

Answer (A) is incorrect because the deflection would have to decrease (not stay constant) to signify the airplane is getting closer to the radial. Answer (B) is incorrect because with the OBS set on the reciprocal heading, the CDI would give reverse indications (not stay constant).

ALL
4557. (Refer to Figure 95.) What is the lateral displacement of the aircraft in NM from the radial selected on the No. 1 NAV?

A— 5.0 NM.
B— 7.5 NM.
C— 10.0 NM.

Aircraft displacement from course is approximately 200 feet per dot per nautical mile. For example, at 30 NM from the station, a one dot deflection indicates approximately a 1 NM displacement of the aircraft from the course centerline. In this case, the CDI is displaced 2.5 dots at 60 NM, which equals 2 NM times 2.5 dots, or approximately a 5 NM displacement from course centerline. (H576) — FAA-H-8083-3, Chapter 11

Answer (B) is incorrect because 7.5 NM would be indicated by a 3/4 needle deflection. Answer (C) is incorrect because 10 NM would be indicated by full needle deflection.

ALL
4558. (Refer to Figure 95.) On which radial is the aircraft as indicated by the No. 1 NAV?

A— R-175.
B— R-165.
C— R-345.

The TO/FROM indicator is the triangular shaped pointer(s) (▲). When the indicator points to the head of the course arrow, it indicates that the course selected, if properly intercepted and flown, will take the aircraft TO the selected facility, and vice versa. In this case, the indicator is pointing to the TAIL of the course arrow, so if flown, the course selected (350) would take the aircraft away FROM the selected facility. Each dot represents 2°. In this case, the aircraft is two and one-half dots (5°) to the LEFT of the selected course, therefore the aircraft is on R-345. (H831) — FAA-H-8083-15, Chapter 7

ALL
4559. (Refer to Figure 95.) Which OBS selection on the No. 1 NAV would center the CDI and change the ambiguity indication to a TO?

A— 175°.
B— 165°.
C— 345°.

Answers

4553 [B]   4554 [C]   4556 [C]   4557 [A]   4558 [C]   4559 [B]

Currently, the course selector is set to 350 (the arrow head of the needle is pointing to 350), with a FROM indication (the triangular shaped pointers are pointing to the tail of the needle), and a left 2.5 dot needle deflection (5° left of course). This means the aircraft is on R-345 (350 – 5). To center the CDI and change the ambiguity indication to a TO, the OBS should be turned to the reciprocal of 345° (345 – 180), or 165°. (H831) — FAA-H-8083-15, Chapter 7

ALL
4560. (Refer to Figure 95.) What is the lateral displacement in degrees from the desired radial on the No. 2 NAV?

A— 1°.
B— 2°.
C— 4°.

Full-scale deflection is 10°, so each dot represents 2°. The CDI is displaced two dots, therefore the lateral displacement is 4°. (H831) — FAA-H-8083-15, Chapter 7

ALL
4561. (Refer to Figure 95.) Which OBS selection on the No. 2 NAV would center the CDI?

A— 174°.
B— 166°.
C— 335°.

Full-scale deflection is 10°, so each dot represents 2°. The CDI is displaced two dots (4°). The OBS is set at 170° with a FROM indication, and the aircraft is 4° to the RIGHT of course (or on the 174° radial). Simply rotating the OBS to 174° would center the CDI. (H831) — FAA-H-8083-15, Chapter 7

ALL
4562. (Refer to Figure 95.) Which OBS selection on the No. 2 NAV would center the CDI and change the ambiguity indication to a TO?

A— 166°.
B— 346°.
C— 354°.

Rotating the OBS until the reciprocal of 170° (350°) is shown under the course arrow will cause the ambiguity indication to change to TO. The CDI now indicates that the aircraft is 4° to the right of course (or on the 354° course TO the selected facility). Simply rotating the OBS to 354° will center the CDI. (H831) — FAA-H-8083-15, Chapter 7

ALL
4601. (Refer to Figure 106.) The course selector of each aircraft is set on 360°. Which aircraft would have a FROM indication on the ambiguity meter and the CDI pointing left of center?

A— 1.
B— 2.
C— 3.

A FROM indication, with a 360° course selection, means the aircraft is in the northern hemisphere. A left CDI needle means the aircraft is in the right quadrant. Airplane 2 fits this description. Remember that the VOR works without regard to the aircraft heading. (H831) — FAA-H-8083-15, Chapter 7

Answer (A) is incorrect because aircraft 1 would have the CDI pointing right (not left) of center. Answer (C) is incorrect because aircraft 3 would have a TO (not FROM) indication, and the CDI would be pointing right (not left) of center.

ALL
4666. Full scale deflection of a CDI occurs when the course deviation bar or needle

A— deflects from left side of the scale to right side of the scale.
B— deflects from the center of the scale to either far side of the scale.
C— deflects from half scale left to half scale right.

Full-scale deflection of the CDI represents 10° or more or a five-dot deflection to one side or the other from center. This will be indicated when the needle moves from the center of the scale to either far side of the scale. (H831) — FAA-H-8083-15, Chapter 7

ALL
4338. (Refer to Figures 65 and 66.) What is your position relative to GRICE intersection?

A— Right of V552 and approaching GRICE intersection.
B— Right of V552 and past GRICE intersection.
C— Left of V552 and approaching GRICE intersection.

The #1 VOR (set to TIBBY VOR) shows you should fly left to get back on course (heading eastbound on V552). Consequently, you are right of course. The #2 NAV indicator is tuned to a localizer, which is why there is no TO/FROM indication. The CDI position indicates that if you were inbound on the localizer, the course center-

Continued

line would be to your left. This corresponds to a position approaching GRICE intersection. (J35) — Enroute Low Altitude Chart

Answer (B) is incorrect because the CDI on the #2 NAV indicator would be to the right if you had passed GRICE. Answer (C) is incorrect because the CDI on the #1 NAV indicator would be to the right if you were left of course.

ALL
4347. (Refer to Figures 71 and 71A.) What is your position relative to the Flosi intersection Northbound on V213?

A—West of V213 and approaching the Flosi intersection.
B—East of V213 and approaching the Flosi intersection.
C—West of V213 and past the Flosi intersection.

The #1 VOR is tuned to Kingston VORTAC (IGN). With the OBS set to 265, a FROM indication, and the CDI deflected to the right, the aircraft is south of the 265 radial and approaching the Flosi intersection. The #2 VOR is tuned to Sparta VORTAC (SAX). With the OBS set to 029, a FROM indication, and the CDI deflected to the right, the aircraft is west of the 029 radial. (J35) — Enroute Low Altitude Chart

Answer (B) is incorrect because the #2 VOR would have a left deflection of the CDI needle, if the aircraft was east of V213. Answer (C) is incorrect because the #1 VOR would have a left-deflected CDI needle, if the aircraft was past the Flosi intersection.

ALL
4495. (Refer to Figures 87 and 88.) What is your position with reference to FALSE intersection (V222) if your VOR receivers indicate as shown?

A—South of V222 and east of FALSE intersection.
B—North of V222 and east of FALSE intersection.
C—South of V222 and west of FALSE intersection.

The #1 VOR is tuned to Beaumont VORTAC (BPT). With the OBS set to 264, FROM, and the CDI needle deflected right, the aircraft is south of V222. The #2 VOR is tuned to Daisetta VORTAC (DAS). With the OBS set to 139, FROM, and the CDI needle deflected right, the aircraft is right (east) of the DAS 139 radial. Since the 139 radial is east of the 142 radial (which defines FALSE intersection), the aircraft is east of FALSE intersection. (J17) — AIM ¶5-3-4

Answer (B) is incorrect because the #1 VOR would have a left CDI deflection if it was north of V222. Answer (C) is incorrect because #2 VOR would have a left CDI deflection if it was west of FALSE intersection.

ALL
4507. (Refer to Figures 89 and 90.) What is your relationship to the airway while en route from BCE VORTAC to HVE VORTAC on V8?

A—Left of course on V8.
B—Left of course on V382.
C—Right of course on V8.

The OBS #1 indicates you are to the right of V382 (033° radial FROM BCE) which coincides with OBS #2's indication that you to the left of V8 (046° radial TO HVE). (J35) — Enroute Low Altitude Chart

ALL
4606. (Refer to Figure 109.) In which general direction from the VORTAC is the aircraft located?

A—Northeast.
B—Southeast.
C—Southwest.

The course selected is 180° and the TO/FROM indicator is pointing to the course, which means the aircraft is north of the course. The CDI needle is deflected to the right, which means the aircraft is left (or east) of course. Therefore, the aircraft is northeast of the VORTAC. See the figure below. (H831) — FAA-H-8083-15, Chapter 7

Answer (B) is incorrect because if the aircraft was southeast there would be a FROM indication. Answer (C) is incorrect because if the aircraft was southwest there would be a FROM indication, and the CDI needle would be deflected to the left.

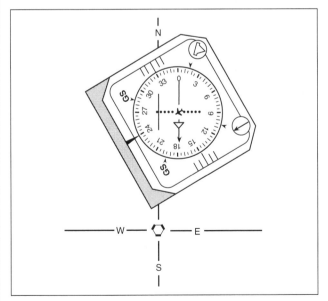

Question 4606

Answers
4347 [A]   4495 [A]   4507 [A]   4606 [A]

ALL
4607. (Refer to Figure 110.) In which general direction from the VORTAC is the aircraft located?

A—Southwest.
B—Northwest.
C—Northeast.

The course selected is 060° and the TO/FROM indicator is pointing away from the course, which means the aircraft is northeast of the course. The CDI needle is deflected right, which means the aircraft is left (or north) of course. Therefore, the aircraft is northeast of the VORTAC. See the figure below. (H831) — FAA-H-8083-15, Chapter 7

Answer (A) is incorrect because if the aircraft was southwest there would be a TO indication. Answer (B) is incorrect because if the aircraft was northwest the CDI needle would be fully deflected to the right.

ALL
4608. (Refer to Figure 111.) In which general direction from the VORTAC is the aircraft located?

A—Northeast.
B—Southeast.
C—Northwest.

The course selected is 360° and the TO/FROM indicator is pointing away from the course, which means the aircraft is north of the course. The CDI needle is deflected to the right, which means the aircraft is left (or west) of course. Therefore, the aircraft is northwest of the VORTAC. See the figure below. (H831) — FAA-H-8083-15, Chapter 7

Answer (A) is incorrect because if the aircraft was northeast the CDI needle would be deflected to the left. Answer (B) is incorrect because if the aircraft was southeast there would be a TO indication.

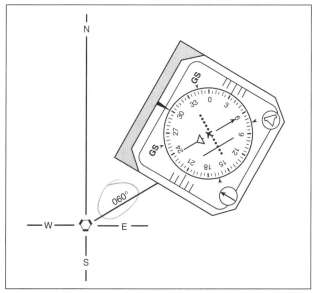

Question 4607

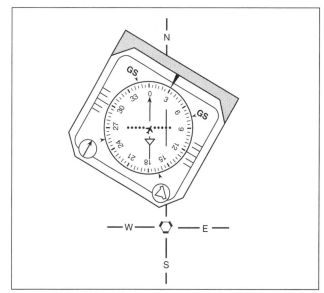

Question 4608

Answers

4607 [C]   4608 [C]

Chapter 4 **Navigation**

HSI

The Horizontal Situation Indicator (HSI) is a combination of two instruments: the heading indicator and the VOR. *See* Figure 4-5.

The aircraft heading displayed on the rotating azimuth card under the upper lubber line in Figure 4-5 is 330°. The course indicating arrowhead that is shown is set to 300°. The tail of the course indicating arrow indicates the reciprocal, or 120°.

The course deviation bar operates with a VOR/LOC navigation receiver to indicate either left or right deviations from the course that is selected with the course indicating arrow. It moves left or right to indicate deviation from the centerline in the same manner that the angular movement of a conventional VOR/LOC needle indicates deviation from course.

The desired course is selected by rotating the course indicating arrow in relation to the azimuth card by means of the course set knob. This gives the pilot a pictorial presentation. The fixed aircraft symbol and the course deviation bar display the aircraft relative to the selected course as though the pilot was above the aircraft looking down.

The TO/FROM indicator is a triangular-shaped pointer. When this indicator points to the head of the course arrow, it indicates that the course selected, if properly intercepted and flown, will take the aircraft TO the selected facility, and vice versa.

The glide slope deviation pointer indicates the relationship of the aircraft to the glide slope. When the pointer is below the center position, the aircraft is above the glide slope and an increased rate of descent is required.

To orient where the aircraft is in relation to the facility, first determine which radial is selected (look at the arrowhead). Next, determine whether the aircraft is flying to or away from the station (look at the TO/FROM indicator) to find which hemisphere the aircraft is in. Next, determine how far from the selected course the aircraft is (look at the deviation bar) to find which quadrant the aircraft is in. Last, consider the aircraft heading (under the lubber line) to determine the aircraft's position within the quadrant. Note that you will have reverse sensing if you're flying the back course. You will know if you're on the back course if the HSI is tuned to the reciprocal of where the localizer is positioned. (For example: if the HSI is tuned to 090 but the localizer is on the 270 extension, then you are on the back course and will have reverse sensing.)

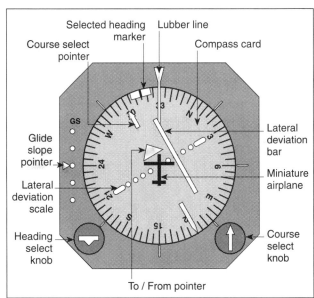

Figure 4-5. Horizontal Situation Indicator (HSI)

Chapter 4 **Navigation**

ALL
4563. (Refer to Figures 96 and 97.) To which aircraft position(s) does HSI presentation "A" correspond?

A—9 and 6.
B—9 only.
C—6 only.

HSI Indicator "A" is set up with the head of the arrow pointing to 270°. The Course Deviation Indicator is centered; therefore, the aircraft is on the extended centerline of RW #9 and #27. With a heading of 360°, Indicator "A" represents an airplane at position #6 or #9. See the figure below. (H831) — FAA-H-8083-15, Chapter 7

Answers (B) and (C) are incorrect because HSI indicator A corresponds to both aircraft positions 6 and 9.

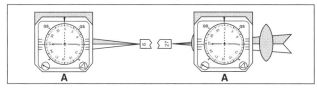

Question 4563

ALL
4564. (Refer to Figures 96 and 97.) To which aircraft position(s) does HSI presentation "B" correspond?

A—11.
B—5 and 13.
C—7 and 11.

HSI Indicator "B" is set up with the head of the arrow pointing to 090°. The CDI indication is reversed and the aircraft is actually to the south of the extended centerline. Indicator "B" then, with the aircraft flying on a heading of 090°, could be at position #13 and #5. Remember that the localizer receiver does not know where you are in relation to the antenna site. Note that you're flying the backcourse (the HSI is turned to 090 but the localizer is on the 270 extension as depicted in Figure 96) so you have reverse sensing. See the figure below. (H831) — FAA-H-8083-15, Chapter 7

Answers (A) and (C) are incorrect because aircrafts 11 and 7 have a 270° heading (not 90°).

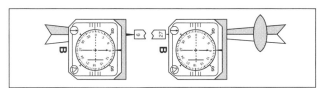

Question 4564

ALL
4565. (Refer to Figures 96 and 97.) To which aircraft position does HSI presentation "C" correspond?

A—9.
B—4.
C—12.

HSI Indicator "C" is set up with the head of the arrow pointing to 090°. With the CDI centered, the aircraft is obviously on the extended centerline. With a heading of 090°, position #12 is the only one which would have that indication. See the figure below. (H831) — FAA-H-8083-15, Chapter 7

Answer (A) is incorrect because aircraft 9 has a 360° heading (not 090°). Answer (B) is incorrect because aircraft 4 has a 270° heading (not 090°).

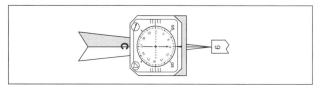

Question 4565

ALL
4566. (Refer to Figures 96 and 97.) To which aircraft position does HSI presentation "D" correspond?

A—1.
B—10.
C—2.

HSI Indicator "D" is set up with the head of the arrow pointing to 090°. The aircraft is to the south of course, on a heading of 310° with position #2 being the only possible choice. See figure below. — (H831) — FAA-H-8083-15, Chapter 7

Answer (A) is incorrect because aircraft 1 has a heading of 225° (not 310°). Answer (B) is incorrect because aircraft 10 has a heading of 135° (not 310°).

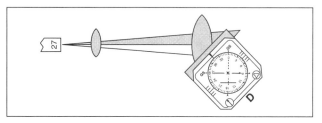

Question 4566

Answers

4563 [A]   4564 [B]   4565 [C]   4566 [C]

Chapter 4 **Navigation**

ALL
4567. (Refer to Figures 96 and 97.) To which aircraft position(s) does HSI presentation "E" correspond?

A—8 only.
B—3 only.
C—8 and 3.

HSI Indicator "E" is set up with the head of the arrow pointing to 090°. The aircraft is to the south of the extended centerline, with position #8 or #3 being the only possible answer for an aircraft on the heading of 045°. See the figure below. (H831) — FAA-H-8083-15, Chapter 7

Answers (A) and (B) are incorrect because both aircraft 8 and 3 are positioned south of the course with a 045° heading.

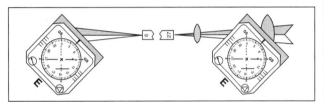

Question 4567

ALL
4568. (Refer to Figures 96 and 97.) To which aircraft position does HSI presentation "F" correspond?

A—4.
B—11.
C—5.

HSI Indicator "F" is set up with the head of the arrow pointing to 270°, and the CDI centered so the aircraft is on the extended centerline. Position #4 is the only one that would be correct for an aircraft on a heading of 270°. See the figure below. (H831) — FAA-H-8083-15, Chapter 7

Answer (B) is incorrect because aircraft 11 has a left course deviation bar. Answer (C) is incorrect because aircraft 5 has a 090° heading (not 270°).

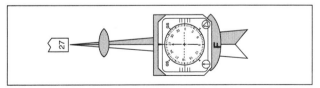

Question 4568

ALL
4569. (Refer to Figures 96 and 97.) To which aircraft position(s) does HSI presentation "G" correspond?

A—7 only.
B—7 and 11.
C—5 and 13.

HSI Indicator "G" is set up with the head of the arrow pointing to 270°. The CDI is deflected to the left, so the aircraft is to the north on a heading of 270°. Again, the indicator does not know which side of the antenna you are on, so position #7 or #11 would be appropriate selections. See the figure below. (H831) — FAA-H-8083-15, Chapter 7

Answer (A) is incorrect because aircraft 11 would also have the indication of HSI "G". Answer (C) is incorrect because aircraft 5 and 13 have a 090° heading (not 270°), and are south (not north) of the course.

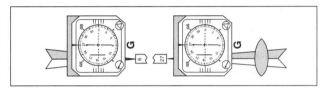

Question 4569

ALL
4570. (Refer to Figures 96 and 97.) To which aircraft position does HSI presentation "H" correspond?

A—8.
B—1.
C—2.

HSI Indicator "H" is set up with the head of the arrow pointing to 270°. The CDI is deflected to the left, so the aircraft is to the north on a heading of 215°. Position #1 is the only position corresponding to the CDI presentation. See the figure below. (H831) — FAA-H-8083-15, Chapter 7

Answer (A) is incorrect because aircraft 8 has a 045° heading (not 215°). Answer (C) is incorrect because aircraft 2 has a 315° heading (not 215°).

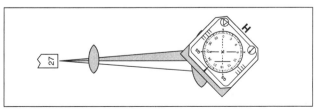

Question 4570

Answers

4567 [C]   4568 [A]   4569 [B]   4570 [B]

ALL
4571. (Refer to Figures 96 and 97.) To which aircraft position does HSI presentation "I" correspond?

A—4.
B—12.
C—11.

HSI Indicator "I" is set up with the head of the arrow pointing to 090°. The aircraft is to the north of the extended centerline on a heading of 270°. The CDI is using reverse sensing, so position #7 or #11 would be appropriate. See the figure below. (H831) — FAA-H-8083-15, Chapter 7

Answer (A) is incorrect because aircraft 4 has a centered CDI, so it is on the extended centerline (not to the north). Answer (B) is incorrect because aircraft 12 heading a 090° heading (not 270°) and is on the extended centerline (not to the north).

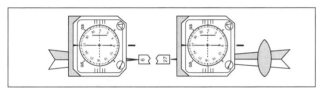

Question 4571

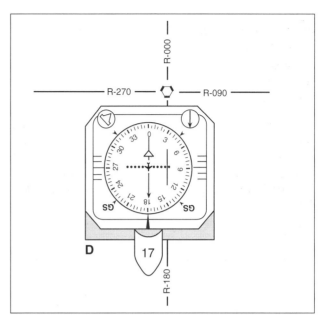

Question 4572

ALL
4572. (Refer to Figures 98 and 99.) To which aircraft position does HSI presentation "D" correspond?

A—4.
B—15.
C—17.

HSI Indicator "D" has a course selection of 180°, and the TO/FROM indicator is pointing to the tail of the course arrow. So the aircraft is flying away FROM the station, and is south of R-270 and R-090. The CDI bar is to the left, which means the aircraft is west of R-180. The aircraft heading is 180°, which describes aircraft 17. See the figure at upper right. (H831) — FAA-H-8083-15, Chapter 7

Answer (A) is incorrect because aircraft 4 would have a TO (not FROM) indication since it is north of R-270 and R-090. Answer (B) is incorrect because aircraft 15 is on a 360° heading (not 180°), and the CDI would be centered (not to the left).

ALL
4573. (Refer to Figures 98 and 99.) To which aircraft position does HSI presentation "E" correspond?

A—5.
B—6.
C—15.

HSI Indicator "E" has a course selection of 360°, and the TO/FROM indicator is pointing to the tail of the course arrow. So the aircraft is flying away FROM the station, and is north of R-270 and R-090. The CDI bar is to the left, which means the aircraft is east of R-000. The aircraft heading is 360°, which describes aircraft 6. See the figure on the next page. (H831) — FAA-H-8083-15, Chapter 7

Answer (A) is incorrect because aircraft 5 has a 180° heading (not 360°), and the CDI would be centered (not deflected to the left). Answer (C) is incorrect because aircraft 15 would have a TO indication (not FROM) and the CDI would be centered (not deflected to the left).

Answers

4571 [C]   4572 [C]   4573 [B]

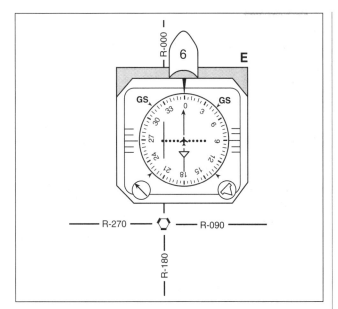

Question 4573

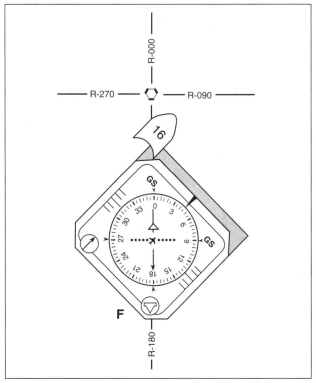

Question 4574

ALL
4574. (Refer to Figures 98 and 99.) To which aircraft position does HSI presentation "F" correspond?

A—10.
B—14.
C—16.

HSI Indicator "F" has a course selection of 180°, and the TO/FROM indicator is pointing to the tail of the course arrow. So the aircraft is flying away FROM the station, and is south of R-270 and R-090. The CDI bar is centered, which means the aircraft is on the extended centerline. The aircraft heading is 045°, which describes aircraft 16. See the figure at upper right. (H831) — FAA-H-8083-15, Chapter 7

Answer (A) is incorrect because aircraft 10 would have a CDI bar deflection to the left (not centered) and a TO (not FROM) indication. Answer (B) is incorrect because aircraft 14 would have a CDI bar deflection to the left (not centered).

ALL
4575. (Refer to Figures 98 and 99.) To which aircraft position does HSI presentation "A" correspond?

A—1.
B—8.
C—11.

HSI Indicator "A" has a course selection of 090°, and the TO/FROM indicator is pointing to the head of the course arrow. So the aircraft is flying TO the station, and is west R-000 and R-180. The CDI is deflected to the right, which means the aircraft is north of R-270. The aircraft heading is 205°, which describes aircraft 1. See the following figure. (H831) — FAA-H-8083-15, Chapter 7

Answer (B) is incorrect because aircraft 8 would have a FROM (not TO) indication. Answer (C) is incorrect because aircraft 11 would have FROM (not TO) indication and the CDI bar would be deflected to the left (not right).

Answers

4574 [C]   4575 [A]

Chapter 4 **Navigation**

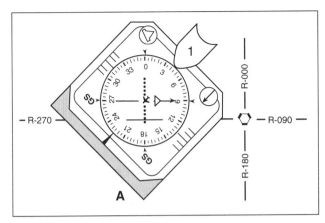

Question 4575

ALL
4576. (Refer to Figures 98 and 99.) To which aircraft position does HSI presentation "B" correspond?

A—9.
B—13.
C—19.

HSI Indicator "B" has a course selection of 270°, and the TO/FROM indicator is pointing to the tail of the course arrow. So the aircraft is flying FROM the station, and is west of R-000 and R-180. The CDI is deflected right, which means the aircraft is south of R-270. The aircraft heading is 135°, which describes aircraft 19. See the figure below. (H831) — FAA-H-8083-15, Chapter 7

Answer (A) is incorrect because aircraft 9 would have a TO (not FROM) indication, and the CDI bar would be deflected to the left (not right). Answer (B) is incorrect because aircraft 13 would have a TO (not FROM) indication.

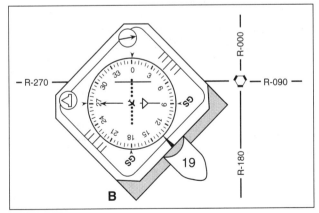

Question 4576

ALL
4577. (Refer to Figures 98 and 99.) To which aircraft position does HSI presentation "C" correspond?

A—6.
B—7.
C—12.

HSI Indicator "C" has a course selection of 360°, and the TO/FROM indicator is pointing to the head of the course arrow. So the aircraft is flying TO the station, and is south of R-270 and R-090. The CDI is deflected left, which means the aircraft is east of R-180. The aircraft heading is 310°, which describes aircraft 12. See the figure below. (H831) — FAA-H-8083-15, Chapter 7

Answer (A) is incorrect because aircraft 6 would have a 360° (not 310°) heading, and a FROM (not TO) indication. Answer (B) is incorrect because aircraft 7 would have a FROM (not TO) indication.

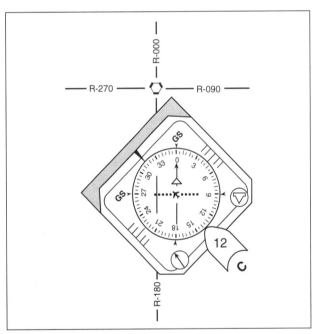

Question 4577

Answers

4576 [C]   4577 [C]

ADF

ADF equipment consists of a receiver that receives in the low- and medium-frequency bands and an instrument needle that points to the station. The ADF may be used to either home or track to a station. Homing is flying the aircraft on any heading required to keep the azimuth needle on 0 until the station is reached. Tracking is following a straight geographic path by establishing a heading that will maintain the desired track, taking into consideration the effects of the wind.

The azimuth needle, pointing to the selected station, indicates the angular difference between the aircraft heading and the direction to the station, measured clockwise from the nose of the aircraft. This angular difference is the relative bearing to the station, and may be read directly from the fixed-scale indicator. To determine magnetic bearing from an aircraft to a selected station, add the magnetic heading (from the heading indicator) to the relative bearing to the station (from the head of the needle on the ADF) to equal the magnetic bearing to the station. Or:

MH + RB = Magnetic Bearing TO the Station.

ALL
4578. (Refer to Figure 101.) What is the magnetic bearing TO the station?

A—060°.
B—260°.
C—270°.

The magnetic bearing to the station is the sum of the Magnetic (compass) Heading and the Relative Bearing, or the formula: MH + RB = MB TO the station.

In this case, MH (350) + RB (270) = 620.

Since there are only 360° in a complete circle, subtract 360 from 620, which equals 260°, which is the MB TO the station. (H830) — FAA-H-8083-15, Chapter 7

ALL
4583. (Refer to instruments in Figure 102.) On the basis of this information, the magnetic bearing TO the station would be

A—175°.
B—255°.
C—355°.

Magnetic Heading plus Relative Bearing equals the Magnetic Bearing TO the Station. Using the information from the FAA Figure, the Magnetic Heading is 215°, the Relative Bearing is 140°, therefore:

215° + 140° = 355° Magnetic Bearing TO the Station

(H830) — FAA-H-8083-15, Chapter 7

ALL
4584. (Refer to instruments in Figure 102.) On the basis of this information, the magnetic bearing FROM the station would be

A—175°.
B—255°.
C—355°.

Magnetic Heading plus Relative Bearing equals the Magnetic Bearing TO the Station. To find Magnetic Bearing FROM the Station, you add or subtract 180° from the Magnetic Bearing TO the Station. Using the information from the FAA Figure, the Magnetic Heading is 215°, the Relative Bearing is 140°, therefore:

215° + 140° = 355° Magnetic Bearing TO the Station
355° – 180° = 175° Magnetic Bearing FROM the Station

(H830) — FAA-H-8083-15, Chapter 7

Answers
4578 [B]   4583 [C]   4584 [A]

ALL
4585. (Refer to instruments in Figure 103.) On the basis of this information, the magnetic bearing FROM the station would be

A—030°.
B—060°.
C—240°.

Magnetic Heading plus Relative Bearing equals the Magnetic Bearing TO the Station. To find Magnetic Bearing FROM the Station, you add or subtract 180° from the Magnetic Bearing TO the Station. Using the information from the FAA Figure, the Magnetic Heading is 330°, the Relative Bearing is 270°, therefore:

330° + 270° = 600° Magnetic Bearing TO the Station
600° − 360° = 240° Magnetic Bearing TO the Station
240° − 180° = 060° Magnetic Bearing FROM the Station

(H830) — FAA-H-8083-15, Chapter 7

ALL
4586. (Refer to instruments in Figure 103.) On the basis of this information, the magnetic bearing TO the station would be

A—060°.
B—240°.
C—270°.

Magnetic Heading plus Relative Bearing equals the Magnetic Bearing TO the Station. Using the information from the FAA Figure, the Magnetic Heading is 330°, the Relative Bearing is 270°, therefore:

330° + 270° = 600° Magnetic Bearing TO the Station
600° − 360° = 240° Magnetic Bearing TO the Station

(H830) — FAA-H-8083-15, Chapter 7

ALL
4591. (Refer to Figure 105.) If the magnetic heading shown for aircraft 7 is maintained, which ADF illustration would indicate the aircraft is on the 120° magnetic bearing FROM the station?

A—2.
B—4.
C—5.

Aircraft 7 is on a magnetic heading of 270°. To determine which ADF would indicate the aircraft is on the 120° magnetic bearing FROM the station, we must calculate the relative bearing, and find the ADF which illustrates this relative bearing. For this problem, we must first convert the magnetic bearing FROM the station to TO the station by adding 180°: 120° + 180° = 300° magnetic bearing TO the station. The following formula is used to find the relative bearing:

Magnetic Heading + Relative Bearing =
Bearing TO the Station

270° + ? = 300° Magnetic Bearing TO the Station
? = 30° Relative Bearing

ADF 5 indicates a 30° relative bearing, which with the magnetic heading of 270° found in aircraft 7, would indicate the aircraft is on the 120° magnetic bearing FROM the station. (H830) — FAA-H-8083-15, Chapter 7

Answer (A) is incorrect because ADF 2 indicates a 060° relative bearing, which would indicate the aircraft is on the 150° (not 120°) MB FROM the station. Answer (B) is incorrect because ADF 4 indicates a 210° relative bearing, which would indicate the aircraft is on the 300° (not 120°) MB FROM the station.

ALL
4592. (Refer to Figure 105.) If the magnetic heading shown for aircraft 5 is maintained, which ADF illustration would indicate the aircraft is on the 210° magnetic bearing FROM the station?

A—2.
B—3.
C—4.

Aircraft 5 is on a magnetic heading of 180°. To determine which ADF would indicate the aircraft is on the 210° magnetic bearing FROM the station, we must calculate the relative bearing, and find the ADF which illustrates this relative bearing. For this problem, we must first convert the magnetic bearing FROM the station to TO the station by subtracting 180°: 210° − 180° = 30° magnetic bearing TO the station. The following formula is used to find the relative bearing:

Magnetic Heading + Relative Bearing =
Bearing TO the Station

180° + ? = 30° Magnetic Bearing TO the Station
? = 210° Relative Bearing

ADF 4 indicates a 210° relative bearing, which with the magnetic heading of 180° found in aircraft 5, would indicate the aircraft is on the 210° magnetic bearing FROM the station. (H830) — FAA-H-8083-15, Chapter 7

Answer (A) is incorrect because ADF 2 indicates a 060° relative bearing, which would indicate the aircraft is on the 060° (not 210°) MB FROM the station. Answer (B) is incorrect because ADF 3 indicates a 255° relative bearing, which would indicate the aircraft is on the 255° (not 210°) MB FROM the station.

Answers

| 4585 | [B] | 4586 | [B] | 4591 | [C] | 4592 | [C] |

Chapter 4 **Navigation**

ALL
4593. (Refer to Figure 105.) If the magnetic heading shown for aircraft 3 is maintained, which ADF illustration would indicate the aircraft is on the 120° magnetic bearing TO the station?

A—4.
B—5.
C—8.

Aircraft 3 is on a magnetic heading of 090°. To determine which ADF would indicate the aircraft is on the 120° magnetic bearing TO the station, we must calculate the relative bearing, and find the ADF which illustrates this relative bearing. The following formula is used to find the relative bearing:

Magnetic Heading + Relative Bearing =
Bearing TO the Station

090° + ? = 120° Magnetic Bearing TO the Station
? = 030° Relative Bearing

ADF 5 indicates a 030° relative bearing, which with the magnetic heading of 090° found in aircraft 3, would indicate the aircraft is on the 120° magnetic bearing TO the station. (H830) — FAA-H-8083-15, Chapter 7

Answer (A) is incorrect because ADF 4 indicates a 210° relative bearing, which would indicate the aircraft is on the 300° (not 120°) MB TO the station. Answer (C) is incorrect because ADF 8 indicates a 135° relative bearing, which would indicate the aircraft is on the 225° (not 120°) MB TO the station.

ALL
4594. (Refer to Figure 105.) If the magnetic heading shown for aircraft 1 is maintained, which ADF illustration would indicate the aircraft is on the 060° magnetic bearing TO the station?

A—2.
B—4.
C—5.

Aircraft 1 is on a magnetic heading of 360°. To determine which ADF would indicate the aircraft is on the 060° magnetic bearing TO the station, we must calculate the relative bearing, and find the ADF which illustrates this relative bearing. The following formula is used to find the relative bearing:

Magnetic Heading + Relative Bearing =
Bearing TO the Station

360° + ? = 060° Magnetic Bearing TO the Station
? = 060° Relative Bearing

ADF 2 indicates a 060° relative bearing, which with the magnetic heading of 360° found in aircraft 1, would indicate the aircraft is on the 060° magnetic bearing TO the station. (H830) — FAA-H-8083-15, Chapter 7

Answer (B) is incorrect because ADF 4 indicates a 210° relative bearing, which would indicate the aircraft is on the 210° (not 060°) MB TO the station. Answer (C) is incorrect because ADF 5 indicates a 030° relative bearing, which would indicate the aircraft is on the 030° (not 060°) MB TO the station.

ALL
4595. (Refer to Figure 105.) If the magnetic heading shown for aircraft 2 is maintained, which ADF illustration would indicate the aircraft is on the 255° magnetic bearing TO the station?

A—2.
B—4.
C—5.

Aircraft 2 is on a magnetic heading of 045°. To determine which ADF would indicate the aircraft is on the 255° magnetic bearing TO the station, we must calculate the relative bearing, and find the ADF which illustrates this relative bearing. The following formula is used to find the relative bearing:

Magnetic Heading + Relative Bearing =
Bearing TO the Station

045° + ? = 255° Magnetic Bearing TO the Station
? = 210° Relative Bearing

ADF 4 indicates a 210° relative bearing, which with the magnetic heading of 045° found in aircraft 2, would indicate the aircraft is on the 255° magnetic bearing TO the station. (H830) — FAA-H-8083-15, Chapter 7

Answer (A) is incorrect because ADF 2 indicates a 060° relative bearing, which would indicate the aircraft is on the 105° (not 255°) MB TO the station. Answer (C) is incorrect because ADF 5 indicates a 030° relative bearing, which would indicate the aircraft is on the 075° (not 255°) MB TO the station.

Answers

4593 [B]   4594 [A]   4595 [B]

ALL
4596. (Refer to Figure 105.) If the magnetic heading shown for aircraft 4 is maintained, which ADF illustration would indicate the aircraft is on the 135° magnetic bearing TO the station?

A—1.
B—4.
C—8.

Aircraft 4 is on a magnetic heading of 135°. To determine which ADF would indicate the aircraft is on the 135° magnetic bearing TO the station, we must calculate the relative bearing, and find the ADF which illustrates this relative bearing. The following formula is used to find the relative bearing:

Magnetic Heading + Relative Bearing =
Bearing TO the Station

135° + ? = 135° Magnetic Bearing TO the Station
? = 000° Relative Bearing
(or 360°)

ADF 1 indicates a 360° relative bearing, which with the magnetic heading of 135° found in aircraft 4, would indicate the aircraft is on the 135° magnetic bearing TO the station. (H830) — FAA-H-8083-15, Chapter 7

Answer (B) is incorrect because ADF 4 indicates a 210° relative bearing, which would indicate the aircraft is on the 345° (not 135°) MB TO the station. Answer (C) is incorrect because ADF 8 indicates a 135° relative bearing, which would indicate the aircraft is on the 270° (not 135°) MB TO the station.

ALL
4597. (Refer to Figure 105.) If the magnetic heading shown for aircraft 6 is maintained, which ADF illustration would indicate the aircraft is on the 255° magnetic bearing FROM the station?

A—2.
B—4.
C—5.

Aircraft 6 is on a magnetic heading of 225°. To determine which ADF would indicate the aircraft is on the 255° magnetic bearing FROM the station, we must calculate the relative bearing, and find the ADF which illustrates this relative bearing. For this problem, we must first convert the magnetic bearing FROM the station to TO the station by subtracting 180°: 255° – 180° = 075° magnetic bearing TO the station. The following formula is used to find the relative bearing:

Magnetic Heading + Relative Bearing =
Bearing TO the Station

225° + ? = 075° Magnetic Bearing TO the Station
? = 210° Relative Bearing

ADF 4 indicates a 210° relative bearing, which with the magnetic heading of 225° found in aircraft 6, would indicate the aircraft is on the 255° magnetic bearing FROM the station. (H830) — FAA-H-8083-15, Chapter 7

Answer (A) is incorrect because ADF 2 indicates a 060° relative bearing, which would indicate the aircraft is on the 105° (not 255°) MB FROM the station. Answer (C) is incorrect because ADF 5 indicates a 030° relative bearing, which would indicate the aircraft is on the 075° (not 255°) MB FROM the station.

ALL
4598. (Refer to Figure 105.) If the magnetic heading shown for aircraft 8 is maintained, which ADF illustration would indicate the aircraft is on the 090° magnetic bearing FROM the station?

A—3.
B—4.
C—6.

Aircraft 8 is on a magnetic heading of 315°. To determine which ADF would indicate the aircraft is on the 090° magnetic bearing FROM the station, we must calculate the relative bearing, and find the ADF which illustrates this relative bearing. For this problem, we must first convert the magnetic bearing FROM the station to TO the station by adding 180°: 090° + 180° = 270° magnetic bearing TO the station. The following formula is used to find the relative bearing:

Magnetic Heading + Relative Bearing =
Bearing TO the Station

315° + ? = 270° Magnetic Bearing TO the Station
? = 315° Relative Bearing

ADF 6 indicates a 315° relative bearing, which with the magnetic heading of 315° found in aircraft 8, would indicate the aircraft is on the 090° magnetic bearing FROM the station. (H830) — FAA-H-8083-15, Chapter 7

Answer (A) is incorrect because ADF 3 indicates a 255° relative bearing, which would indicate the aircraft is on the 030° (not 090°) MB FROM the station. Answer (B) is incorrect because ADF 4 indicates a 210° relative bearing, which would indicate the aircraft is on the 345° (not 090°) MB FROM the station.

Answers

4596 [A]   4597 [B]   4598 [C]

Chapter 4 **Navigation**

ALL
4599. (Refer to Figure 105.) If the magnetic heading shown for aircraft 5 is maintained, which ADF illustration would indicate the aircraft is on the 240° magnetic bearing TO the station?
A—2.
B—3.
C—4.

Aircraft 5 is on a magnetic heading of 180°. To determine which ADF would indicate the aircraft is on the 240° magnetic bearing TO the station, we must calculate the relative bearing, and find the ADF which illustrates this relative bearing. The following formula is used to find the relative bearing:

*Magnetic Heading + Relative Bearing =
Bearing TO the Station*

*180° + ? = 240° Magnetic Bearing TO the Station
? = 060° Relative Bearing*

ADF 2 indicates a 060° relative bearing, which with the magnetic heading of 180° found in aircraft 5, would indicate the aircraft is on the 240° magnetic bearing TO the station. (H830) — FAA-H-8083-15, Chapter 7

Answer (B) is incorrect because ADF 3 indicates a 255° relative bearing, which would indicate the aircraft is on the 075° (not 240°) MB TO the station. Answer (C) is incorrect because ADF 4 indicates a 210° relative bearing, which would indicate the aircraft is on the 030° (not 240°) MB TO the station.

ALL
4600. (Refer to Figure 105.) If the magnetic heading shown for aircraft 8 is maintained, which ADF illustration would indicate the aircraft is on the 315° magnetic bearing TO the station?
A—3.
B—4.
C—1.

Aircraft 8 is on a magnetic heading of 315°. To determine which ADF would indicate the aircraft is on the 315° magnetic bearing TO the station, we must calculate the relative bearing, and find the ADF which illustrates this relative bearing. The following formula is used to find the relative bearing:

*Magnetic Heading + Relative Bearing =
Bearing TO the Station*

*315° + ? = 315° Magnetic Bearing TO the Station
? = 000° Relative Bearing
  (or 360°)*

ADF 1 indicates a 360° relative bearing, which with the magnetic heading of 315° found in aircraft 8, would indicate the aircraft is on the 315° magnetic bearing TO the station. (H830) — FAA-H-8083-15, Chapter 7

Answer (A) is incorrect because ADF 3 indicates a 255° relative bearing, which would indicate the aircraft is on the 210° (not 315°) MB TO the station. Answer (B) is incorrect because ADF 4 indicates a 210° relative bearing, which would indicate the aircraft is on the 165° (not 315°) MB TO the station.

Answers

4599 [A]   4600 [C]

RMI

The Radio Magnetic Indicator (RMI) is a slaved compass with two needles. One or both needles can point to either a VOR or NDB. The compass card rotates as the aircraft turns, displaying the magnetic heading of the aircraft under the index at the top of the instrument. See Figure 4-6.

When the needle is pointing to a VOR, the tail of the needle shows the radial the aircraft is on. Thus, if the VOR is tuned to a VOT, the RMI needle will point to 180°, showing the aircraft on the 360° radial. If a VOT is used to check accuracy of dual VORs, each RMI needle must point to 180° ± 4°. When dual VOR receivers are checked against each other, the RMI needles must be within 4° of each other when both VOR receivers are tuned to the same station.

When the needle points to an NDB, the tail of the needle shows the magnetic bearing FROM the station.

To orient where the aircraft is in relation to the facility, first determine which radial is selected to find which quadrant you are in (look at the tail of the needle; if you are trying to orient yourself relative to the VOR, make sure you are using the VOR needle). Next, consider the aircraft heading (under the lubber line) to determine the aircraft's position within the quadrant.

Wind orientation is important when using the RMI to fly a DME arc. With an RMI, in a no-wind condition, the arc should theoretically be flown by keeping the RMI needle on the wingtip reference point. If a crosswind is drifting the aircraft away from the station, the aircraft must be turned until the bearing pointer is ahead of the wingtip reference. If a crosswind is drifting the aircraft toward the facility, the aircraft must turned until the bearing pointer is behind the wingtip reference. As a guide in making range corrections, change the relative bearing 10° to 20° for each 1/2 mile of deviation from the desired arc.

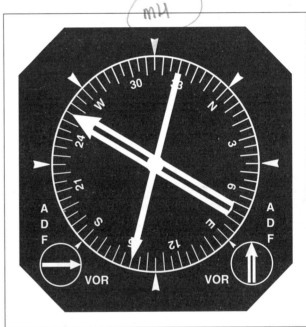

Figure 4-6. Radio Magnetic Indicator (RMI)

ALL
4579. (Refer to Figure 100.) Which RMI illustration indicates the aircraft to be flying outbound on the magnetic bearing of 235° FROM the station? (Wind 050° at 20 knots.)

A—2.
B—3.
C—4.

The magnetic heading of the aircraft is always directly under the index at the top of the instrument (assuming no compass deviation error). The bearing pointer displays bearings TO the selected station, the tail displays bearings FROM the station. RMI 3 depicts the magnetic bearing of 235° FROM the station since that is where the tail is pointing. The aircraft heading is also 235° indicating it is tracking outbound on the 235° radial. Since the wind is blowing from 050° at 20 knots (which is a direct tailwind) a wind correction angle would not be required. See the figure on the next page. (H831) — FAA-H-8083-15, Chapter 7

Answer (A) is incorrect because RMI 2 is indicating a 235° magnetic bearing TO (not FROM) the station. Answer (C) is incorrect because RMI 4 is indicating a strong wind correction to the right.

Answers
4579 [B]

Chapter 4 Navigation

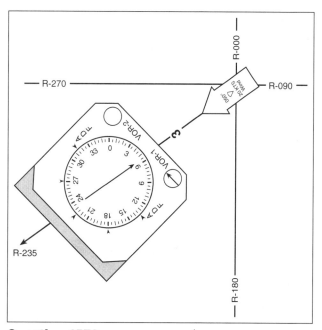

Question 4579

ALL
4580. (Refer to Figure 100.) What is the magnetic bearing TO the station as indicated by illustration 4?

A—285°.
B—055°.
C—235°.

The bearing pointer displays bearing TO the selected station. In RMI 4, the needle is pointing to 055°, which is the magnetic bearing TO the station. (H831) — FAA-H-8083-15, Chapter 7

Answer (A) is incorrect because 285° is the magnetic heading (not the magnetic bearing TO the station). Answer (C) is incorrect because 235° is the magnetic bearing FROM (not TO) the station.

ALL
4581. (Refer to Figure 100.) Which RMI illustration indicates the aircraft is southwest of the station and moving closer TO the station?

A—1.
B—2.
C—3.

An aircraft southwest of the station and moving closer TO the station would have both the heading and the bearing pointer indicating northeast. This describes RMI 1 which has a heading, and magnetic bearing to the station, of 055°. (H831) — FAA-H-8083-15, Chapter 7

Answer (B) is incorrect because RMI 2 shows the aircraft northeast (not southwest) of the station and moving away FROM (not TO) the station (heading 055° and station behind the airplane). Answer (C) is incorrect because RMI 3 shows the aircraft heading southwest and moving away FROM the station.

ALL
4582. (Refer to Figure 100.) Which RMI illustration indicates the aircraft is located on the 055° radial of the station and heading away from the station?

A—1.
B—2.
C—3.

The radial, or magnetic bearing FROM the station, is determined from the tail of the bearing pointer. The aircraft is located on the radial when the aircraft heading matches the radial. RMI 2 is indicating 055° with the tail of the needle, and the aircraft located on that radial with a 055° heading. (H831) — FAA-H-8083-15, Chapter 7

Answer (A) is incorrect because RMI 1 indicates the aircraft is on the 235° radial flying TO the station. Answer (C) is incorrect because RMI 3 indicates the aircraft is on the 235° radial flying FROM the station.

ALL
4587. (Refer to Figure 104.) If the radio magnetic indicator is tuned to a VOR, which illustration indicates the aircraft is on the 115° radial?

A—1.
B—2.
C—3.

VOR radials are FROM the station, and are indicated on a radio magnetic indicator (RMI) by the tail of the bearing indicator. Illustration 1 shows the tail of the indicator on 115°, which indicates the aircraft is on the 115° radial. (H831) — FAA-H-8083-15, Chapter 7

Answer (B) is incorrect because RMI 2 indicates the aircraft is on the 315° radial. Answer (C) is incorrect because RMI 3 indicates the aircraft is on the 010° radial.

ALL
4588. (Refer to Figure 104.) If the radio magnetic indicator is tuned to a VOR, which illustration indicates the aircraft is on the 335° radial?

A—2.
B—3.
C—4.

VOR radials are FROM the station, and are indicated on a radio magnetic indicator (RMI) by the tail of the bearing indicator. Illustration 4 shows the tail of the indicator on

Answers

| 4580 [B] | 4581 [A] | 4582 [B] | 4587 [A] | 4588 [C] |

335°, which indicates the aircraft is on the 335° radial. (H831) — FAA-H-8083-15, Chapter 7

Answer (A) is incorrect because RMI 2 indicates the aircraft is on the 315° radial. Answer (B) is incorrect because RMI 3 indicates the aircraft is on the 010° radial.

ALL
4589. (Refer to Figure 104.) If the radio magnetic indicator is tuned to a VOR, which illustration indicates the aircraft is on the 315° radial?

A—2.
B—3.
C—4.

VOR radials are FROM the station, and are indicated on a radio magnetic indicator (RMI) by the tail of the bearing indicator. Illustration 2 shows the tail of the indicator on 315°, which indicates the aircraft is on the 315° radial. (H831) — FAA-H-8083-15, Chapter 7

Answer (B) is incorrect because RMI 3 indicates the aircraft is on the 010° radial. Answer (C) is incorrect because RMI 4 indicates the aircraft is on the 335° radial.

ALL
4590. (Refer to Figure 104.) If the radio magnetic indicator is tuned to a VOR, which illustration indicates the aircraft is on the 010° radial?

A—1.
B—2.
C—3.

VOR radials are FROM the station, and are indicated on a radio magnetic indicator (RMI) by the tail of the bearing indicator. Illustration 3 shows the tail of the indicator on 010°, which indicates the aircraft is on the 010° radial. (H831) — FAA-H-8083-15, Chapter 7

Answer (A) is incorrect because RMI 1 indicates the aircraft is on the 115° radial. Answer (B) is incorrect because RMI 2 indicates the aircraft is on the 315° radial.

ALL
4602. (Refer to Figure 107.) Where should the bearing pointer be located relative to the wing-tip reference to maintain the 16 DME range in a right-hand arc with a right crosswind component?

A—Behind the right wing-tip reference for VOR-2.
B—Ahead of the right wing-tip reference for VOR-2.
C—Behind the right wing-tip reference for VOR-1.

The #2 (double-barred) needle is pointing toward the right wing-tip reference, so it is evidently tuned to the facility being used for a right-hand arc. The general rule is if a crosswind is blowing you away from the facility (as it is in this case), turn until the bearing pointer is ahead of the wing-tip reference. (H831) — FAA-H-8083-15, Chapter 7

Answer (A) is incorrect because it would only be appropriate for the bearing pointer to be behind the right wing-tip reference for a left-hand crosswind during a right-hand arc. Answer (C) is incorrect because VOR-1 is only appropriate for a left-hand arc.

ALL
4603. (Refer to Figure 108.) Where should the bearing pointer be located relative to the wing-tip reference to maintain the 16 DME range in a left-hand arc with a left crosswind component?

A—Ahead of the left wing-tip reference for the VOR-2.
B—Ahead of the right wing-tip reference for the VOR-1.
C—Behind the left wing-tip reference for the VOR-2.

The #2 (double-barred) needle is pointing toward the left wing-tip reference, so it is evidently tuned to the facility being used for a left-hand arc. The general rule is if a crosswind is blowing you away from the facility (as it is in this case), turn until the bearing pointer is ahead of the wing-tip reference. (H831) — FAA-H-8083-15, Chapter 7

Answer (B) is incorrect because VOR is only appropriate for a right-hand arc. Answer (C) is incorrect because it would only be appropriate for the bearing pointer to be behind the left wing-tip reference for a right-hand crosswind during a left-hand arc.

ALL
4331. (Refer to Figures 60A and 61.) What is your position relative to the PLATS intersection, glide slope, and the localizer course?

A—Past PLATS, below the glide slope, and right of the localizer course.
B—Approaching PLATS, above the glide slope, and left of the localizer course.
C—Past PLATS, above the glide slope, and right of the localizer course.

The standard LOC/GS indicator on the right side of FAA Figure 61 shows the aircraft to the right of course and high. The RMI shows a heading of 030° direct to the LOM (TUTTE). The large arrow shows the aircraft on the 300° radial from Scholes VORTAC. Therefore, the aircraft is past PLATS intersection. (J42) — Instrument Approach Procedures

Answers

| 4589 | [A] | 4590 | [C] | 4602 | [B] | 4603 | [A] | 4331 | [C] |

Chapter 4 **Navigation**

ALL

4367. (Refer to Figures 78 and 79.) What is your position relative to the VOR COP southeast bound on V86 between the BOZEMAN and LIVINGSTON VORTACs? The No. 1 VOR is tuned to 116.1 and the No. 2 VOR is tuned to 112.2.

A—Past the LVM R-246 and west of the BZN R-110.
B—Approaching the LVM R-246 and west of the BZN R-110.
C—Past the LVM R-246 and east of the BZN R-110.

The aircraft is proceeding SE on a heading of 130° (indicated under the index at the top of the instrument). The #1 needle (thin needle) is tuned to Livingston VORTAC (116.1) and on the 239° radial (indicated by the tail of the needle). This means the aircraft is past the 246° radial of LVM. The #2 needle (double needle) is tuned to Bozeman (112.2) and the aircraft is on the 102° radial (indicated by the tail of the needle). This means the aircraft is to the left (east) of the 110° radial of BZN. (J35) — Enroute Low Altitude Chart

Answer (A) is incorrect because the aircraft is east (not west) of the BZN R-110. Answer (B) is incorrect because the aircraft is past (not approaching) the LVM R-246 and east (not west) of the BZN R-110.

ALL

4269. (Refer to Figure 30.) During the arc portion of the instrument departure procedure (GNATS1.MOURN), a left crosswind is encountered. Where should the bearing pointer of an RMI be referenced relative to the wing-tip to compensate for wind drift and maintain the 15 DME arc?

A—Behind the right wing-tip reference point.
B—On the right wing-tip reference point.
C—Behind the left wing-tip reference point.

If a crosswind is drifting you toward the facility, turn until the RMI bearing pointer is behind the wing tip. With a crosswind from the left on a right-hand arc, the right wing-tip reference point will be behind the wing tip. (H832) — FAA-H-8083-15, Chapter 7

RNAV and LORAN

RNAV is a system of navigation which allows a pilot to fly a selected course without the need to overfly ground-based navigation facilities. The system is based on the existing VORTACs. To fly either an RNAV route or to execute an RNAV approach under instrument flight rules, the aircraft must have an approved RNAV receiver.

A waypoint is a predetermined geographical position used for route definition and/or progress reporting purposes.

Long Range Navigation (LORAN) is a pulsed hyperbolic system, operating in the 90-110 kHz frequency band. The system is based upon measurement of the difference in time of arrival of pulses of radio-frequency energy radiated by a group, or a chain of transmitters. Some installed LORAN equipment are approved for IFR enroute operations. This can be confirmed by checking the supplement section of the Airplane Flight Manual or an FAA Form 337.

ALL

4069. What is a way point when used for an IFR flight?

A—A predetermined geographical position used for an RNAV route or an RNAV instrument approach.
B—A reporting point defined by the intersection of two VOR radials.
C—A location on a victor airway which can only be identified by VOR and DME signals.

Waypoint: A predetermined geographical position used for route/instrument approach definition, or progress reporting purposes, that is defined relative to a VORTAC station, or in terms of latitude/longitude coordinates. (H862) — FAA-H-8083-15, Glossary

Answer (B) is incorrect because it describes an intersection, not a waypoint. Answer (C) is incorrect because it describes a DME fix, not a waypoint.

Answers

4367 [C]   4269 [A]   4069 [A]

ALL
4684. (Refer to Figure 129.) What minimum airborne equipment is required to be operative for RNAV RWY 36 approach at Adams Field?

A—An approved RNAV receiver that provides both horizontal and vertical guidance.
B—A transponder and an approved RNAV receiver that provides both horizontal and vertical guidance.
C—Any approved RNAV receiver.

An approved RNAV receiver is required for RNAV approaches. (J42) — Instrument Approach Procedures

Answer (A) is incorrect because RNAV approaches do not require RNAV equipment with vertical guidance capability. Answer (B) is incorrect because ATC may authorize continued flight with an inoperative transponder, and vertical guidance capability is not required.

ALL
4665. By which means may a pilot determine if a Loran C equipped aircraft is approved for IFR operations?

A—Not necessary; Loran C is not approved for IFR.
B—Check aircraft logbook.
C—Check the Airplane Flight Manual Supplement.

Some installed aircraft LORAN-C are approved for IFR enroute operations. This can be confirmed by checking the supplement section of the Airplane Flight Manual or an FAA Form 337. (J01) — AIM ¶1-1-16

Answer (A) is incorrect because some LORAN-C receivers are approved for IFR by the FAA. Answer (B) is incorrect because the approval may be found in the aircraft maintenance records, not necessarily in the aircraft logbook.

ILS

An Instrument Landing System (ILS) consists of a localizer, a glide slope, marker beacons, and approach lights. *See* **Figure 4-7**. The localizer provides azimuth information, furnishing the pilot with course guidance to the runway centerline. The approach course is called the front course and the localizer signal is transmitted from approximately 1,000 feet past the far end of the runway. At the runway threshold, the course width is adjusted to 700 feet. When tracking inbound on the localizer, drift corrections should be small and should be reduced as the course narrows. When the outer marker is reached, enough drift correction should be established so as to allow completion of the approach with heading corrections no greater than 2°. The localizer identification, which consists of a three-letter identifier preceded by the letter "I", is transmitted in Morse Code on the localizer frequency.

The glide slope transmitter is automatically tuned when the localizer frequency is selected. Offset from the runway approximately 1,000 feet from the approach end, it projects a beam 1.4° from full-scale. The glide slope angle is normally adjusted to 3° above horizontal so that it intersects the middle marker approximately 200 feet AGL, and is usable to a distance of about 10 NM. **Figures 4-8** and **4-9** on the next page depict typical indications that a pilot would see during localizer and/or glide slope operations.

Low-powered transmitters called marker beacons are located along the ILS approach course. These beacons transmit their signals vertically upward in a very narrow beam across the localizer course. When an aircraft flies through the beam a receiver is activated which informs the pilot of the aircraft's position relative to the runway. The marker beacon farthest from the runway is the Outer Marker (OM).

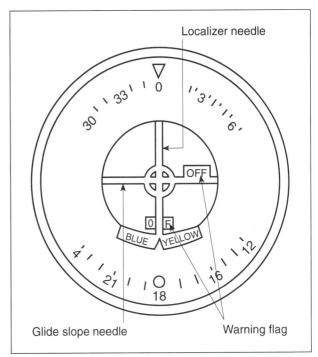

Figure 4-7. Localizer and glide slope needles

Answers
4684 [C]   4665 [C]

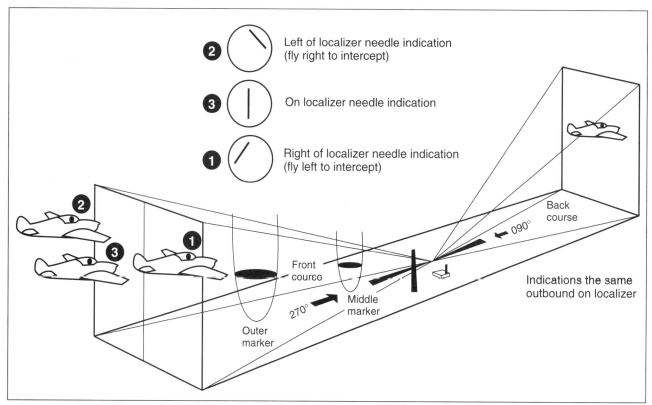

Figure 4-8. Localizer course indications

Passing over the OM the pilot will hear a series of dashes, and a blue (or purple) light will illuminate on the instrument panel. Located 4 to 7 miles from the runway, the OM indicates a position at which an aircraft at the appropriate altitude on the localizer course will intercept the glide slope. Approximately 3,500 feet from the landing threshold, an aircraft on the glide slope will cross the Middle Marker (MM) at an altitude of approximately 200 feet. The pilot will hear alternating dots and dashes, and an amber light will illuminate on the instrument panel. At some locations, an Inner Marker (IM) is installed between the MM and the runway threshold. The IM is identified by dots transmitted at the rate of 6 per second and the illumination of a white light on the instrument panel. Frequently, low-powered non-directional beacons called Compass Locators are co-located with outer and middle markers. For identification, a locator outer marker (LOM) transmits the first two letters of the localizer identification group, and the locator middle marker (LMM) transmits the last two letters of the localizer identifier.

A Localizer-Type Directional Aid (LDA) approach is of comparable utility and accuracy to a localizer. Like the localizer, it has a course width of approximately 5°, but it is not part of a complete ILS system and it is not aligned with the runway. It may have straight-in landing minimums.

The Simplified Directional Facility (SDF) provides a final approach course similar to that of the ILS localizer. It has no glide slope, and it may or may not be aligned with the runway. The course width will be fixed at either 6° or 12°, and the antenna may be offset from the runway centerline.

Chapter 4 **Navigation**

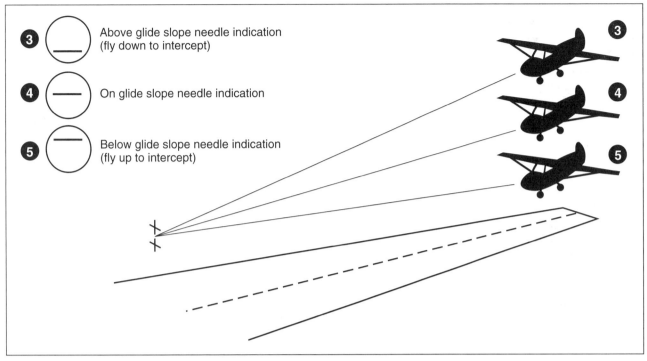

Figure 4-9. Glide slope indications

ALL
4353. (Refer to Figure 73.) Which sequence of marker beacon indicator lights, and their respective codes, will you receive on the ILS RWY 6 approach procedure to the MAP?

A—Blue—alternate dots and dashes; amber—dashes.
B—Amber—alternate dots and dashes; blue—dashes.
C—Blue—dashes; amber—alternate dots and dashes.

The OM is modulated at 400 Hz and identified with continuous dashes at the rate of two dashes per second and a blue marker beacon light. The MM is modulated at 1300 Hz and identified with alternate dots and dashes keyed at the rate of 95 dot/dash combinations per minute and an amber marker beacon light. The IM is modulated at 3000 Hz and identified with continuous dots keyed at the rate of six dots per second and a white marker beacon light (but in this case, the IM is passed after the MAP). (J01) — AIM ¶1-1-10

ALL
4702. What is a difference between an SDF and an LDA facility?

A—The SDF course width is either 6° or 12° while the LDA course width is approximately 5°.
B—The SDF course has no glide slope guidance while the LDA does.
C—The SDF has no marker beacons while the LDA has at least an OM.

An LDA is of comparable utility and accuracy to a localizer but is not part of a complete ILS. The LDA usually provides a more precise approach course than the SDF, which may have a course width of 6° or 12°. The LDA course widths are from 3° to 6°. (J01) — AIM ¶1-1-10 and ¶1-1-11

Answer (B) is incorrect because neither the SDF or the LDA are equipped with glide slope guidance. Answer (C) is incorrect because both the SDF or LDA may be equipped with marker beacons.

Answers

4353 [C]   4702 [A]

Chapter 4 Navigation

ALL
4703. What is the difference between a Localizer-Type Directional Aid (LDA) and the ILS localizer?

A—The LDA is not aligned with the runway.
B—The LDA uses a course width of 6° or 12°, while an ILS uses only 5°.
C—The LDA signal is generated from a VOR-type facility and has no glide slope.

An LDA is of comparable utility and accuracy to a localizer but is not part of a complete ILS. The LDA usually provides a more precise approach course than the SDF, which may have a course width of 6° or 12°. The LDA is not aligned with the runway. (J01) — AIM ¶1-1-10

Answer (B) is incorrect because the course width of a LDA and an ILS are the same (3° to 6°). Answer (C) is incorrect because an LDA is a localizer-type signal (not a VOR-type).

ALL
4704. How wide is an SDF course?

A—Either 3° or 6°.
B—Either 6° or 12°.
C—Varies from 5° to 10°.

The SDF signal emitted from the transmitter is either 6° or 12° wide as needed to provide maximum flyability and optimum course quality. (J01) — AIM ¶1-1-11

Answer (A) is incorrect because 3° or 6° is half the course width for the SDF. Answer (C) is incorrect because a course of 5° to 10° is not related to an SDF, LDA, or an ILS.

ALL
4705. What are the main differences between the SDF and the localizer of an ILS?

A—The useable off-course indications are limited to 35° for the localizer and up to 90° for the SDF.
B—The SDF course may not be aligned with the runway and the course may be wider.
C—The course width for the localizer will always be 5° while the SDF course will be between 6° and 12°.

The SDF course may not be aligned with the runway and the course may be wider than an ILS localizer course. The ILS localizer course width varies between 3° and 6° and is tailored to provide 700 feet at the threshold (full-scale limits). (J01) — AIM ¶1-1-11

Answer (A) is incorrect because both approaches are limited to 35° for off-course indications. Answer (C) is incorrect because the course width for a localizer is usually between 3° and 6° while the SDF is either 6° or 12° (not between).

ALL
4729. Which range facility associated with the ILS is identified by the last two letters of the localizer identification group?

A—Inner marker.
B—Outer marker.
C—Middle compass locator.

Compass locators transmit two-letter identification groups. The outer locator transmits the first two letters of the localizer identification group, and the middle locator transmits the last two letters. (J01) — AIM ¶1-1-10

Answers (A) and (B) are incorrect because marker beacons are not identified by letters unless they are compass locations.

ALL
4730. Which range facility associated with the ILS can be identified by a two-letter coded signal?

A—Middle marker.
B—Outer marker.
C—Compass locator.

Compass locators transmit two-letter identification groups. The outer locator transmits the first two letters of the localizer identification group, and the middle locator transmits the last two letters. (J01) — AIM ¶1-1-10

Answers (A) and (B) are incorrect because marker beacons are not identified by letters unless they are compass locations.

ALL
4747. Which indications will a pilot receive where an IM is installed on a front course ILS approach?

A—One dot per second and a steady amber light.
B—Six dots per second and a flashing white light.
C—Alternate dashes and a blue light.

The Inner Marker (IM), where installed, is located on the front course between the middle marker and the landing threshold. It indicates the point at which an aircraft is at a designated Decision Height (DH) on the glidepath. The IM is modulated at 3000 Hz and identified with continuous dots at the rate of 6 per second, and a white marker beacon light. (J01) — AIM ¶1-1-10

Answer (A) is incorrect because a steady amber light identifies the MM (not the IM) with 95 dot-dash combinations. Answer (C) is incorrect because a blue light and continuous flashes identify the outer marker (not IM).

Answers

| 4703 | [A] | 4704 | [B] | 4705 | [B] | 4729 | [C] | 4730 | [C] | 4747 | [B] |

ALL
4753. Approximately what height is the glide slope centerline at the MM of a typical ILS?
A—100 feet.
B—200 feet.
C—300 feet.

The MM indicates a position at which an aircraft is approximately 3,500 feet from the landing threshold. This will also be the position at which an aircraft on the glidepath will be at an altitude of approximately 200 feet above the elevation of the touchdown zone. (J01) — AIM ¶1-1-10

ALL
4773. When tracking inbound on the localizer, which of the following is the proper procedure regarding drift corrections?
A—Drift corrections should be accurately established before reaching the outer marker and completion of the approach should be accomplished with heading corrections no greater than 2°.
B—Drift corrections should be made in 5° increments after passing the outer marker.
C—Drift corrections should be made in 10° increments after passing the outer marker.

Drift corrections should be small and reduced proportionately as the course narrows. By the time you reach the outer marker, your drift correction should be established accurately enough on a well-executed approach to permit completion of the approach with heading corrections no greater than 2°. (H837) — FAA-H-8083-15, Chapter 7

Answers (B) and (C) are incorrect because drift corrections should be no greater than 2° (not 5° or 10°).

ALL
4824. (Refer to Figures 139 and 140.) Which displacement from the localizer and glide slope at the 1.9 NM point is indicated?
A—710 feet to the left of the localizer centerline and 140 feet below the glide slope.
B—710 feet to the right of the localizer centerline and 140 feet above the glide slope.
C—430 feet to the right of the localizer centerline and 28 feet above the glide slope.

With the CDI two dots to the left, we are to the right of course, and according to FAA Figure 139, at the 1.9 NM location 710 feet to the right of course. With the glide slope indication two dots below, we are above the glide slope and at the 1.9 NM mark on FAA Figure 139, that would put you 140 feet above the glide slope. (H837) — FAA-H-8083-15, Chapter 7

ALL
4825. (Refer to Figures 139 and 141.) Which displacement from the localizer centerline and glide slope at the 1,300-foot point from the runway is indicated?
A—21 feet below the glide slope and approximately 320 feet to the right of the runway centerline.
B—28 feet above the glide slope and approximately 250 feet to the left of the runway centerline.
C—21 feet above the glide slope and approximately 320 feet to the left of the runway centerline.

With the CDI a dot and one-half to the right, we are to the left of course, and according to FAA Figure 139, at the 1,300-foot location (you must interpolate between 500 feet and 1,500 feet AGL), we are 322.5 feet to the left of course. The glide slope indication is a dot and a half above the glide slope and at the 1,300-foot location; that would put you 21 feet above the glide slope. (H837) — FAA-H-8083-15, Chapter 7

ALL
4826. (Refer to Figures 139 and 142.) Which displacement from the localizer and glide slope at the outer marker is indicated?
A—1,550 feet to the left of the localizer centerline and 210 feet below the glide slope.
B—1,550 feet to the right of the localizer centerline and 210 feet above the glide slope.
C—775 feet to the left of the localizer centerline and 420 feet below the glide slope.

With the CDI two dots to the right, we are to the left of course, and according to FAA Figure 139, at the 5.6 NM location 1,550 feet to the left of course. The glide slope indication is one dot below the glide slope and at the 5.6 NM mark, that would put you 210 feet below the glide slope. (H837) — FAA-H-8083-15, Chapter 7

Answers
4753 [B]   4773 [A]   4824 [B]   4825 [C]   4826 [A]

Chapter 4 **Navigation**

ALL
4685. (Refer to Figure 130.) How does an LDA facility, such as the one at Roanoke Regional, differ from a standard ILS approach facility?

A—The LOC is wider.
B—The LOC is offset from the runway.
C—The GS is unusable beyond the MM.

The LDA is not aligned with the runway. Straight-in minimums may be published where alignment does not exceed 30° between the course and runway. Circling minimums only are published where this alignment exceeds 30°. (H837) — FAA-H-8083-15, Chapter 7

MLS and GPS

The Microwave Landing System (MLS) provides precision navigation guidance for exact alignment and descent of aircraft on an approach to, and landing on, a runway. It provides azimuth and elevation angle guidance and range information, all of which is interpreted by the aircraft receiver to determine the aircraft's position.

The MLS identification consists of a three-letter Morse code identifier preceded by the Morse code letter "M" which is two dashes (– –). An MLS example would be "M-STP." The "M" distinguishes this system from an ILS, which is preceded by the Morse code letter "I" which is two dots (• •); for example "I-STP."

The Global Positioning System (GPS) is a satellite-based radio navigational, positioning, and time transfer system. The GPS receiver verifies the integrity (usability) of the signals received from the GPS satellites through receiver autonomous integrity monitoring (RAIM) to determine if a satellite is providing corrupted information. Without RAIM capability, the pilot has no assurance of the accuracy of the GPS position. If RAIM is not available, another type of navigation and approach system must be used, another destination selected, or the trip delayed until RAIM is predicted to be available on arrival.

The database may not contain all of the transitions or departures from all runways and some GPS receivers do not contain DPs in the data base. It is necessary that helicopter procedures be flown at 70 knots or less since helicopter departure procedures and missed approaches use a 20:1 obstacle clearance surface (OCS), which is double the fixed-wing OCS, and turning areas are based on this speed as well. Any required alternate airport must have an approved instrument approach procedure other than GPS, which is anticipated to be operational and available at the estimated time of arrival and which the aircraft is equipped to fly.

Answers
4685 [B]

ALL

4798. What international Morse Code identifier is used to identify a specific interim standard microwave landing system?

A— A two letter Morse Code identifier preceded by the Morse Code for the letters "IM."
B— A three letter Morse Code identifier preceded by the Morse Code for the letter "M."
C— A three letter Morse Code identifier preceded by the Morse Code for the letters "ML."

The identification consists of a three-letter Morse Code identifier preceded by the Morse Code for "M" (— —) (i.e., M-STP). The "M" distinguishes this system from ILS which is preceded by the Morse Code for "I" (• •) (i.e., I-STP). (J01) — AIM ¶1-1-12

ALL

4799. If Receiver Autonomous Integrity Monitoring (RAIM) is not available when setting up a GPS approach, the pilot should

A— select another type of navigation and approach system.
B— continue to the MAP and hold until the satellites are recaptured.
C— continue the approach, expecting to recapture the satellites before reaching the FAF.

If RAIM is not available, another type of navigation and approach system must be used, another destination selected, or the trip delayed until RAIM is predicted to be available on arrival. (J01) — AIM ¶1-1-21

RTC

4800. For a helicopter GPS instrument approach to be practical in a metropolitan area, airspeed must be limited to

A— 90 knots.
B— 70 knots.
C— 60 knots.

It is necessary that helicopter procedures be flown at 70 knots or less since helicopter departure procedures and missed approaches use a 20:1 obstacle clearance surface. (J01) — AIM ¶1-1-21

ALL

4801. When using GPS for navigation and instrument approaches, any required alternate airport must have

A— authorization to fly approaches under IFR using GPS avionics systems.
B— a GPS approach that is anticipated to be operational and available at the ETA.
C— an approved operational instrument approach procedure other than GPS.

Any required alternate airport must have an approved instrument approach procedure other than GPS, which is anticipated to be operational and available at the estimated time of arrival and which the aircraft is equipped to fly. (J01) — AIM ¶1-1-21

Answers

4798 [B]   4799 [A]   4800 [B]   4801 [C]

Chapter 5
Regulations and Procedures

Requirements for Instrument Rating   *5–3*

Instrument Currency Requirements   *5–6*

Equipment Requirements   *5–10*

Inspection Requirements   *5–15*

Oxygen Requirements   *5–16*

Logbook Requirements   *5–18*

Preflight Requirements   *5–19*

Airspace   *5–20*

Cloud Clearance and Visibility Requirements   *5–24*

Aircraft Accident/Incident Reporting and NOTAMs   *5–26*

Spatial Disorientation   *5–27*

Optical Illusions   *5–29*

Cockpit Lighting and Scanning   *5–31*

Altitude and Course Requirements   *5–32*

Communication Reports   *5–34*

Lost Communication Requirements   *5–36*

Chapter 5 **Regulations and Procedures**

Requirements for Instrument Rating

Although "FAR" is used as the acronym for "Federal Aviation Regulations," and found throughout the regulations themselves and hundreds of other publications, the FAA is now actively discouraging its use. "FAR" also means "Federal Acquisition Regulations." To eliminate any possible confusion, the FAA cites the federal aviation regulations with reference to Title 14 of the Code of Federal Regulations. For example, "FAR Part 91.3" is referenced as "14 CFR Part 91 Section 3."

Federal Aviation Regulations Part 61 stipulates that no person may act as pilot-in-command (PIC) of a civil aircraft under IFR or in weather conditions less than the minimums prescribed for visual flight rules (VFR) unless the pilot holds an instrument rating. The rating must be for the category of aircraft to be flown; e.g., airplane or rotorcraft.

In addition, any flight in Class A airspace (from 18,000 feet MSL to and including FL600) requires an instrument rating. VFR flight is not allowed in Class A airspace.

Commercial airplane pilots who carry passengers for hire at night, or on cross-country flights of more than 50 nautical miles (NM), are also required to hold an instrument rating.

ALL
4024. When are you required to have an instrument rating for flight in VMC?

A—Flight through an MOA.
B—Flight into an ADIZ.
C—Flight into Class A airspace.

Each person operating an aircraft in Class A airspace must conduct that operation under instrument flight rules (IFR). (A20) — 14 CFR §61.3(e) and §91.135

Answer (A) is incorrect because an instrument rating is not required for VMC flight through an MOA (Military Operations Area). Answer (B) is incorrect because an instrument rating is not required for VMC flight into an ADIZ (Air Defense Identification Zone).

ALL
4025. The pilot in command of a civil aircraft must have an instrument rating only when operating

A—under IFR in positive control airspace.
B—under IFR, in weather conditions less than the minimum for VFR flight, and in a Class A airspace.
C—in weather conditions less than the minimum prescribed for VFR flight.

No person may act as pilot-in-command of a civil aircraft under instrument flight rules, or in weather conditions less than the minimums prescribed for VFR flight unless:

1. *In the case of an airplane, the pilot holds an Instrument Rating or an Airline Transport Pilot Certificate with an airplane category rating on it; or*

2. *In the case of a helicopter, the pilot holds a helicopter instrument rating or an Airline Transport Pilot Certificate with a rotorcraft category and helicopter class rating not limited to VFR.*

Also, each person operating an aircraft in Class A airspace must conduct that operation under instrument flight rules (IFR). (A20) — 14 CFR §61.3(e) and §91.135

Answer (A) is incorrect because it doesn't address the requirement for an instrument rating if flying in IFR conditions in uncontrolled airspace. Answer (C) is incorrect because it doesn't address the option of flying in VFR conditions with an IFR clearance.

ALL
4028. A certificated commercial pilot who carries passengers for hire at night or in excess of 50 NM is required to have at least

A—an associated type rating if the airplane is of the multiengine class.
B—a First-Class Medical Certificate.
C—an instrument rating in the same category and class of aircraft.

A Commercial pilot applicant must hold an instrument rating, or the Commercial Pilot Certificate that is issued is endorsed with a limitation prohibiting the carriage of passengers for hire in aircraft on cross-country flights of more than 50 nautical miles (NM), or at night. (A20) — 14 CFR §61.133

Answer (A) is incorrect because regardless of whether the aircraft needs a type rating, an instrument rating is required to carry passengers for hire at night. Answer (B) is incorrect because a commercial pilot does not need a first-class medical certificate, but rather a second-class medical certificate.

Answers

4024 [C]   4025 [B]   4028 [C]

Chapter 5 **Regulations and Procedures**

AIR

4029. You intend to carry passengers for hire on a night VFR flight in a single-engine airplane within a 25-mile radius of the departure airport. You are required to possess at least which rating(s)?

A—A Commercial Pilot Certificate with a single-engine land rating.
B—A Commercial Pilot Certificate with a single-engine and instrument (airplane) rating.
C—A Private Pilot Certificate with a single-engine land and instrument airplane rating.

A Commercial Certificate, with single-engine airplane class, is required to carry passengers for hire in that class of aircraft. In addition, the applicant must hold an instrument rating (airplane), or the Commercial Pilot Certificate is endorsed with a limitation prohibiting the carriage of passengers for hire in airplanes on cross-country flights of more than 50 nautical miles (NM), or at night. (A20) — 14 CFR §61.133

Answer (A) is incorrect because in order to carry passengers for hire at night, an instrument rating is required in addition to a commercial pilot certificate. Answer (C) is incorrect because a commercial (not private) certificate is required to carry passengers for hire.

ALL

4031. Under which condition must the pilot in command of a civil aircraft have at least an instrument rating?

A—When operating in Class E airspace.
B—For a flight in VFR conditions while on an IFR flight plan.
C—For any flight above an altitude of 1,200 feet AGL, when the visibility is less than 3 miles.

No person may act as pilot-in-command of a civil aircraft under instrument flight rules, or in weather conditions less than the minimums prescribed for VFR flight unless:

1. *In the case of an airplane, the pilot holds an Instrument Rating or an Airline Transport Pilot Certificate with an airplane category rating on it; or*
2. *In the case of a helicopter, the pilot holds a helicopter Instrument Rating or an Airline Transport Pilot Certificate with a rotorcraft category and helicopter class rating not limited to VFR.*

Also, each person operating an aircraft in Class A airspace must conduct that operation under instrument flight rules (IFR). (A20) — 14 CFR §61.3

Answer (A) is incorrect because an instrument rating is required for flight in Class A airspace, not in the Continental Control Area. Answer (C) is incorrect because VFR is allowed in uncontrolled airspace during the day with as little as 1 SM visibility when more than 1,200 feet AGL but less than 10,000 feet MSL.

AIR

4002. What limitation is imposed on a newly certificated commercial airplane pilot if that person does not hold an instrument pilot rating?

A—The carrying of passengers or property for hire on cross-country flights at night is limited to a radius of 50 nautical miles (NM).
B—The carrying of passengers for hire on cross-country flights is limited to 50 NM for night flights, but not limited for day flights.
C—The carrying of passengers for hire on cross-country flights is limited to 50 NM and the carrying of passengers for hire at night is prohibited.

The commercial pilot applicant must hold an instrument rating (airplane), or the Commercial Pilot Certificate that is issued is endorsed with a limitation prohibiting the carriage of passengers for hire in airplanes on cross-country flights of more than 50 NM, or at night. (A24) — 14 CFR §61.133

Answer (A) is incorrect because the carriage of any property is not prohibited at night. Answer (B) is incorrect because passengers may not be carried during night and flight is limited to 50 NM for flights without an instrument rating.

AIR

4034. Which limitation is imposed on the holder of a Commercial Pilot Certificate if that person does not hold an instrument rating?

A—That person is limited to private pilot privileges at night.
B—The carrying of passengers or property for hire on cross-country flights at night is limited to a radius of 50 NM.
C—The carrying of passengers for hire on cross-country flights is limited to 50 NM and the carrying of passengers for hire at night is prohibited.

The commercial pilot applicant must hold an instrument rating (airplane), or the Commercial Pilot Certificate that is issued is endorsed with a limitation prohibiting the carriage of passengers for hire in airplanes on cross-country flights of more than 50 NM, or at night. (A20) — 14 CFR §61.133

Answer (A) is incorrect because the pilot may exercise commercial pilot privileges at night, but with limitations imposed. Answer (B) is incorrect because no passengers may be carried at night without an instrument rating.

Answers

4029 [B]   4031 [B]   4002 [C]   4034 [C]

AIR
4035. To carry passengers for hire in an airplane on cross-country flights of more than 50 NM from the departure airport, the pilot in command is required to hold at least

A—a Category II pilot authorization.
B—a First-Class Medical certificate.
C—a Commercial Pilot Certificate with an instrument rating.

In order to fly for hire, the pilot-in-command must hold a Commercial Pilot Certificate. In addition commercial pilots must hold an instrument rating (airplane), or they will be prohibited to carry passengers for hire in airplanes on cross-country flights of more than 50 NM, or at night. (A20) — 14 CFR §61.133

Answer (A) is incorrect because Category II pilot authorization is in reference to an authorization for reduced ILS approach minimums. Answer (B) is incorrect because a First-Class Medical Certificate is only required for airline transport pilots.

RTC
4018. Under which condition may you act as pilot in command of a helicopter under IFR?

Your certificates and ratings: Private Pilot Certificate with AMEL and Airplane instrument, rotorcraft category rating, and helicopter class rating.

A—If a certificated helicopter instrument flight instructor is on board.
B—If you meet the recent helicopter IFR experience requirements.
C—If you acquire a helicopter instrument rating and meet IFR currency requirements.

No person may act as pilot-in-command of a helicopter unless he/she has a valid pilot certificate with rotorcraft category and helicopter class ratings. In addition, to operate IFR, the pilot-in-command must have either a helicopter instrument rating or an Airline Transport Pilot Certificate with rotorcraft category and helicopter class rating not limited to VFR. In addition to having the appropriate ratings, a pilot must have the required recent experience set forth in 14 CFR §61.57. (A20) — 14 CFR §61.57

Answer (A) is incorrect because the pilot must have the appropriate category, class, and instrument ratings in order to operate as PIC of an instrument flight, regardless of who is accompanying the flight. Answer (B) is incorrect because the pilot must have the helicopter instrument rating and meet the currency requirements in order to operate as PIC of an instrument flight.

RTC
4030-1. Do regulations permit you to act as pilot in command of a helicopter in IMC if you hold a Private Pilot Certificate with ASEL, airplane instrument rating, rotorcraft category, and helicopter class rating?

A—Yes, if you comply with the recent IFR experience requirements for a helicopter.
B—No, you must hold either an unrestricted Airline Transport Pilot-Helicopter Certificate or a helicopter instrument rating.
C—No, however, you may do so if you hold an Airline Transport Pilot-Helicopter Certificate, limited to VFR.

No person may act as pilot-in-command of a helicopter unless the pilot has a valid pilot certificate with rotorcraft category and helicopter class ratings. In addition, to operate IFR, the pilot-in-command must have either an instrument-helicopter rating or an Airline Transport Pilot Certificate with rotorcraft category, and helicopter class rating not limited to VFR. Along with the appropriate ratings, a pilot must have the required recent experience set forth in 14 CFR §61.57. (A20) — 14 CFR §61.3 and §61.57

Answer (A) is incorrect because the pilot must hold the appropriate certificate, in addition to meeting the currency requirements. Answer (C) is incorrect because even with the ATP-Helicopter certificate, it must be unrestricted, to allow flight in IMC.

AIR
4030-2. Do regulations permit you to act as pilot in command of an airplane in IMC if you hold a Private Pilot Certificate with ASEL, Rotorcraft category, with helicopter class rating and instrument helicopter rating?

A—No, however, you may do so if you hold an Airline Transport Pilot-Airplane Certificate, limited to VFR.
B—No, you must hold either an unrestricted Airline Transport Pilot-Airplane Certificate or an airplane instrument rating.
C—Yes, if you comply with the recent IFR experience requirements for a helicopter.

No person may act as pilot-in-command of an airplane unless the pilot has a valid pilot certificate with airplane category. In addition, to operate IFR, the pilot-in-command must have either an instrument-airplane rating or an Airline Transport Pilot Certificate with airplane category not limited to VFR. Along with the appropriate ratings, a pilot must have the required recent experience set forth in 14 CFR §61.57. (A20) — 14 CFR §61.3 and §61.57

Answer (A) is incorrect because even with the ATP-Helicopter certificate, it must be unrestricted, to allow flight in IMC. Answer (C) is incorrect because the pilot must hold the appropriate certificate, in addition to meeting the currency requirements.

Answers

4035 [C]　　　4018 [C]　　　4030-1 [B]　　　4030-2 [B]

Chapter 5 **Regulations and Procedures**

Instrument Currency Requirements

No person may act as pilot in command under IFR or in weather conditions less than the minimums prescribed for VFR, unless within the preceding 6 calendar months, that person has performed and logged under actual or simulated instrument conditions:

1. At least 6 instrument approaches;
2. Holding procedures; and
3. Intercepting and tracking courses through the use of navigation systems.

Satisfactory accomplishment (within the last 6 calendar months) of an instrument competency check in the category of aircraft to be flown will also meet this recency requirements.

If the recent instrument experience for 6 calendar months is not met, the pilot may not act as PIC under IFR until an instrument competency check in the category of aircraft involved has been passed. This check must be given by an approved FAA examiner, instrument instructor, or FAA inspector.

ALL
4012. To meet the minimum instrument experience requirements, within the last 6 calendar months you need

A—six instrument approaches, holding procedures, and intercepting and tracking courses in the appropriate category of aircraft.
B—six hours in the same category aircraft.
C—six hours in the same category aircraft, and at least 3 of the 6 hours in actual IFR conditions.

To act as pilot-in-command under IFR, a pilot must have logged, in the past 6 calendar months, at least 6 instrument approaches, holding procedures, and intercepting and tracking courses through the use of navigation systems, in the appropriate category of aircraft for the instrument privileges sought. (A20) — 14 CFR §61.57

Answers (B) and (C) are incorrect because pilots are not required to fly a specified amount of flight time under IFR conditions.

ALL
4013. After your recent IFR experience lapses, how much time do you have before you must pass an instrument competency check to act as pilot in command under IFR?

A—6 months.
B—90 days.
C—12 months.

A pilot who does not meet the recent instrument experience requirements during the prescribed time, has 6 months thereafter to pass an instrument competency check in the category of aircraft involved. (A20) — 14 CFR §61.57

Answer (B) is incorrect because "90 days" refers to the takeoff and landing currency requirements to carry passengers. Answer (C) is incorrect because 12 months is the time from when you gain IFR currency to when you need another instrument competency check (assuming you have not maintained currency).

ALL
4014. An instrument rated pilot, who has not logged any instrument time in 1 year or more, cannot serve as pilot in command under IFR, unless the pilot

A—completes the required 6 hours and six approaches, followed by an instrument proficiency check given by an FAA-designated examiner.
B—passes an instrument proficiency check in the category of aircraft involved, given by an approved FAA examiner, instrument instructor, or FAA inspector.
C—passes an instrument proficiency check in the category of aircraft involved, followed by 6 hours and six instrument approaches, 3 of those hours in the category of aircraft involved.

A pilot who does not meet the recent instrument experience requirements during the prescribed time, has 6 months thereafter to pass an instrument proficiency check in the category of aircraft involved, given by an approved FAA Examiner, instrument instructor, or FAA inspector. (A20) — 14 CFR §61.57

Answers (A) and (C) are incorrect because an instrument proficiency check alone provides currency. Additional hours and approaches are not required.

Answers
4012 [A]   4013 [A]   4014 [B]

Chapter 5 **Regulations and Procedures**

ALL
4015. A pilot's recent IFR experience expires on July 1 of this year. What is the latest date the pilot can meet the IFR experience requirement without having to take an instrument proficiency check?

A—December 31, this year.
B—June 30, next year.
C—July 31, this year.

A pilot who does not meet the recent instrument experience requirements during the prescribed time, has 6 months thereafter to pass an instrument proficiency check in the category of aircraft involved. If a pilot's IFR currency expires on July 1 of this year, that pilot would have 6 months, or until December 31, this year, to gain the IFR experience before an instrument proficiency check would be required. (A20) — 14 CFR §61.57

Answer (B) is incorrect because this is a 12-month period, not the required 6 months. Answer (C) is incorrect because this is 1 month, not the required 6 months.

ALL
4017. What minimum conditions are necessary for the instrument approaches required for IFR currency?

A—The approaches may be made in an aircraft, approved instrument ground trainer, or any combination of these.
B—At least three approaches must be made in the same category of aircraft to be flown.
C—At least three approaches must be made in the same category and class of aircraft to be flown.

The minimum conditions necessary for the instrument approaches is to complete 6 within the past 6 months, to be completed in an aircraft, approved instrument ground trainer, or any combination of these. (A20) — 14 CFR §61.57

Answers (B) and (C) are incorrect because it is not necessary to complete 3 approaches in the same category and class of aircraft; they can be in an approved instrument ground trainer, as well.

ALL
4020. How may a pilot satisfy the recent flight experience requirement necessary to act as pilot in command in IMC in powered aircraft? Within the previous 6 calendar months, logged

A—six instrument approaches and 3 hours under actual or simulated IFR conditions within the last 6 months; three of the approaches must be in the category of aircraft involved.
B—six instrument approaches, holding procedures, and intercepting and tracking courses using navigational systems.
C—6 hours of instrument time under actual or simulated IFR conditions within the last 3 months, including at least six instrument approaches of any kind. Three of the 6 hours must be in flight in any category aircraft.

To act as pilot-in-command under IFR, a pilot must have logged, in the past 6 calendar months, at least 6 instrument approaches, holding procedures, and intercepting and tracking courses through the use of navigation systems, in the appropriate category of aircraft for the instrument privileges sought. (A20) — 14 CFR §61.57

ALL
4021. How long does a pilot meet the recency of experience requirements for IFR flight after successfully completing an instrument competency check if no further IFR flights are made?

A—90 days.
B—6 calendar months.
C—12 calendar months.

A pilot who does not meet the recent instrument experience requirements during the prescribed time, may not serve as pilot-in-command under IFR until that pilot passes an instrument competency check in the category of aircraft involved. If no other instrument flight is conducted, that instrument competency check will remain current for 6 months. (A20) — 14 CFR §61.57

Answer (A) is incorrect because 90 days refers to the takeoff and landing requirements for carrying passengers. Answer (C) is incorrect because 12 months is the time after which another instrument competency check will be required.

Answers

4015 [A]   4017 [A]   4020 [B]   4021 [B]

Chapter 5 **Regulations and Procedures**

AIR
4023. What recent instrument flight experience requirements must be met before you may act as pilot in command of an airplane under IFR?

A—A minimum of six instrument approaches in an airplane, or an approved simulator (airplane) or ground trainer, within the preceding 6 calendar months.
B—A minimum of six instrument approaches, at least three of which must be in an aircraft within the preceding 6 calendar months.
C—A minimum of six instrument approaches in an aircraft, at least three of which must be in the same category within the preceding 6 calendar months.

The minimum conditions necessary for the instrument approaches is to complete 6 within the past 6 months, to be completed in an aircraft, approved instrument ground trainer, or any combination of these. (A20) — 14 CFR §61.57

Answers (B) and (C) are incorrect because all 6 approaches can be in an approved instrument ground trainer, and all must be in the same category of aircraft.

AIR
4026. What additional instrument experience is required for you to meet the recent flight experience requirements to act as pilot in command of an airplane under IFR?

Your present instrument experience within the preceding 6 calendar months is:

1. 3 hours with holding, intercepting and tracking courses in an approved airplane flight simulator.
2. two instrument approaches in an airplane.

A—Three hours of simulated or actual instrument flight time in a helicopter, and two instrument approaches in an airplane or helicopter.
B—Three instrument approaches in an airplane.
C—Four instrument approaches in an airplane, or an approved airplane flight simulator or training device.

To act as pilot-in-command under IFR, a pilot must have logged, in the past 6 calendar months, at least 6 instrument approaches, holding procedures, and intercepting and tracking courses through the use of navigation systems in the appropriate category of aircraft for the instrument privileges sought. (A20) — 14 CFR §61.57

ALL
4027. To meet the minimum required instrument flight experience to act as pilot in command of an aircraft under IFR, you must have logged within the preceding 6 calendar months in the same category of aircraft: six instrument approaches,

A—holding procedures, intercepting and tracking courses through the use of navigation systems.
B—and 6 hours of instrument time in any aircraft.
C—three of which must be in the same category and class of aircraft to be flown, and 6 hours of instrument time in any aircraft.

To act as pilot-in-command under IFR, a pilot must have logged, in the past 6 calendar months, at least 6 instrument approaches, holding procedures, and intercepting and tracking courses through the use of navigation systems in the appropriate category of aircraft for the instrument privileges sought. (A20) — 14 CFR §61.57

ALL
4001. No pilot may act as pilot-in-command of an aircraft under IFR or in weather conditions less than the minimums prescribed for VFR unless that pilot has, within the preceding 6 calendar months, completed at least

A—three instrument approaches and logged 3 hours.
B—six instrument flights under actual IFR conditions.
C—six instrument approaches, holding procedures, intercepting and tracking courses using navigational systems, or passed an instrument proficiency check.

To act as pilot-in-command under IFR, a pilot must have logged, in the past 6 calendar months, at least 6 instrument approaches, holding procedures, and intercepting and tracking courses through the use of navigation systems, in the appropriate category of aircraft for the instrument privileges sought. (A20) — 14 CFR §61.57

Chapter 5 Regulations and Procedures

RTC
4016. What additional instrument approaches, if any, must you perform to meet the recent flight experience requirements for IFR operation in a helicopter?

Within the preceding 6 calendar months, you have accomplished:

One approach in a helicopter.
Two approaches in an airplane.
Two approaches in an approved airplane simulator.

A—One approach in an airplane, helicopter, or approved simulator.
B—None.
C—Five approaches in a helicopter or an approved rotorcraft simulator.

To act as pilot-in-command under IFR, a pilot must have logged, in the past 6 calendar months, at least 6 instrument approaches, holding procedures, and intercepting and tracking courses through the use of navigation systems, in the appropriate category of aircraft for the instrument privileges sought. (A20) — 14 CFR §61.57

RTC
4019. What additional flight hours within the preceding 6 calendar months are required to maintain IFR currency in a helicopter, if you already have 3 hours in an instrument simulator?

A—None, but 6 instrument approaches, holding procedures and tracking courses must be accomplished.
B—None, but three instrument approaches must also be accomplished.
C—3 hours of actual or simulated instrument time in the same type helicopter.

To act as pilot-in-command under IFR, a pilot must have logged, in the past 6 calendar months, at least 6 instrument approaches, holding procedures, and intercepting and tracking courses through the use of navigation systems, in the appropriate category of aircraft for the instrument privileges sought. (A20) — 14 CFR §61.57

Answer (B) is incorrect because 6 approaches (not 3) must be accomplished, along with holding procedures and tracking courses. Answer (C) is incorrect because pilots are not required to fly a specified amount of flight time under IFR conditions.

RTC
4022. Which additional instrument experience is required before you may act as pilot-in-command of a helicopter under IFR?

Your instrument experience within the preceding 6 calendar months is:

2 hours and one instrument approach in an approved helicopter simulator with holding, intercepting and tracking procedures, and one instrument approach in an airplane.

A—Three instrument approaches in an airplane or helicopter.
B—1 hour of simulated instrument flight time and two instrument approaches in a helicopter.
C—Five instrument approaches in a helicopter or an approved helicopter simulator.

To act as pilot-in-command under IFR, a pilot must have logged, in the past 6 calendar months, at least 6 instrument approaches, holding procedures, and intercepting and tracking courses through the use of navigation systems, in the appropriate category of aircraft for the instrument privileges sought. (A20) — 14 CFR §61.57

Answers (A) and (B) are incorrect because a total of 6 approaches must be accomplished all in the aircraft category for the instrument privileges sought, and pilots are not required to fly a specified amount of flight time under IFR conditions.

Answers

4016 [C]   4019 [A]   4022 [C]

Chapter 5 **Regulations and Procedures**

Equipment Requirements

For IFR flight, the following instruments and equipment are required:

1. All VFR day and night equipment.
2. Two-way radio and navigational equipment appropriate to the ground facilities to be used.
3. Gyroscopic rate-of-turn indicator.
4. Slip-skid indicator.
5. Gyroscopic attitude indicator.
6. Gyroscopic heading indicator.
7. Sensitive altimeter.
8. Clock with sweep-second hand or digital display.
9. Generator of adequate capacity.
10. Mode C transponder with encoding altimeter (above 10,000 feet MSL in controlled airspace).
11. Distance measuring equipment (DME) when at or above FL240 using VORs for navigation. Should the DME fail at or above 24,000 feet MSL, the PIC shall report the failure to ATC immediately and may then continue at and above 24,000 feet MSL to the next airport of intended landing.

ALL
4037. In the 48 contiguous states, excluding the airspace at or below 2,500 feet AGL, an operable coded transponder equipped with Mode C capability is required in all controlled airspace at and above

A—12,500 feet MSL.
B—10,000 feet MSL.
C—Flight level (FL) 180.

A coded transponder with altitude reporting (Mode C) is required for all operations within the 48 contiguous states and the District of Columbia above an altitude of 10,000 feet MSL. Operations in airspace below 2,500 feet AGL are excluded from this rule. (B11) — 14 CFR §91.215(b)

Answer (A) is incorrect because 12,500 feet MSL is not an altitude which defines airspace. Answer (C) is incorrect because FL180 is the floor of Class A airspace.

ALL
4038. A coded transponder equipped with altitude reporting capability is required in all controlled airspace

A—at and above 10,000 feet MSL, excluding at and below 2,500 feet AGL.
B—at and above 2,500 feet above the surface.
C—below 10,000 feet MSL, excluding at and below 2,500 feet AGL.

A coded transponder with altitude reporting (Mode C) is required for all operations within the 48 contiguous states and the District of Columbia above an altitude of 10,000 feet MSL. Operations in airspace below 2,500 feet AGL are excluded from this rule. (B11) — 14 CFR §91.215(b)

ALL
4051. An aircraft operated under 14 CFR Part 91 IFR is required to have which of the following?

A—Radar altimeter.
B—Dual VOR system.
C—Gyroscopic direction indicator.

For IFR flight, the following instruments and equipment are required:

1. *Instruments and equipment specified for VFR flight (and those required for night flight, if applicable).*
2. *Two-way radio communications system and navigational equipment appropriate to the ground facilities to be used.*
3. *Gyroscopic rate-of-turn indicator (with certain exceptions).*
4. *Slip-skid indicator.*
5. *Sensitive altimeter adjustable for barometric pressure.*
6. *A clock displaying hours, minutes and seconds with a sweep-second pointer or digital presentation.*

Answers

4037 [B]   4038 [A]   4051 [C]

7. Generator or alternator of adequate capacity.
8. Gyroscopic pitch and bank indicator (artificial horizon).
9. Gyroscopic direction indicator (directional gyro or equivalent).

(B11) — 14 CFR §91.205(d)

Answer (A) is incorrect because a sensitive (not radar) altimeter is required for IFR flight. Answer (B) is incorrect because, if VOR navigation is being used, only a single VOR is required.

ALL
4055. What minimum navigation equipment is required for IFR flight?

A—VOR/LOC receiver, transponder, and DME.
B—VOR receiver and, if in ARTS III environment, a coded transponder equipped for altitude reporting.
C—Navigation equipment appropriate to the ground facilities to be used.

Two-way radio communications system and navigational equipment appropriate to the ground facilities to be used is required for IFR flight. (B11) — 14 CFR §91.205(d)

Answer (A) is incorrect because a VOR/LOC receiver and DME are only required if that is the primary means of navigation used for this flight (and a transponder is not a navigation system). Answer (B) is incorrect because a VOR is only required if that is the primary means of navigation (and a transponder is not a navigation system).

ALL
4050. Where is DME required under IFR?

A—At or above 24,000 feet MSL if VOR navigational equipment is required.
B—In positive control airspace.
C—Above 18,000 feet MSL.

DME is required if VOR is used for navigation on flights at or above 24,000 feet MSL. (B11) — 14 CFR §91.205(e)

Answers (B) and (C) are incorrect because Class A airspace begins at 18,000 feet MSL and DME is required for flights at or above 24,000 feet MSL, if using VOR for navigation.

ALL
4448. What action should you take if your DME fails at FL240?

A—Advise ATC of the failure and land at the nearest available airport where repairs can be made.
B—Notify ATC that it will be necessary for you to go to a lower altitude, since your DME has failed.
C—Notify ATC of the failure and continue to the next airport of intended landing where repairs can be made.

For flight at and above 24,000 feet MSL, if VOR navigational equipment is required, no person may operate a U.S. registered civil aircraft within the 50 states, and the District of Columbia, unless that aircraft is equipped with approved distance measuring equipment (DME). When DME fails at and above FL240, the pilot-in-command of the aircraft shall notify ATC immediately, and may then continue operations at and above FL240 to the next airport of intended landing at which repairs or replacement of the equipment can be made. (B08) — 14 CFR §91.205(e)

Answer (A) is incorrect because it is not necessary to land at the nearest airport; the pilot may continue the flight to the destination. Answer (B) is incorrect because it is not necessary to descend; the pilot may continue the flight at or above FL240.

ALL
4459. What is the procedure when the DME malfunctions at or above 24,000 feet MSL?

A—Notify ATC immediately and request an altitude below 24,000 feet.
B—Continue to your destination in VFR conditions and report the malfunction.
C—After immediately notifying ATC, you may continue to the next airport of intended landing where repairs can be made.

For flight at and above 24,000 feet MSL, if VOR navigational equipment is required, no person may operate a U.S.-registered civil aircraft within the 50 states, and the District of Columbia, unless that aircraft is equipped with approved distance measuring equipment (DME). When DME fails at and above FL240, the pilot-in-command of the aircraft shall notify ATC immediately, and may then continue operations at and above FL240 to the next airport of intended landing at which repairs or replacement of the equipment can be made. (B08) — 14 CFR §91.205(e)

Answer (A) is incorrect because it is not necessary to descent to a lower altitude; the pilot may continue the flight at or above 24,000 feet. Answer (B) is incorrect because it is not necessary to fly only in VFR conditions; the pilot may continue the flight as normal.

Answers

4055 [C]	4050 [A]	4448 [C]	4459 [C]

Chapter 5 **Regulations and Procedures**

ALL
4007. If the aircraft's transponder fails during flight within Class B airspace,

A—the pilot should immediately request clearance to depart the Class B airspace.
B—ATC may authorize deviation from the transponder requirement to allow aircraft to continue to the airport of ultimate destination.
C—aircraft must immediately descend below 1,200 feet AGL and proceed to destination.

ATC may authorize deviations from transponder requirements to allow an aircraft with an inoperative transponder to continue to the airport of ultimate destination, including any intermediate stops, or to proceed to a place where suitable repairs can be made. (B11) — 14 CFR §91.215(d)(1)

Answer (A) is incorrect because ATC can authorize a deviation from the transponder requirement and does not require the pilot to request clearance to depart Class B airspace. Answer (C) is incorrect because a pilot may only descend after a clearance from ATC is obtained.

ALL
4438. When an aircraft is not equipped with a transponder, what requirement must be met before ATC will authorize a flight within Class B airspace?

A—A request for the proposed flight must be made to ATC at least 1 hour before the flight.
B—The proposed flight must be conducted when operating under instrument flight rules.
C—The proposed flight must be conducted in visual meteorological conditions (VMC).

ATC may authorize deviations on a continuing or individual basis for operations of aircraft without a transponder, in which case the request for a deviation must be submitted to the ATC facility having jurisdiction over the airspace concerned at least 1 hour before the proposed operation. (J08) — 14 CFR §91.215(d)(3)

ALL
4043. Aircraft being operated under IFR are required to have, in addition to the equipment required for VFR and night, at least

A—distance measuring equipment.
B—dual VOR receivers.
C—a slip skid indicator.

For IFR flight, the following instruments and equipment are required:

1. *Instruments and equipment specified for VFR flight (and those required for night flight, if applicable).*
2. *Two-way radio communications system and navigational equipment appropriate to the ground facilities to be used.*
3. *Gyroscopic rate-of-turn indicator (with certain exceptions).*
4. *Slip-skid indicator.*
5. *Sensitive altimeter adjustable for barometric pressure.*
6. *A clock displaying hours, minutes and seconds with a sweep-second pointer or digital presentation.*
7. *Generator or alternator of adequate capacity.*
8. *Gyroscopic pitch and bank indicator (artificial horizon).*
9. *Gyroscopic direction indicator (directional gyro or equivalent).*

(B11) — 14 CFR §91.205(d)

Answers (A) and (B) are incorrect because distance measuring equipment and dual VOR receivers are not mandatory for instrument flight, although they may be helpful.

ALL
4057. To meet the requirements for flight under IFR, an aircraft must be equipped with certain operable instruments and equipment. One of those required is

A—a clock with sweep-second pointer or digital presentation.
B—a radar altimeter.
C—a transponder with altitude reporting capability.

For IFR flight, the following instruments and equipment are required:

1. *Instruments and equipment specified for VFR flight (and those required for night flight, if applicable)*
2. *Two-way radio communications system and navigational equipment appropriate to the ground facilities to be used.*
3. *Gyroscopic rate-of-turn indicator (with certain exceptions).*
4. *Slip-skid indicator.*
5. *Sensitive altimeter adjustable for barometric pressure.*
6. *A clock displaying hours, minutes and seconds with a sweep-second pointer or digital presentation.*

Answers

| 4007 | [B] | 4438 | [A] | 4043 | [C] | 4057 | [A] |

Chapter 5 **Regulations and Procedures**

7. Generator or alternator of adequate capacity.
8. Gyroscopic pitch and bank indicator (artificial horizon).
9. Gyroscopic direction indicator (directional gyro or equivalent).

(B11) — 14 CFR §91.205(d)

Answer (B) is incorrect because a sensitive altimeter (not radar) is required. Answer (C) is incorrect because a transponder is not required for instrument flight, but rather for flight at specified altitudes and airspaces.

ALL
4653. (Refer to Figure 123.) What minimum navigation equipment is required to complete the VOR/DME-A procedure?

A—One VOR receiver.
B—One VOR receiver and DME.
C—Two VOR receivers and DME.

VOR/DME procedure number means that both operative VOR and DME receivers and ground equipment in normal operation are required to use the procedure. In the VOR/DME procedure, when either the VOR or DME is inoperative, the procedure is not authorized. (J42) — Instrument Approach Procedures

ALL
4375. The aircraft's transponder fails during flight within Class D airspace.

A—The pilot should immediately request clearance to depart the Class D airspace.
B—No deviation is required because a transponder is not required in Class D airspace.
C—Pilot must immediately request priority handling to proceed to destination.

ATC may authorize deviations from transponder requirements immediately to allow an aircraft with an inoperative transponder to continue to the airport of ultimate destination, including any intermediate stops, or to proceed to a place where suitable repairs can be made. Within Class D airspace there are no transponder requirements, therefore if the aircraft's transponder fails while in Class D airspace, no deviation will be necessary. (J08) — 14 CFR §91.215

Answers (A) and (C) are incorrect because there are no transponder requirements within Class D airspace, therefore no deviation or priority handling will be necessary.

ALL
4695. (Refer to Figure 131.) Other than VOR/DME RNAV, what additional navigation equipment is required to conduct the VOR/DME RNAV RWY 4R approach at BOS?

A—None.
B—VNAV.
C—Transponder with altitude encoding and Marker Beacon.

No additional equipment is necessary to conduct the VOR/DME RNAV RWY 4R approach at BOS. (J42) — Instrument Approach Procedures

Answer (B) is incorrect because the chart would be annotated with "VNAV" in the upper left part of the chart if this was required. Answer (C) is incorrect because marker beacons are not used on this approach.

ALL
4426. In addition to a VOR receiver and two-way communications capability, which additional equipment is required for IFR operation in Class B airspace?

A—DME and an operable coded transponder having Mode C capability.
B—Standby communications receiver, DME, and coded transponder.
C—An operable coded transponder having Mode C capability.

No person may operate an aircraft within Class B airspace unless that aircraft is equipped with—

1. *For IFR operation. An operable VOR or TACAN receiver; and*
2. *For all operations. An operable two-way radio capable of communications with ATC on appropriate frequencies for that Class B airspace area and a transponder having Mode C capability.*

(J08) — 14 CFR §91.131(c)

Answer (A) is incorrect because DME is not required, and the transponder must have Mode C capability. Answer (B) is incorrect because a standby radio receiver and DME are not required, and the transponder must have Mode C capability.

Answers

| 4653 | [B] | 4375 | [B] | 4695 | [A] | 4426 | [C] |

Chapter 5 **Regulations and Procedures**

ALL
4439. Prior to operating an aircraft not equipped with a transponder in Class B airspace, a request for a deviation must be submitted to the

A—FAA Administrator at least 24 hours before the proposed operation.
B—nearest FAA General Aviation District Office 24 hours before the proposed operation.
C—controlling ATC facility at least 1 hour before the proposed flight.

ATC may authorize deviations on a continuing or individual basis for operation of an aircraft that is not equipped with a transponder. The request must be made at least one hour before the proposed operation. (J08) — 14 CFR §91.215(d)(3)

Answers (A) and (B) are incorrect because the ATC facility having jurisdiction over Class B airspace (not the FAA Administrator or the nearest GADO) can authorize deviations from the transponder requirements, and the request must be made at least one hour before the proposed operation.

ALL
4440. Which of the following is required equipment for operating an aircraft within Class B airspace?

A—A 4096 code transponder with automatic pressure altitude reporting equipment.
B—A VOR receiver with DME.
C—A 4096 code transponder.

No person may operate an aircraft in Class B airspace unless the aircraft is equipped with—

1. *For IFR operation. An operable VOR or TACAN receiver, and*
2. *For all operations. An operable two-way radio capable of communications with ATC on appropriate frequencies for that Class B airspace area, and a transponder with Mode C capability.*

(J08) — 14 CFR §91.131(c),(d)

Answer (B) is incorrect because DME is not required and a VOR receiver is only required for IFR operations. Answer (C) is incorrect because the transponder must also have Mode C capability.

ALL
4004. The use of certain portable electronic devices is prohibited on aircraft that are being operated under

A—IFR.
B—VFR.
C—DVFR.

With certain exceptions, no person may operate, nor may any operator or pilot in command of an aircraft allow the operation of any portable electronic device on U.S. registered civil aircraft operated in IFR conditions. (B07) — 14 CFR §91.21(a)(2)

ALL
4284. How can a pilot determine if a Global Positioning System (GPS) installed in an aircraft is approved for IFR enroute and IFR approaches?

A—Flight manual supplement.
B—GPS operator's manual.
C—Aircraft owner's handbook.

An appropriate Airplane or Rotorcraft Flight Manual Supplement (or, for aircraft without an FAA Approved Flight Manual, a Supplemental Flight Manual) containing the limitations and operating procedures applicable to the equipment installed, should be provided for each installation of GPS navigation equipment for IFR approval. A Flight Manual Supplement may be necessary for installations limited to VFR use only, depending upon the complexity of the installation and the need to identify necessary limitations and operating procedures. (K26) — AC 20-138, page 21

RTC
4329. (Refer to Figure 59.) Unless otherwise authorized by ATC, what is the minimum equipment for navigation of helicopters on an IFR cross-country flight when in the immediate vicinity of the HUMBLE VORTAC?

A—VOR receiver, transponder with Mode C capability and two-way communications.
B—Transponder with Mode C capability and two-way communications.
C—VOR (or TACAN) and two-way communications.

At HUMBLE VORTAC the helicopter would be in Class B airspace. Unless otherwise authorized by ATC, no person may operate an aircraft within a Class B airspace unless that aircraft is equipped with an operable two-way radio, and the applicable operating transponder and automatic altitude reporting equipment. Because the question states "on an IFR cross-country flight," VOR is also required. (B11) — 14 CFR §91.131(c),(d) and §91.205(d)(2)

Answers

4439 [C] 4440 [A] 4004 [A] 4284 [A] 4329 [A]

Inspection Requirements

The altimeter, transponder, static system, and encoder must be checked every 24 calendar months.

The emergency locator transmitter (ELT) batteries must be replaced when 50% of their shelf life has expired, or after 1 hour of cumulative use. Check the date on the outside of the transmitter case.

The VOR receiver(s) must be checked within 30 days and found to be within limits.

AIR
4036. When must an operational check on the aircraft VOR equipment be accomplished when used to operate under IFR?

A—Within the preceding 10 days or 10 hours of flight time.
B—Within the preceding 30 days or 30 hours of flight time.
C—Within the preceding 30 days.

No person may operate a civil aircraft under IFR using the VOR system of radio navigation unless the VOR equipment of that aircraft:

1. *Is maintained, checked, and inspected under an approved procedure; or*
2. *Has been operationally checked within the preceding 30 days and was found to be within the limits of the permissible indicated bearing error.*

(B10) — 14 CFR §91.171(a)

Answer (A) is incorrect because the transponder must be checked every 30 (not 10) days. Answer (B) is incorrect because there is no requirement for regarding hours in operation.

ALL
4047. Your aircraft had the static pressure system and altimeter tested and inspected on January 5, of this year, and was found to comply with FAA standards. These systems must be reinspected and approved for use in controlled airspace under IFR by

A—January 5, next year.
B—January 5, 2 years hence.
C—January 31, 2 years hence.

No person may operate an airplane or helicopter in controlled airspace under IFR unless within the preceding 24 calendar months, each static pressure system, each altimeter instrument, and each automatic pressure altitude reporting system has been tested and inspected. (B13) — 14 CFR §91.411(a)(1)

Answer (A) is incorrect because it is 24 months (not 12). Answer (B) is incorrect because the test is valid until the end of the month.

ALL
4048. Which checks and inspections of flight instruments or instrument systems must be accomplished before an aircraft can be flown under IFR?

A—VOR within 30 days, altimeter systems within 24 calendar months, and transponder within 24 calendar months.
B—ELT test within 30 days, altimeter systems within 12 calendar months, and transponder within 24 calendar months.
C—VOR within 24 calendar months, transponder within 24 calendar months, and altimeter system within 12 calendar months.

No person may operate a civil aircraft under IFR using the VOR system of radio navigation unless the VOR equipment of that aircraft has been operationally checked within the preceding 30 days and was found to be within the limits of the permissible indicated bearing error.

No person may operate an airplane in controlled airspace under IFR unless within the preceding 24 calendar months, each static pressure system, each altimeter instrument, and each automatic pressure altitude reporting system has been tested and inspected.

No person may use an ATC transponder unless, within the preceding 24 calendar months, that ATC transponder has been tested and inspected. (B10) — 14 CFR §91.171(a)(2), §91.413(a) and §91.411(a)(1)

Answer (B) is incorrect because the altimeter must be checked every 24 (not 12) month, and ELTs are maintained in accordance with the manufacturer's requirements (only ELT batteries must be replaced on a specified time interval). Answer (C) is incorrect because VORs must be checked within 30 days (not 24 months), and altimeter systems within 24 months (not 12).

Answers

4036 [C] 4047 [C] 4048 [A]

Chapter 5 Regulations and Procedures

ALL
4049. An aircraft altimeter system test and inspection must be accomplished within

A—12 calendar months.
B—18 calendar months.
C—24 calendar months.

No person may operate an airplane or helicopter in controlled airspace under IFR unless the aircraft altimeter system test and inspection has been accomplished within the preceding 24 calendar months. (B13) — 14 CFR §91.411(a)(1)

Answers (A) and (B) are incorrect because the altimeter system must be checked every 24 calendar months (not 18 or 12).

ALL
4039. Who is responsible for determining that the altimeter system has been checked and found to meet 14 CFR Part 91 requirements for a particular instrument flight?

A—Owner.
B—Operator.
C—Pilot-in-command.

The pilot-in-command of an aircraft is directly responsible for, and is the final authority as to, the operation of that aircraft. (B07) — 14 CFR §91.3(a)

Answers (A) and (B) are incorrect because the owner or operator is primarily responsible for maintaining the aircraft, but the pilot-in-command is responsible for determining that the aircraft is airworthy.

Oxygen Requirements

Pilots are encouraged to use supplemental oxygen above 5,000 feet MSL at night. 14 CFR §91.211 stipulates that the required flight crew must use oxygen after 30 minutes at cabin pressure altitudes above 12,500 feet MSL, and at all times at cabin pressure altitudes above 14,000 feet MSL. Every occupant of the aircraft must be provided with supplemental oxygen at cabin pressure altitudes above 15,000 feet MSL.

Hypoxia is a state of oxygen deficiency, and impairs functions of the brain and other organs. Headache, drowsiness, dizziness, and euphoria are all symptoms of hypoxia. Hyperventilation, a deficiency of carbon dioxide within the body, can be the result of rapid or extra deep breathing due to emotional tension, anxiety, or fear. Symptoms will subside after the rate and depth of breathing are brought under control. Anxiety is an uneasiness over an impending or anticipated ill, or a fearful concern.

ALL
4042. If an unpressurized aircraft is operated above 12,500 feet MSL, but not more than 14,000 feet MSL, for a period of 2 hours 20 minutes, how long during that time is the minimum flightcrew required to use supplemental oxygen?

A—2 hours 20 minutes.
B—1 hour 20 minutes.
C—1 hour 50 minutes.

No person may operate a civil aircraft of U.S. registry at cabin pressure altitudes above 12,500 feet MSL up to and including 14,000 feet MSL unless the required minimum flight crew is provided with and uses supplemental oxygen for that part of the flight at those altitudes that is of more than 30 minutes duration.

Solution:

```
  2 hours 20  minutes
  -       30  minutes
  ─────────────────────
  1 hour  50  minutes
```

(B11) — 14 CFR §91.211(a)

Answer (A) is incorrect because a pilot can fly for 30 minutes before using oxygen between these altitudes. Answer (B) is incorrect because a pilot can fly for 30 minutes (not 1 hour) before using oxygen between these altitudes.

ALL
4045. What is the maximum cabin pressure altitude at which a pilot can fly for longer than 30 minutes without using supplemental oxygen?

A—10,500 feet.
B—12,000 feet.
C—12,500 feet.

No person may operate a civil aircraft of U.S. registry at cabin pressure altitudes above 12,500 feet MSL up to and including 14,000 feet MSL unless the required minimum flight crew is provided with and uses supplemental oxygen for that part of the flight at those altitudes that is of more than 30 minutes duration. (B11) — 14 CFR §91.211(a)(1)

Answers (A) and (B) are incorrect because 12,500 is the highest altitude at which the pilot can fly for longer than 30 minutes without using supplemental oxygen.

Answers

4049 [C]   4039 [C]   4042 [C]   4045 [C]

ALL
4052. What is the maximum IFR altitude you may fly in an unpressurized aircraft without providing passengers with supplemental oxygen?

A— 12,500 feet.
B— 14,000 feet.
C— 15,000 feet.

No person may operate a civil aircraft of U.S. registry at cabin pressure altitudes above 15,000 feet (MSL) unless each occupant of the aircraft is provided with supplemental oxygen. (B11) — 14 CFR §91.211(a)(3)

Answer (A) is incorrect because between cabin pressure altitudes 12,500 feet and 14,000 feet MSL, only the flight crew must use supplemental oxygen, and only for that part of the flight more than 30 minutes duration. Answer (B) is incorrect because at cabin pressure altitudes 14,000 feet MSL and above, only the flight crew is required to use supplemental oxygen.

ALL
4053. What is the oxygen requirement for an unpressurized aircraft at 15,000 feet?

A— All occupants must use oxygen for the entire time at this altitude.
B— Crew must start using oxygen at 12,000 feet and passengers at 15,000 feet.
C— Crew must use oxygen for the entire time above 14,000 feet and passengers must be provided supplemental oxygen only above 15,000 feet.

No person may operate a civil aircraft of U.S. registry at cabin pressure altitudes above 15,000 feet (MSL) unless each occupant of the aircraft is provided with supplemental oxygen. (B11) — 14 CFR §91.211(a)(3)

Answer (A) is incorrect because all occupants must only be provided with oxygen, and crew must use oxygen above 14,000 feet MSL. Answer (B) is incorrect because the crew must begin using oxygen at cabin pressure altitudes at 14,000 feet MSL or at 12,500 feet MSL for those parts of the flight more than 30 minutes duration.

ALL
4503. (Refer to Figure 89.) What are the oxygen requirements for an IFR flight northeast bound from Bryce Canyon on V382 at the lowest appropriate altitude in an unpressurized aircraft?

A— The required minimum crew must be provided and use supplemental oxygen for that part of the flight of more than 30 minutes.
B— The required minimum crew must be provided and use supplemental oxygen for that part of the flight of more than 30 minutes, and the passengers must be provided supplemental oxygen.
C— The required minimum crew must be provided and use supplemental oxygen, and all occupants must be provided supplemental oxygen for the entire flight above 15,000 feet.

The lowest appropriate altitude northeast bound on V382 is 16,000 feet. At altitudes above 14,000 feet MSL, the minimum flight crew must be provided with and use supplemental oxygen during the entire flight time at those altitudes. At altitudes above 15,000 feet MSL, each occupant of the aircraft must be provided with supplemental oxygen. (B11) — 14 CFR §91.211

ALL
4513. (Refer to Figure 91.) What are the oxygen requirements for an IFR flight eastbound on V520 from DBS VORTAC in an unpressurized aircraft at the MEA?

A— The required minimum crew must be provided and use supplemental oxygen for that part of the flight of more than 30 minutes.
B— The required minimum crew must be provided and use supplemental oxygen for that part of the flight of more than 30 minutes, and the passengers must be provided supplemental oxygen.
C— The required minimum crew must be provided and use supplemental oxygen.

Eastbound V520 is in the bottom, center of FAA Figure 91. The MEA for V520 eastbound from DBS VORTAC is 15,000 feet; therefore the required crew must be provided with, and use, supplemental oxygen. (B11) — 14 CFR §91.211(a)

Answer (A) is incorrect because this implies the oxygen requirement for flights between 12,500 feet MSL and 14,000 feet MSL. Answer (B) is incorrect because the crew must use supplemental oxygen at all times above 14,000 feet MSL, and passengers must be provided with oxygen only above 15,000 feet MSL.

Answers

| 4052 | [C] | 4053 | [C] | 4503 | [C] | 4513 | [C] |

Chapter 5 **Regulations and Procedures**

ALL
4809. Why is hypoxia particularly dangerous during flights with one pilot?

A—Night vision may be so impaired that the pilot cannot see other aircraft.
B—Symptoms of hypoxia may be difficult to recognize before the pilot's reactions are affected.
C—The pilot may not be able to control the aircraft even if using oxygen.

The effects of hypoxia are usually difficult to recognize, especially when they occur gradually. (J31) — AIM ¶8-1-2(a)(5)

Answer (A) is incorrect because hypoxia affects more than just night vision. Answer (C) is incorrect because hypoxia will not occur if oxygen is properly used.

ALL
4816. What action should be taken if hyperventilation is suspected?

A—Breathe at a slower rate by taking very deep breaths.
B—Consciously breathe at a slower rate than normal.
C—Consciously force yourself to take deep breaths and breathe at a faster rate than normal.

The symptoms of hyperventilation subside within a few minutes after the rate and depth of breathing are consciously brought back under control. The buildup of carbon dioxide in the body can be hastened by controlled breathing in and out with a paper bag held over the nose and mouth. (J31) — AIM ¶8-1-3(b)

Answers (A) and (C) are incorrect because taking very deep breaths and breathing at a faster rate will further aggravate hyperventilation.

Logbook Requirements

Pilots are authorized to log both simulated and actual instrument time, but only for that period during which the pilot operates the aircraft solely by reference to instruments. An instrument instructor may log instrument time when instruction is given during actual instrument weather conditions. Each logbook entry must include the place and type of each instrument approach, and for each flight in simulated instrument conditions, the name of the safety pilot.

A safety pilot must occupy the other control seat when the PIC is operating the aircraft solely by reference to instruments. The safety pilot is required to be an appropriately-rated pilot, but there is no requirement to have an instrument rating.

ALL
4008. To meet instrument experience requirements of 14 CFR part 61, section 61.57(c), a pilot enters the condition of flight in the pilot logbook as simulated instrument conditions, what qualifying information must also be entered?

A—Location and type of each instrument approach completed and name of safety pilot.
B—Number and type of instrument approaches completed and route of flight.
C—Name and pilot certificate number of safety pilot and type of approaches completed.

Each instrument flight entry must include the place and type of each instrument approach completed, and the name of the safety pilot for each simulated instrument flight. (A20) — 14 CFR §61.51

Answer (B) is incorrect because the number of instrument approaches completed and the route of flight is not required in the logbook entry. Answer (C) is incorrect because the pilot certificate number of the safety pilot is not required in the logbook entry.

ALL
4009. What portion of dual instruction time may a certificated instrument flight instructor log as instrument flight time?

A—All time during which the instructor acts as instrument instructor, regardless of weather conditions.
B—All time during which the instructor acts as instrument instructor in actual instrument weather conditions.
C—Only the time during which the instructor flies the aircraft by reference to instruments.

Instrument flight instructors may log as instrument flight time that time during which they act as instrument flight instructor in actual instrument weather conditions. (A20) — 14 CFR §61.51

Answers (A) and (C) are incorrect because the flight conditions must be IMC for the instructor to enter flight instruction as instrument time.

Answers

4809 [B]   4816 [B]   4008 [A]   4009 [B]

ALL
4010. Which flight time may be logged as instrument time when on an instrument flight plan?

A—All of the time the aircraft was not controlled by ground references.
B—Only the time you controlled the aircraft solely by reference to flight instruments.
C—Only the time you were flying in IFR weather conditions.

A pilot may log as instrument flight time, only that time during which the pilot operates the aircraft solely by reference to instruments under actual or simulated instrument flight conditions. (A20) — 14 CFR §61.51

Answer (A) is incorrect because VFR-On-Top can be conducted without reference to ground reference. The regulations specifically state flight must be conducted solely by reference to the instruments. Answer (C) is incorrect because flight conducted solely by reference to the instruments can be done in actual as well as simulated IFR weather conditions.

ALL
4011. What are the minimum qualifications for a person who occupies the other control seat as safety pilot during simulated instrument flight?

A—Private pilot certificate with appropriate category and class ratings for the aircraft.
B—Private pilot.
C—Private pilot with appropriate category, class, and instrument rating.

No person may operate a civil aircraft in simulated instrument flight unless the other control seat is occupied by a safety pilot who possesses at least a Private Pilot Certificate with category and class ratings appropriate to the aircraft being flown. (B08) — 14 CFR §91.109(b)

Answer (B) is incorrect because a private, helicopter pilot may not act as safety pilot in an airplane. The safety pilot's certificate must have the appropriate category and class rating. Answer (C) is incorrect because a safety pilot is not required to be instrument rated.

Preflight Requirements

Before beginning a flight, the PIC is required to become familiar with all available information concerning that flight. This must include:

1. Weather reports and forecasts.
2. Fuel requirements.
3. Alternatives available if the flight cannot be completed as planned.
4. Any known traffic delays.
5. Runway lengths.
6. Expected takeoff and landing distances.

ALL
4003. Before beginning any flight under IFR, the pilot in command must become familiar with all available information concerning that flight including:

A—all instrument approaches at the destination airport.
B—an alternate airport and adequate takeoff and landing performance at the destination airport.
C—the runway lengths at airports of intended use, and the aircraft's takeoff and landing data.

Each pilot-in-command shall, before beginning a flight, become familiar with all available information concerning that flight. This information must include:

1. *For a flight under IFR or a flight not in the vicinity of an airport, weather reports and forecasts, fuel requirements, alternatives available if the planned flight cannot be completed, and any known traffic delays of which the pilot has been advised by ATC.*

2. *For any flight, runway lengths at airports of intended use, and takeoff and landing distance information.*

3. *For civil aircraft for which an approved airplane or rotorcraft flight manual containing takeoff and landing distance data is required, the takeoff and landing distance data contained therein.*

(B08) — 14 CFR §91.103

Answer (A) is incorrect because it is not a required preflight action to know what approaches are available (but it is a good idea to know what a pilot's options will be). Answer (B) is incorrect because listing an alternate airport is only required when the destination is forecast to have ceilings below 2,000 feet and visibility less than 3 SM.

Answers

4010 [B]   4011 [A]   4003 [C]

Chapter 5 **Regulations and Procedures**

ALL
4033. Before beginning any flight under IFR, the pilot in command must become familiar with all available information concerning that flight. In addition, the pilot must

A—list an alternate airport on the flight plan and become familiar with the instrument approaches to that airport.
B—list an alternate airport on the flight plan and confirm adequate takeoff and landing performance at the destination airport.
C—be familiar with the runway lengths at airports of intended use, and the alternatives available if the flight cannot be completed.

Each pilot-in-command shall, before beginning a flight, become familiar with all available information concerning the flight. This information must include:

1. *For a flight under IFR or a flight not in the vicinity of an airport, weather reports and forecasts, fuel requirements, alternatives available if the planned flight cannot be completed, and any known traffic delays of which the pilot has been advised by ATC.*
2. *For any flight, runway lengths at airports of intended use.*
3. *For civil aircraft for which an approved airplane or rotorcraft flight manual containing takeoff and landing distance data is required, the takeoff and landing distance data contained therein.*

(B08) — 14 CFR §91.103

Answers (A) and (B) are incorrect because listing an alternate airport is not required for all IFR flight, but only when the destination is forecast to have ceilings below 2,000 feet AGL and visibility less than 3 SM.

Airspace

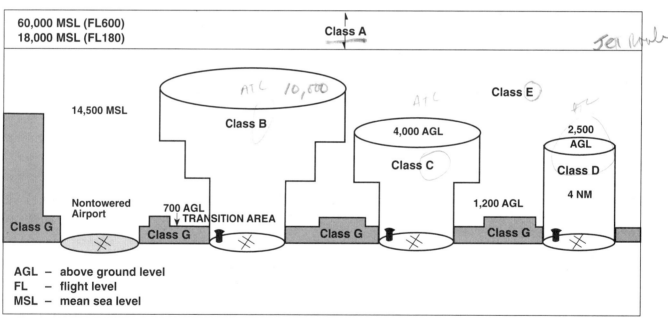

Figure 5-1

Class A—Class A airspace extends from 18,000 feet MSL to and including FL600. No VFR flight, including VFR-On-Top, is authorized in Class A airspace.

Class B—Class B airspace consists of controlled airspace within which all aircraft are subject to certain operating rules as well as pilot and equipment requirements. Each location will contain at least one primary airport. ATC clearance is required prior to operating within Class B airspace. In addition, each aircraft must be equipped with: two-way radio with appropriate ATC frequencies, a VOR or TACAN receiver unless flying VFR, and a Mode C transponder (there are some local exceptions when

Answers
4033 [C]

operating to non-primary airports). ATC may authorize deviations from the transponder requirements. Requests for deviation must be submitted to the controlling ATC facility at least 1 hour prior to the proposed operation. A pilot landing or taking off from an airport within Class B airspace must hold at least a Private Pilot Certificate, or meet stringent student pilot requirements. When operating to or from the primary airport, large turbine-powered airplanes must remain at or above the floor of Class B airspace.

Class C—Class C airspace is controlled airspace surrounding designated airports within which ATC provides radar vectoring and sequencing for all IFR and VFR aircraft. Two-way radio is required, and communication with ATC must be maintained while flying within Class C airspace. A Mode C transponder is required within, and up to 10,000 feet MSL over Class C airspace.

Class D—Class D airspace exists only when and where an airport traffic control tower is in operation. It usually extends for a 5 statute mile radius (4 nautical miles) from the center of the airport and from the surface up to, but not including, 2,500 feet AGL. The actual dimensions may be different, as needed. ATC authorization is required for all operations within the airspace. When the control tower is not operating, the Class D airspace becomes Class E or Class G, as appropriate.

Class E—Class E airspace is controlled airspace that has not been designated Class A, B, C, or D.

Class G—Class G airspace is the portion of airspace that has not been designated Class A, B, C, D, or E airspace. It is uncontrolled; ATC has neither the authority nor the responsibility for exercising control over air traffic in these areas.

Transition Area—Class E airspace which begins at 700 feet AGL or at 1,200 feet AGL and is used as a transition to/from the terminal environment.

Alert Area—contains a high volume of pilot training or other unusual aerial activity.

Military Operations Area (MOA)—designated to separate or segregate certain military activities from IFR traffic and to let VFR traffic know where these activities are taking place.

Prohibited and Restricted Areas—denote the presence of unusual, often invisible, hazards to flight.

Warning Areas—contain the same sort of hazardous activities as those in Restricted Areas, but are located in international airspace.

ALL
4526. (Refer to Figure 93.) What is the floor of Class E airspace when designated in conjunction with an airway?

A—700 feet AGL.
B—1,200 feet AGL.
C—1,500 feet AGL.

The floor of Class E airspace when designated in conjunction with an airway is 1,200 feet AGL. (J08) — AIM ¶3-2-1

Answer (A) is incorrect because 700 feet AGL is the floor of Class E airspace when designated in conjunction with an airport which has an approved instrument approach procedure. Answer (C) is incorrect because 1,200 (not 1,500) feet AGL is the floor of Class E airspace when designated in conjunction with an airway.

ALL
4528. (Refer to Figure 93.) What is the floor of Class E airspace when designated in conjunction with an airport which has an approved IAP?

A—500 feet AGL.
B—700 feet AGL.
C—1,200 feet AGL.

Assuming Class E airspace has not been designated to the surface, the floor of Class E airspace is 700 feet AGL when designated in conjunction with an airport which has an approved instrument approach procedure. (J08) — AIM ¶3-2-6

Answer (A) is incorrect because the floor is 700 (not 500) feet. Answer (C) is incorrect because 1,200 feet AGL is the floor of Class E airspace when designated in conjunction with an airway.

Answers

4526 [B]   4528 [B]

Chapter 5 Regulations and Procedures

ALL
4530. (Refer to Figure 93.) What is the maximum altitude that Class G airspace will exist? (Does not include airspace less than 1,500 feet AGL.)

A—18,000 feet MSL.
B—14,500 feet MSL.
C—14,000 feet MSL.

The maximum altitude Class G airspace exists is 14,500 feet MSL. (J06) — AIM ¶3-2-1

Answer (A) is incorrect because 18,000 feet MSL is where Class A airspace begins, not where Class G airspace ends. Answer (C) is incorrect because 14,000 feet MSL does not define an airspace.

ALL
4531. (Refer to Figure 93.) What is generally the maximum altitude for Class B airspace?

A—4,000 feet MSL.
B—10,000 feet MSL.
C—14,500 feet MSL.

The shape of Class B airspace is individually tailored to fit the specific requirements of the terrain and traffic flows. However, the vertical dimension is generally from the surface to 10,000 feet MSL. (J08) — AIM ¶3-2-3

ALL
4532. (Refer to Figure 93.) What are the normal lateral limits for Class D airspace?

A—8 NM.
B—5 NM.
C—4 NM.

The normal lateral limits for Class D airspace is 4 NM. (J08) — AIM ¶3-2-5

ALL
4533. (Refer to Figure 93.) What is the floor of Class A airspace?

A—10,000 feet MSL.
B—14,500 feet MSL.
C—18,000 feet MSL.

Class A airspace exists from 18,000 feet MSL up to and including FL600. (J08) — AIM ¶3-2-2

ALL
4434. MOAs are established to

A—prohibit all civil aircraft because of hazardous or secret activities.
B—separate certain military activities from IFR traffic.
C—restrict civil aircraft during periods of high-density training activities.

Military Operations Areas (MOAs) consist of airspace of defined vertical and lateral limits established for the purpose of separating certain military training activities from IFR traffic. (J09) — AIM ¶3-4-5

Answer (A) is incorrect because a prohibited area (not MOA) would prohibit all aircraft because of hazardous or secret activities. Answer (C) is incorrect because a restricted area (not MOA) would be used to restrict aircraft during period of high-density training activities.

ALL
4473. Which airspace is defined as a transition area when designated in conjunction with an airport which has a prescribed IAP?

A—The Class E airspace extending upward from 700 feet or more above the surface and terminating at the base of the overlying controlled airspace.
B—That Class D airspace extending from the surface and terminating at the base of the continental control area.
C—The Class C airspace extending from the surface to 700 or 1,200 feet AGL, where designated.

A Transition Area is Class E airspace starting from 700 feet or more above the surface when designated in conjunction with an airport for which an instrument approach procedure has been prescribed, and extends up to the overlying controlled airspace. (J08) — AIM ¶3-2-6(e)

ALL
4529. (Refer to Figure 93.) Which altitude is the upper limit for Class A airspace?

A—14,500 feet MSL.
B—18,000 feet MSL.
C—60,000 feet MSL.

Class A airspace is that airspace from 18,000 feet MSL up to and including FL600, including that airspace overlying the waters within 12 nautical miles of the coast of the 48 contiguous states and Alaska. (J08) — AIM ¶3-2-2

Answers

| 4530 | [B] | 4531 | [B] | 4532 | [C] | 4533 | [C] | 4434 | [B] | 4473 | [A] |
| 4529 | [C] | | | | | | | | | | |

Chapter 5 **Regulations and Procedures**

ALL
4474. The vertical extent of the Class A airspace throughout the conterminous U.S. extends from

A—18,000 feet to and including FL450.
B—18,000 feet to and including FL600.
C—12,500 feet to and including FL600.

Class A airspace includes airspace within the continental U.S. from 18,000 feet MSL up to and including FL600. (J08) — AIM ¶3-2-2

ALL
4475. Class G airspace is that airspace where

A—ATC does not control air traffic.
B—ATC controls only IFR flights.
C—the minimum visibility for VFR flight is 3 miles.

Class G airspace is that portion of the airspace that has not been designated as Class A, B, C, D, or E airspace. If flight is conducted outside of these areas, ATC does not exercise control over that flight. (J07) — AIM ¶3-3-1

ALL
4476. What are the vertical limits of a transition area that is designated in conjunction with an airport having a prescribed IAP?

A—Surface to 700 feet AGL.
B—1,200 feet AGL to the base of the overlying controlled airspace.
C—700 feet AGL or more to the base of the overlying controlled airspace.

A Transition Area is Class E airspace starting from 700 feet or more above the surface when designated in conjunction with an airport for which an instrument approach procedure has been prescribed, and extends up to the overlying controlled airspace. (J08) — AIM ¶3-2-6(e)

ALL
4485. Unless otherwise specified on the chart, the minimum en route altitude along a jet route is

A—18,000 feet MSL.
B—24,000 feet MSL.
C—10,000 feet MSL.

Jet Route is a route designed to serve aircraft operations from 18,000 feet MSL up to, and including flight level 450. (J17) — AIM ¶5-3-4(a)(2)

ALL
4508. (Refer to Figure 89.) What type airspace exists above Bryce Canyon Airport from the surface to 1,200 feet AGL?

A—Class D airspace.
B—Class E airspace.
C—Class G airspace.

Class G airspace is that portion of the airspace that has not been designated as Class A through E. On the Enroute Low Altitude Chart, open areas (white) indicate controlled airspace, while shaded areas (brown) indicate uncontrolled airspace up to 14,500. The Federal Airways are in Class E airspace areas, and unless otherwise specified, extend upward from 1,200 feet to but not including 18,000 feet MSL. Since Bryce Canyon is an uncontrolled field in a white area, the airspace is Class G from the surface to 1,200 feet AGL. (J35) — Enroute Low Altitude Chart Legend

ALL
4527. (Refer to Figure 93.) Which altitude is the normal upper limit for Class D airspace?

A—1,000 feet AGL.
B—2,500 feet AGL.
C—4,000 feet AGL.

Class D airspace is that airspace from the surface to 2,500 feet above the airport elevation (charted in MSL) surrounding those airports that have an operational control tower. (J08) — AIM ¶3-2-5

Answers

| 4474 | [B] | 4475 | [A] | 4476 | [C] | 4485 | [A] | 4508 | [C] | 4527 | [B] |

Chapter 5 **Regulations and Procedures**

ALL
4539. What minimum aircraft equipment is required for operation within Class C airspace?

A—Two-way communications and Mode C transponder.
B—Two-way communications.
C—Transponder and DME.

A Mode C transponder and two-way communications are required in order to operate within Class C airspace. (J08) — AIM ¶3-2-4(d)(2)

Answer (B) is incorrect because a Mode C transponder is also required in Class C airspace. Answer (C) is incorrect because DME is not required, and two-way radio is required, to operate in Class C airspace.

Cloud Clearance and Visibility Requirements

When operating under VFR, or when operating with a VFR-On-Top clearance, the pilot must maintain the required visibility and cloud clearance appropriate to the altitude flown. *See* Figures 5-2 and 5-3.

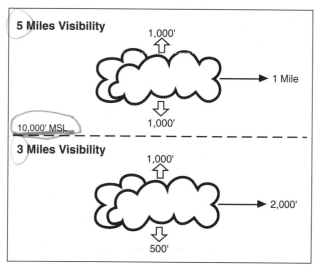

Figure 5-2. Visibility and cloud clearance

Airspace	Flight Visibility	Distance from Clouds
Class A	Not applicable	Not applicable
Class B	3 statute miles	Clear of clouds
Class C	3 statute miles	500 feet below 1,000 feet above 2,000 feet horizontal
Class D	3 statute miles	500 feet below 1,000 feet above 2,000 feet horizontal
Class E: Less than 10,000 feet MSL	3 statute miles	500 feet below 1,000 feet above 2,000 feet horizontal
At or above 10,000 feet MSL	5 statute miles	1,000 feet below 1,000 feet above 1 statute mile horizontal
Class G: 1,200 feet or less above the surface (regardless of MSL altitude)		
Day except as provided in § 91.155(b)	1 statute mile	Clear of clouds
Night except as provided in § 91.155(b)	3 statute miles	500 feet below 1,000 feet above 2,000 feet horizontal
More than 1,200 feet above the surface but less than 10,000		
Day	1 statute mile	500 feet below 1,000 feet above 2,000 feet horizontal
Night	3 statute miles	500 feet below 1,000 feet above 2,000 feet horizontal
More than 1,200 feet above the surface and at or above 10,000 feet MSL	5 statute miles	1,000 feet below 1,000 feet above 1 statute mile horizontal

Figure 5-3. Basic VFR weather minimums

ALL
4518. What is the minimum flight visibility and distance from clouds for flight at 10,500 feet with a VFR-On-Top clearance during daylight hours? (Class E airspace.)

A—3 SM, 1,000 feet above, 500 feet below, and 2,000 feet horizontal.
B—5 SM, 1,000 feet above, 1,000 feet below, and 1 mile horizontal.
C—5 SM, 1,000 feet above, 500 feet below, and 1 mile horizontal.

In Class E airspace, at or above 10,000 feet MSL, the VFR weather minimums are flight visibility 5 statute miles, and cloud clearance of 1,000 feet below, 1,000 feet above, and 1 statute mile horizontal. (B09) — 14 CFR §91.155

Answers
4539 [A]   4518 [B]

ALL
4519. What is the required flight visibility and distance from clouds if you are operating in Class E airspace at 9,500 feet MSL with a VFR-On-Top clearance during daylight hours?

A—3 SM, 1,000 feet above, 500 feet below, and 2,000 feet horizontal.
B—5 SM, 500 feet above, 1,000 feet below, and 2,000 feet horizontal.
C—3 SM, 500 feet above, 1,000 feet below, and 2,000 feet horizontal.

In Class E airspace, less than 10,000 feet MSL, the VFR weather minimums are flight visibility 3 statute miles, and cloud clearance of 500 feet below, 1,000 feet above, and 2,000 feet horizontal. (B09) — 14 CFR §91.155

ALL
4520. (Refer to Figure 92.) What is the minimum in-flight visibility and distance from clouds required for a VFR-On-Top flight at 9,500 feet MSL (above 1,200 feet AGL) during daylight hours for area 3?

A—2,000 feet; (E) 1,000 feet; (F) 2,000 feet; (H) 500 feet.
B—5 miles; (E) 1,000 feet; (F) 2,000 feet; (H) 500 feet.
C—3 miles; (E) 1,000 feet; (F) 2,000 feet; (H) 500 feet.

In Class E airspace, less than 10,000 feet MSL, the VFR weather minimums are flight visibility 3 statute miles, and cloud clearance of 500 feet below, 1,000 feet above, and 2,000 feet horizontal (regardless of whether it is day or night). (B09) — 14 CFR §91.155

ALL
4521. (Refer to Figure 92.) A flight is to be conducted in VFR-On-Top conditions at 12,500 feet MSL (above 1200 feet AGL). What is the in-flight visibility and distance from clouds required for operation in Class E airspace during daylight hours for area 1?

A—5 miles; (A) 1,000 feet; (B) 2,000 feet; (D) 500 feet.
B—5 miles; (A) 1,000 feet; (B) 1 mile; (D) 1,000 feet.
C—3 miles; (A) 1,000 feet; (B) 2,000 feet; (D) 1,000 feet.

In Class E airspace, at or above 10,000 feet MSL, the VFR weather minimums are flight visibility 5 statute miles, and cloud clearance of 1,000 feet below, 1,000 feet above, and 1 statute mile horizontal (regardless of whether it is day or night). (B09) — 14 CFR §91.155

ALL
4522. (Refer to Figure 92.) What is the minimum in-flight visibility and distance from clouds required in VFR conditions above clouds at 13,500 feet MSL (above 1,200 feet AGL) in Class G airspace during daylight hours for area 2?

A—5 miles; (A) 1,000 feet; (C) 2,000 feet; (D) 500 feet.
B—3 miles; (A) 1,000 feet; (C) 1 mile; (D) 1,000 feet.
C—5 miles; (A) 1,000 feet; (C) 1 mile; (D) 1,000 feet.

In Class G airspace, more than 1,200 feet AGL and at or above 10,000 feet MSL, during daylight hours, the VFR weather minimums are flight visibility 5 miles, and cloud clearance of 1,000 feet above, 1,000 feet below, and 1 statute mile horizontal. (B09) — 14 CFR §91.155

ALL
4523. (Refer to Figure 92.) What in-flight visibility and distance from clouds is required for a flight at 8,500 feet MSL (above 1,200 feet AGL) in Class G airspace in VFR conditions during daylight hours in area 4?

A—1 mile; (E) 1,000 feet; (G) 2,000 feet; (H) 500 feet.
B—3 miles; (E) 1,000 feet; (G) 2,000 feet; (H) 500 feet.
C—5 miles; (E) 1,000 feet; (G) 1 mile; (H) 1,000 feet.

In Class G airspace, more than 1,200 feet AGL but less than 10,000 feet MSL, during daylight hours, the VFR weather minimums are flight visibility 1 statute mile, and cloud clearance of 1,000 feet above, 500 feet below, and 2,000 feet horizontal. (B09) — 14 CFR §91.155

ALL
4524. (Refer to Figure 92.) What is the minimum in-flight visibility and distance from clouds required for an airplane operating less than 1,200 feet AGL during daylight hours in area 6?

A—3 miles; (I) 1,000 feet; (K) 2,000 feet; (L) 500 feet.
B—1 mile; (I) clear of clouds; (K) clear of clouds; (L) clear of clouds.
C—1 mile; (I) 500 feet; (K) 1,000 feet; (L) 500 feet.

In Class G airspace, less than 1,200 feet AGL, during daylight hours, the VFR weather minimums are flight visibility 1 statute mile, and the pilot must stay clear of clouds. (B09) — 14 CFR §91.155

Answers

4519 [A]   4520 [C]   4521 [B]   4522 [C]   4523 [A]   4524 [B]

Chapter 5 **Regulations and Procedures**

ALL
4525. (Refer to Figure 92.) What is the minimum in-flight visibility and distance from clouds required for an airplane operating less than 1,200 feet AGL under special VFR during daylight hours in area 5?

A— 1 mile; (I) 2,000 feet; (J) 2,000 feet; (L) 500 feet.
B— 3 miles; (I) clear of clouds; (J) clear of clouds; (L) 500 feet.
C— 1 mile; (I) clear of clouds; (J) clear of clouds; (L) clear of clouds.

Special VFR operations below 10,000 feet MSL at night require the pilot to remain clear of clouds, and ground visibility must be at least 1 statute mile, or if ground visibility is not reported, flight visibility must be at least 1 statute mile. (B09) — 14 CFR §91.157

Aircraft Accident/Incident Reporting and NOTAMs

Reports are required following an aircraft accident, if an aircraft is overdue and is believed to have been involved in an accident, or after any of a number of listed incidents. The reporting procedures are set forth in National Transportation Safety Board (NTSB) Part 830.

Notice to Airman (NOTAM) information is time-critical information that could affect a pilot's decision to make a flight. It is either temporary in nature, or not known far enough in advance to be included in publications. There are three NOTAM categories:

1. NOTAM (L) information, distributed locally, advises of taxiway closures, airport beacon outage, equipment near runways, and NOTAM information concerning airports that are not annotated with the NOTAM Service symbol (§) in the Airport/Facility Directory.

2. NOTAM (D) data receives wide dissemination and is appended to hourly weather reports. It includes information pertaining to all airports with approved instrument approach procedures, NAVAIDs that are part of the National Airspace System, and certain VFR airports.

3. FDC (Flight Data Center) NOTAMs are regulatory in nature and primarily advise of safety-of-flight items such as changes to aeronautical charts and instrument approach procedures. FDC NOTAMs are kept on file at FSSs until published or canceled.

ALL
4088. Which publication covers the procedures required for aircraft accident and incident reporting responsibilities for pilots?

A—FAR Part 61.
B—FAR Part 91.
C—NTSB Part 830.

NTSB Part 830 contains regulations pertaining to notification and reporting of aircraft accidents or incidents and overdue aircraft, and preservation of aircraft wreckage, mail, cargo, and records. (G10) — NTSB Part 830

Answer (A) is incorrect because 14 CFR Part 61 contains regulations on certification of pilots. Answer (B) is incorrect because 14 CFR Part 91 contains regulations on general operating and flight rules.

ALL
4079. Which sources of aeronautical information, when used collectively, provide the latest status of airport conditions (e.g., runway closures, runway lighting, snow conditions)?

A— Aeronautical Information Manual, aeronautical charts, and Distant (D) Notice to Airmans (NOTAM's).
B— Airport Facility Directory, FDC NOTAM's, and Local (L) NOTAM's.
C— Airport Facility Directory, Distant (D) NOTAM's, and Local (L) NOTAM's.

The Airport/Facility Directory provides current airport conditions known. The Distant (D) NOTAMs, and Local (L) NOTAMs provide the latest status of airport condi-

Answers

4525 [C]　　　4088 [C]　　　4079 [C]

tions, for those items too current or temporary to be published in the A/FD. (J34) — AIM ¶5-1-3

Answer (A) is incorrect because the Aeronautical Information Manual does not provide specific airport information, and aeronautical charts do not provide current runway conditions. Answer (B) is incorrect because FDC NOTAMs provide information on regulatory changes, not airport conditions.

ALL
4080. What is the purpose of FDC NOTAMs?

A—To provide the latest information on the status of navigation facilities to all FSS facilities for scheduled broadcasts.
B—To issue notices for all airports and navigation facilities in the shortest possible time.
C—To advise of changes in flight data which affect instrument approach procedure (IAP), aeronautical charts, and flight restrictions prior to normal publication.

FDC (Flight Data Center) NOTAMs are regulatory in nature, and contain such things as amendments to published IAPs and other current aeronautical charts. They are also used to advertise temporary flight restrictions caused by such things as natural disasters or large-scale public events that may generate a congestion of air traffic over a site. (J06) — AIM ¶5-1-3

Answer (A) is incorrect because it describes the purpose of NOTAM(D)s. Answer (B) is incorrect because it describes the purpose of both NOTAM(D)s or (L)s.

ALL
4406. From what source can you obtain the latest FDC NOTAM's?

A—Notices to Airmen Publications.
B—FAA AFSS/FSS.
C—Airport/Facility Directory.

FDC NOTAMs are transmitted via Service A and are kept on file at the FSS until published or canceled. (J15) — AIM ¶5-1-3

Answer (A) is incorrect because Class II NOTAMs are only published every two weeks and the FSS may have more current data. Answer (C) is incorrect because the A/FD only includes changes that are permanent and were known at the time of publication.

Spatial Disorientation

Illusions caused by the motion-sensing system of the body are most commonly encountered during instrument flight. The system may be stimulated by motion of the aircraft or by head or body movement. It is not capable of distinguishing between centrifugal force and gravity, nor can it detect small changes in velocity. This system may also produce false sensations, such as interpreting deceleration as a turn in the opposite direction. These illusions and false sensations may lead to spatial disorientation.

The most common form of spatial disorientation is "the leans," resulting from a banked attitude not being perceived by the pilot. Abrupt correction of a banked attitude may stimulate the motion-sensing fluid of the inner ear, creating the sensation of banking in the opposite direction. The pilot may roll the aircraft back to its original attitude until he/she thinks the aircraft is straight and level, or if level flight is maintained, will still feel compelled to align his/her body with the perceived vertical.

The motion-sensing system may lead to a false perception of the true vertical. For example, in a well-coordinated turn without visual reference, the only sensation is that of the body being pressed into the seat, a sensation normally associated with a climb, and the pilot may falsely interpret it as such. On the other hand, recovering from turns reduces pressure on the seat and may lead the pilot to believe the aircraft is descending.

Spatial disorientation can happen to anyone since it is due to the normal function and limitations of the senses of balance. It only becomes dangerous when the pilot fails to suppress the false sensations and place complete reliance on the indications of the flight instruments. Eventually, as instrument flight experience is acquired, the onset of spatial disorientation lessens as trust in the flight instruments is built up.

An abrupt change from climb to straight-and-level flight can create the illusion of tumbling backwards. The disoriented pilot will push the aircraft abruptly into a nose-low attitude, possibly intensifying this illusion. This is called "inversion illusion." A rapid acceleration during takeoff can create the illusion of being in a nose-up attitude. The disoriented pilot will push the aircraft into a nose-low, or dive attitude. This is called "somatogravic illusion."

Answers

4080 [C]   4406 [B]

Chapter 5 Regulations and Procedures

ALL
4810. The sensations which lead to spatial disorientation during instrument flight conditions

A—are frequently encountered by beginning instrument pilots, but never by pilots with moderate instrument experience.
B—occur, in most instances, during the initial period of transition from visual to instrument flight.
C—must be suppressed and complete reliance placed on the indications of the flight instruments.

The sensations which lead to illusions during instrument flight conditions are normal perceptions experienced by normal individuals. These undesirable sensations cannot be completely prevented, but they can and must be ignored or sufficiently suppressed by developing absolute reliance upon what the flight instruments are telling about the attitude of the aircraft. (J31) — AIM ¶8-1-5

Answers (A) and (B) are incorrect because spatial disorientation can occur to anyone at all levels of experience.

ALL
4811. How can an instrument pilot best overcome spatial disorientation?

A—Rely on kinesthetic sense.
B—Use a very rapid cross-check.
C—Read and interpret the flight instruments, and act accordingly.

To overcome spatial orientation, pilots must become proficient in the use of flight instruments and rely upon them. Sight is the only reliable sense during instrument flight. (J31) — AIM ¶8-1-5

Answer (A) is incorrect because kinesthetic senses cause spatial disorientation, and they must be ignored. Answer (B) is incorrect because it is not beneficial to use a very rapid cross-check.

ALL
4814. A pilot is more subject to spatial disorientation if

A—kinesthetic senses are ignored.
B—eyes are moved often in the process of cross-checking the flight instruments.
C—body signals are used to interpret flight attitude.

Illusions that lead to spatial disorientation are created by information received from our motion-sensing system located in each inner ear. In flight, the system may be stimulated by motion of the aircraft alone, or in combination with head and body movement. Sight is the only reliable sense when flying solely by reference to flight instruments. (J31) — AIM ¶8-1-5

Answer (A) is incorrect because ignoring the kinesthetic senses is a method of overcoming spatial disorientation. Answer (B) is incorrect because proper cross-checking of the instruments is a method for overcoming spatial disorientation.

ALL
4815. Which procedure is recommended to prevent or overcome spatial disorientation?

A—Reduce head and eye movements to the extent possible.
B—Rely on the kinesthetic sense.
C—Rely on the indications of the flight instruments.

Spatial disorientation cannot be completely prevented, but it can and must be ignored or sufficiently suppressed by developing absolute reliance upon what the flight instruments are telling about the attitude of the aircraft. (J31) — AIM ¶8-1-5

Answer (A) is incorrect because eye movement is necessary for proper instrument scan. Answer (B) is incorrect because kinesthetic senses cause spatial disorientation and must be ignored.

ALL
4813. How can an instrument pilot best overcome spatial disorientation?

A—Use a very rapid cross-check.
B—Properly interpret the flight instruments and act accordingly.
C—Avoid banking in excess of 30°.

To overcome spatial orientation, pilots must become proficient in the use of flight instruments and rely upon them. Sight is the only reliable sense during instrument flight. (J31) — AIM ¶8-1-5

Answer (A) is incorrect because it is important to properly interpret the instruments. Answer (C) is incorrect because even in banks less than 30°, pilots can still experience spatial disorientation.

ALL
4802. Without visual aid, a pilot often interprets centrifugal force as a sensation of

A—rising or falling.
B—turning.
C—motion reversal.

A pilot often interprets centrifugal force as a sensation of rising or falling. (H800) — FAA-H-8083-15, Chapter 1

Answer (B) is incorrect because centrifugal force is caused by turning. Answer (C) is incorrect because is often interpreted as a vertical movement (not motion reversal).

Answers

4810 [C]   4811 [C]   4814 [C]   4815 [C]   4813 [B]   4802 [A]

ALL
4805. Abrupt head movement during a prolonged constant rate turn in IMC or simulated instrument conditions can cause

A—pilot disorientation.
B—false horizon.
C—elevator illusion.

An abrupt head movement in a prolonged constant rate turn that has ceased stimulating the motion-sensing system can create the illusion of rotation or movement on an entirely different axis. The disoriented pilot will maneuver the aircraft into a dangerous attitude in an attempt to stop rotation. This most overwhelming of all illusions (called the Coriolis illusion) may be prevented by not making any sudden, extreme head movements, particularly during prolonged constant-rate turns under IFR conditions. (J31) — AIM ¶8-1-5(b)

Answer (B) is incorrect because a false horizon is created by sloping cloud formations, an obscured horizon, a dark scene spread with ground lights and stars, or certain geometric patterns of ground light. Answer (C) is incorrect because elevator illusion is caused by an abrupt vertical acceleration, usually an updraft.

ALL
4807. An abrupt change from climb to straight-and-level flight can create the illusion of

A—tumbling backwards.
B—a noseup attitude.
C—a descent with the wings level.

An abrupt change from climb to straight-and-level flight can create the illusion of tumbling backwards. The disoriented pilot will push the aircraft abruptly into a nose-low attitude, possibly intensifying this illusion. This is called "Inversion Illusion." (J31) — AIM ¶8-1-5(b)

ALL
4808. A rapid acceleration during takeoff can create the illusion of

A—spinning in the opposite direction.
B—being in a noseup attitude.
C—diving into the ground.

A rapid acceleration during takeoff can create the illusion of being in a nose-up attitude. The disoriented pilot will push the aircraft into a nose-low, or dive attitude. This is called "Somatogravic Illusion." (J31) — AIM ¶8-1-5(b)

Optical Illusions

Various atmospheric and surface features encountered while landing may create the illusion of incorrect distance from, or height above, the landing runway. Some of the more common illusions are:

Runway Width Illusion—A runway that is narrower-than-usual can create the illusion that the aircraft is at a higher altitude that it actually is, causing the pilot to fly a lower than normal approach. A wider-than-usual runway can have the opposite effect.

Runway and Terrain Slopes Illusion—An upsloping runway or upsloping terrain can create the illusion that the aircraft is at a higher altitude than it actually is, while a down-sloping runway will have the opposite effect.

Atmospheric Illusion—Atmospheric haze can create the illusion of being at a greater distance from the runway. Rain on the windshield can create an illusion of greater height. The pilot who does not recognize these illusions will fly a lower approach.

Sloping cloud formations, an obscured horizon, a dark scene spread with ground lights and stars, or certain geometric patterns of ground light can create illusions of not being aligned correctly with the actual horizon. The disoriented pilot will place the aircraft in a dangerous position. This illusion is called "false horizons."

Answers

4805 [A]   4807 [A]   4808 [B]

Chapter 5 **Regulations and Procedures**

ALL
4806. A sloping cloud formation, an obscured horizon, and a dark scene spread with ground lights and stars can create an illusion known as

A—elevator illusions.
B—autokinesis.
C—false horizons.

Sloping cloud formations, an obscured horizon, a dark scene spread with ground lights and stars, or certain geometric patterns of ground light can create illusions of not being aligned correctly with the actual horizon. The disoriented pilot will place the aircraft in a dangerous position. This illusion is called "False Horizons." (J31) — AIM ¶8-1-5

Answer (A) is incorrect because elevator illusion is caused by an abrupt vertical acceleration, usually an updraft, giving the sensation of being in a climb. Answer (B) is incorrect because autokinesis is caused by staring at a static light for many seconds, which creates the illusion the light is moving.

ALL
4803. Due to visual illusion, when landing on a narrower-than-usual runway, the aircraft will appear to be

A—higher than actual, leading to a lower-than-normal approach.
B—lower than actual, leading to a higher-than-normal approach.
C—higher than actual, leading to a higher-than-normal approach.

A narrower-than-usual runway can create the illusion that the aircraft is at a higher altitude than it actually is. The pilot who does not recognize this illusion will fly a lower approach, with the risk of striking objects along the approach path or landing short. (J31) — AIM ¶8-1-5

Answer (B) is incorrect because wider (not narrower) runways give a lower-than-actual illusion. Answer (C) is incorrect because a narrower runway appears to be higher than actual, which leads to a lower (not higher) than normal approach.

ALL
4804. What visual illusion creates the same effect as a narrower-than-usual runway?

A—An upsloping runway.
B—A wider-than-usual runway.
C—A downsloping runway.

An upsloping runway, upsloping terrain, or both can create the illusion that the aircraft is at a higher altitude than it actually is. The pilot who does not recognize this illusion will fly a lower approach. This is the same effect as a narrower-than-usual runway. (J31) — AIM ¶8-1-5

Answers (B) and (C) are incorrect because wider-than-usual runways and downsloping runways create an illusion that the aircraft is lower than it is. The narrower-than-usual runway creates an illusion that the aircraft is higher than it is.

ALL
4819. What effect does haze have on the ability to see traffic or terrain features during flight?

A—Haze causes the eyes to focus at infinity, making terrain features harder to see.
B—The eyes tend to overwork in haze and do not detect relative movement easily.
C—Haze creates the illusion of being a greater distance than actual from the runway, and causes pilots to fly a lower approach.

Atmospheric haze causes the illusion of being a greater distance from the runway. The pilot who does not recognize this illusion will fly a lower approach. (J31) — AIM ¶8-1-5

Answers

4806 [C]   4803 [A]   4804 [A]   4819 [C]

Cockpit Lighting and Scanning

Dark adaptation, during which vision becomes more sensitive to light, can be achieved to a moderate degree within 20 minutes under dim red cockpit lighting. After that, any exposure to white light, even for a few seconds, will seriously impair night vision.

Only a very small portion of the eye has the ability to send clear messages to the brain. Because the eyes focus on only a narrow viewing area, effective scanning is accomplished with a series of short, regularly-spaced eye movements that bring successive areas of the sky into the central viewing area of the retina. Each movement should not exceed 10°, and each area should be observed for at least 1 second to enable the eyes to detect a moving or contrasting object.

Pilots should execute gentle banks, at a frequency which permits continuous visual scanning of the airspace about them, during climbs and descents in flight conditions which permit visual detection of other traffic.

ALL
4812. Which statement is correct regarding the use of cockpit lighting for night flight?

A—Reducing the lighting intensity to a minimum level will eliminate blind spots.
B—The use of regular white light, such as a flashlight, will impair night adaptation.
C—Coloration shown on maps is least affected by the use of direct red lighting.

In darkness, vision becomes more sensitive to light through a process called dark adaptation. Since any degree of dark adaptation is lost within a few seconds of viewing a bright light, the pilot should avoid it, or close one eye when using white light. (J31) — AIM ¶8-1-6

Answer (A) is incorrect because reducing the lighting intensity to a minimum level will be insufficient to read charts, instruments, etc. Answer (C) is incorrect because red lighting distorts colors on maps.

ALL
4817. Which use of cockpit lighting is correct for night flight?

A—Reducing the interior lighting intensity to a minimum level.
B—The use of regular white light, such as a flashlight, will not impair night adaptation.
C—Coloration shown on maps is least affected by the use of direct red lighting.

In darkness, vision becomes more sensitive to light through a process called dark adaptation. Since any degree of dark adaptation is lost within a few seconds of viewing a bright light, the pilot should avoid it, or close one eye when using white light. (J31) — AIM ¶8-1-6

Answer (B) is incorrect because white light will impair night vision within a few seconds. Answer (C) is incorrect because red light significantly distorts color.

ALL
4818. Which technique should a pilot use to scan for traffic to the right and left during straight-and-level flight?

A—Systematically focus on different segments of the sky for short intervals.
B—Concentrate on relative movement detected in the peripheral vision area.
C—Continuous sweeping of the windshield from right to left.

While the eyes can observe an approximate 200° arc of the horizon at one glance, only a very small center area called the fovea, in the rear of the eye, has the ability to send clear, sharply focused messages to the brain. Because the eyes can focus only on this narrow viewing area, effective scanning is accomplished with a series of short, regularly spaced eye movements that bring successive areas of the sky into the central visual field. Each movement should not exceed 10°, and each area should be observed for at least one second to enable detection. (J31) — AIM ¶8-1-6

Answers

4812 [B]   4817 [A]   4818 [A]

Chapter 5 **Regulations and Procedures**

ALL
4634. What is expected of you as pilot on an IFR flight plan if you are descending or climbing in VFR conditions?

A—If on an airway, climb or descend to the right of the centerline.
B—Advise ATC you are in visual conditions and will remain a short distance to the right of the centerline while climbing.
C—Execute gentle banks, left and right, at a frequency which permits continuous visual scanning of the airspace about you.

During climbs and descents in flight conditions which permit visual detection of other traffic, pilots should execute gentle banks, left and right at a frequency which permits continuous visual scanning of the airspace about them. (J14) — AIM ¶4-4-14(b)

Answers (A) and (B) are incorrect because the pilot is required to remain on the centerline (not to the right) of the airway unless in VFR conditions and maneuvering to avoid other traffic, and the pilot is not required to advise ATC when in visual conditions.

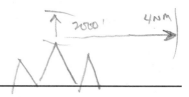

Altitude and Course Requirements

During IFR operations in mountainous terrain, pilots may not operate below 2,000 feet above the highest obstacle, within a horizontal distance of 4 nautical miles from the course to be flown (except for takeoff or landing). During IFR operations in non-mountainous terrain, pilots may not operate below 1,000 feet above the highest obstacle.

VFR cruising altitudes are based on magnetic course.

ALL
4428. Unless otherwise prescribed, what is the rule regarding altitude and course to be maintained during an IFR off-airways flight over mountainous terrain?

A—1,000 feet above the highest obstacle within a horizontal distance of 5 NM of course.
B—7,500 feet above the highest obstacle within a horizontal distance of 3 NM of course.
C—2,000 feet above the highest obstacle within 4 NM of course.

In the case of operations over an area designated as mountainous terrain, no person may operate an aircraft under IFR (except when necessary for takeoff or landing) below an altitude of 2,000 feet above the highest obstacle, within a horizontal distance of 4 nautical miles from the course to be flown. (J07) — 14 CFR §91.177

ALL
4425. Unless otherwise prescribed, what is the rule regarding altitude and course to be maintained during an off-airways IFR flight over nonmountainous terrain?

A—1,000 feet above the highest obstacle within 5 SM of course.
B—2,000 feet above the highest obstacle within 5 SM of course.
C—1,000 feet above the highest obstacle within 3 NM of course.

Except when necessary for takeoff or landing, or unless otherwise authorized by the Administrator, no person may operate an aircraft in nonmountainous terrain under IFR below an altitude 1,000 feet above the highest obstacle within a horizontal distance of 4 NM from the course to be flown (4.6 SM). (J07) — 14 CFR §91.177(a)(2)(ii)

Answers
4634 [C]   4428 [C]   4425 [A]

ALL
4441. Which procedure is recommended while climbing to an assigned altitude on the airway?

A—Climb on the centerline of the airway except when maneuvering to avoid other air traffic in VFR conditions.
B—Climb slightly on the right side of the airway when in VFR conditions.
C—Climb far enough to the right side of the airway to avoid climbing or descending traffic coming from the opposite direction if in VFR conditions.

Unless otherwise authorized by ATC, no person may operate an aircraft within controlled airspace under IFR except as follows:

1. On a Federal airway, along the centerline of that airway.

2. On any other route, along the direct course between the navigational aids or fixes defining that route.

However, this section does not prohibit maneuvering the aircraft to pass well clear of other air traffic or maneuvering the aircraft, in VFR conditions, to clear the intended flight path both before and during climb or descent. (B08) — 14 CFR §91.181

Answers (B) and (C) are incorrect because the pilot is required to maintain the centerline (not on the right side) while operating on a Federal airway.

ALL
4541. In the case of operations over an area designated as a mountainous area where no other minimum altitude is prescribed, no person may operate an aircraft under IFR below an altitude of

A—500 feet above the highest obstacle.
B—1,000 feet above the highest obstacle.
C—2,000 feet above the highest obstacle.

Except when necessary for takeoff or landing, or unless otherwise authorized by the Administrator, no person may operate an aircraft in an area designated as a mountainous area under IFR below an altitude of 2,000 feet above the highest obstacle within a horizontal distance of 4 NM from the course to be flown. (B08) — 14 CFR §91.177(a)(2)(i)

Chapter 5 Regulations and Procedures

ALL
4543. If, while in Class E airspace, a clearance is received to "maintain VFR conditions on top," the pilot should maintain a VFR cruising altitude based on the direction of the

A—true course.
B—magnetic heading.
C—magnetic course.

Each person operating an aircraft under the ATC clearance "VFR Conditions On-Top," shall maintain an altitude or flight level appropriate to the magnetic course. (B08) — 14 CFR §91.159

ALL
4006. Except when necessary for takeoff or landing or unless otherwise authorized by the Administrator, the minimum altitude for IFR flight is

A—3,000 feet over all terrain.
B—3,000 feet over designated mountainous terrain; 2,000 feet over terrain elsewhere.
C—2,000 feet above the highest obstacle over designated mountainous terrain; 1,000 feet above the highest obstacle over terrain elsewhere.

Except when necessary for takeoff or landing, no person may operate an aircraft under IFR below an altitude of 2,000 feet above the highest obstacle within 4 NM for mountainous areas, and in any other case, 1,000 feet above. (B10) — 14 CFR §91.177

Answer (A) is incorrect because the minimum IFR altitude is not 3,000 feet over all terrain, but 2,000 feet while over mountainous terrain, and 1,000 feet over terrain elsewhere. Answer (B) is incorrect because the minimum IFR altitude while over mountainous terrain is 2,000 feet and 1,000 feet over terrain elsewhere.

ALL
4765. In the case of operations over an area designated as a mountainous area, no person may operate an aircraft under IFR below 2,000 feet above the highest obstacle within a horizontal distance of

A—3 SM from the course flown.
B—4 SM from the course flown.
C—4 NM from the course flown.

Except when necessary for takeoff or landing, or unless otherwise authorized by the Administrator, no person may operate an aircraft in a designated mountainous area under IFR below an altitude of 2,000 feet above the highest obstacle within a horizontal distance of 4 nautical miles from the course to be flown. (B10) — 14 CFR §91.177(a)(2)(i)

Answers

| 4441 | [A] | 4541 | [C] | 4543 | [C] | 4006 | [C] | 4765 | [C] |

Communication Reports

The PIC of an aircraft operating under IFR in controlled airspace is required to report the following items as soon as possible:

1. Any unforecast weather.
2. The time and altitude passing each designated reporting point (including fixes used to define direct routes) except when in radar contact. (Resume normal position reporting when advised "Radar Contact Lost" or "Radar Service Terminated").
3. Any other information relating to the safety of flight.

The pilot should also advise ATC immediately should any of the following malfunctions occur in flight:

1. Loss of VOR or ADF capability.
2. Complete or partial panel of ILS receiver capability.
3. Impairment of air/ground communications capability.

Some additional reports should be made without specific request from ATC. These include:

1. When an approach has been missed.
2. When leaving a previously assigned altitude for a newly assigned altitude.
3. When unable to climb or descend at a rate of at least 500 fpm.
4. Changing altitude when operating VFR-On-Top.
5. Change in average true airspeed of ±5% or 10 knots (whichever is greater) from that filed in the flight plan.
6. Time and altitude reaching a holding fix or clearance limit.
7. Leaving holding.

Additionally, when not in radar contact, a report should be made upon leaving the final approach fix inbound on final, and should a previously submitted estimate be made more than 3 minutes in error, a corrected estimate should be given to ATC.

ALL
4456. Which report should be made to ATC without a specific request when not in radar contact?

A—Entering instrument meteorological conditions.
B—When leaving final approach fix inbound on final approach.
C—Correcting an E.T.A. any time a previous E.T.A. is in error in excess of 2 minutes.

The following reports should be made to ATC or FSS facilities without a specific ATC request when not in radar contact:

1. *When leaving final approach fix inbound on final approach.*
2. *A corrected estimate at anytime it becomes apparent that an estimate previously submitted is in error in excess of 3 minutes.*

(J17) — AIM ¶5-3-3(a)(2)

Answer (A) is incorrect because a report should be made to the ATC controller any time (not only in IMC) when leaving any assigned holding fix or point. Answer (C) is incorrect because a report should be made to the ATC controller any time a previous ETA is in error in excess of 3 minutes.

Answers

4456 [B]

ALL
4071. For which speed variation should you notify ATC?

A—When the ground speed changes more than 5 knots.
B—When the average true airspeed changes 5 percent or 10 knots, whichever is greater.
C—Any time the ground speed changes 10 MPH.

The pilot of an IFR flight should report changes in the average true airspeed (at cruising altitude) of 5 percent or 10 knots (whichever is greater) from that filed in the flight plan. (J15) — AIM ¶5-3-3(a)(1)(e)

Answers (A) and (C) are incorrect because the pilot should notify ATC if the true airspeed (not ground speed) changes more than 10 knots (not 5 knots or 10 MPH).

ALL
4379. What does declaring "minimum fuel" to ATC imply?

A—Traffic priority is needed to the destination airport.
B—Emergency handling is required to the nearest useable airport.
C—Merely an advisory that indicates an emergency situation is possible should any undue delay occur.

"Minimum Fuel" indicates that an aircraft's fuel supply has reached a state where, upon reaching the destination, it can accept little or no delay. This is not an emergency situation, but merely indicates an emergency situation is possible should any undue delay occur. (J12) — AIM ¶5-5-15

Answers (A) and (B) are incorrect because priority may be issued only when declaring an emergency, and minimum fuel is not an emergency.

ALL
4380. When ATC has not imposed any climb or descent restrictions and aircraft are within 1,000 feet of assigned altitude, pilots should attempt to both climb and descend at a rate of between

A—500 feet per minute and 1,000 feet per minute.
B—500 feet per minute and 1,500 feet per minute.
C—1,000 feet per minute and 2,000 feet per minute.

When ATC has not used the phrase "at pilot's discretion" nor imposed any climb or descent restrictions, pilots should initiate climb or descent promptly on acknowledgment of the clearance. Descend or climb at an optimum rate consistent with the operating characteristics of the aircraft to 1,000 feet above or below the assigned altitude, and then attempt to descend or climb at a rate of between 500 and 1,500 fpm until the assigned altitude is reached. (J14) — AIM ¶4-4-9(d)

ALL
4078. For IFR planning purposes, what are the compulsory reporting points when using VOR/DME or VORTAC fixes to define a direct route not on established airways?

A—Fixes selected to define the route.
B—There are no compulsory reporting points unless advised by ATC.
C—At the changeover points.

When a pilot is on a direct flight, pilots shall report over each reporting point used in the flight plan to define the route of flight. (J15) — AIM ¶5-3-2(c)(2)

Answer (B) is incorrect because the compulsory points on a direct flight are those fixes used to define the route. Answer (C) is incorrect because some direct flights do not require changeover points; for example, flights using Loran or GPS.

ALL
4460. What action should you take if your No. 1 VOR receiver malfunctions while operating in controlled airspace under IFR? Your aircraft is equipped with two VOR receivers. The No. 1 receiver has VOR/Localizer/Glide Slope capability, and the No. 2 has only VOR/Localizer capability.

A—Report the malfunction immediately to ATC.
B—Continue the flight as cleared; no report is required.
C—Continue the approach and request a VOR or NDB approach.

The pilot-in-command of each aircraft operated in controlled airspace under IFR, shall report as soon as practical to ATC any malfunctions of navigational, approach, or communication equipment occurring in flight. (J17) — 14 CFR §91.187(a)

Answer (B) is incorrect because a report must be made to ATC regarding the situation. Answer (C) is incorrect because ATC will know a VOR or NDB approach will be necessary if you report the ILS receiver inoperative.

Answers

| 4071 | [B] | 4379 | [C] | 4380 | [B] | 4078 | [A] | 4460 | [A] |

ALL
4381. During an IFR flight in IMC, a distress condition is encountered, (fire, mechanical, or structural failure). The pilot should

A—not hesitate to declare an emergency and obtain an amended clearance.
B—wait until the situation is immediately perilous before declaring an emergency.
C—contact ATC and advise that an urgency condition exists and request priority consideration.

Pilots should not hesitate to declare an emergency when faced with a distress condition, such as fire, mechanical failure, or structural damage. A distress condition is defined as being threatened by serious and/or imminent danger and requiring immediate assistance. (J21) — AIM ¶6-1-2

Answer (B) is incorrect because a distress condition is perilous and an emergency condition should be declared. Answer (C) is incorrect because the pilot should contact ATC and advise that an emergency condition exists and obtain an amended clearance.

ALL
4605. During the en route phase of an IFR flight, the pilot is advised "Radar service terminated." What action is appropriate?

A—Set transponder to code 1200.
B—Resume normal position reporting.
C—Activate the IDENT feature of the transponder to re-establish radar contact.

Pilots should resume normal position reporting when ATC advises, "Radar Contact Lost," or "Radar Service Terminated." (J17) — AIM ¶5-3-2(c)(3)

Answer (A) is incorrect because squawking 1200 is for VFR flights only. Answer (C) is incorrect because the IDENT feature of the transponder should only be used upon request by ATC.

Lost Communication Requirements

If unable to contact ATC on a newly assigned frequency, the pilot should go back to the previous frequency. If contact is lost on that frequency also, the pilot should attempt to establish contact with the nearest FSS. If it becomes apparent that two-way communication has been lost, the pilot should change the transponder to code 7600.

If communications are lost in VFR conditions, the pilot should remain VFR and land as soon as practicable.

Should the communication failure occur in IFR conditions, the pilot should proceed to the destination by:

1. *Route:*
   a. The route assigned in the last ATC clearance; or
   b. If being radar vectored, by the direct route to the fix, route, or airway specified in the vector clearance; or
   c. If no route has been assigned, by the route the pilot was told to expect; or
   d. In the absence of any of the above, by the route filed in the flight plan.
2. *Altitude:* At the highest of the following altitudes for the route segment being flown.
   a. The altitude assigned in the last clearance received; or
   b. The altitude ATC has advised may be expected in a further clearance; or
   c. The minimum enroute altitude (MEA).

If given holding instructions or a clearance limit, the pilot should leave that fix at the expect further clearance (EFC) time given. If holding is necessary in the terminal area, it should be depicted, or, if none

Answers
4381 [A]   4605 [B]

is depicted, at the initial approach fix (IAF). If no holding was required en route, the pilot should proceed to the terminal area without delay and hold, maintaining the enroute altitude.

Descent for approach may begin upon arrival at the IAF, but not before the ETA shown on the flight plan, as amended with ATC. If an Expect Approach Clearance (EAC) time has been received, begin descent for the approach at that time.

ALL
4462. During an IFR flight in IMC, you enter a holding pattern (at a fix that is not the same as the approach fix) with an EFC time of 1530. At 1520 you experience complete two-way communications failure. Which procedure should you follow to execute the approach to a landing?

A—Depart the holding fix to arrive at the approach fix as close as possible to the EFC time and complete the approach.
B—Depart the holding fix at the EFC time, and complete the approach.
C—Depart the holding fix at the earliest of the flight planned ETA or the EFC time, and complete the approach.

If the clearance limit is not a fix from which an approach begins, leave the clearance limit at the expected-further-clearance time if one has been received. Then proceed to a fix from which an approach begins, and commence descent, or descent and approach, as close as possible to the estimated time of arrival as calculated from the filed, or amended (with ATC), estimated time en route. (J21) — 14 CFR §91.185(c)

Answer (A) is incorrect because the pilot is to leave for (not arrive at) the approach fix at the EFC time. Answer (C) is incorrect because the EFC time takes priority over the flight plan ETA.

ALL
4463. Which procedure should you follow if you experience two-way communications failure while holding at a holding fix with an EFC time? (The holding fix is not the same as the approach fix.)

A—Depart the holding fix to arrive at the approach fix as close as possible to the EFC time.
B—Depart the holding fix at the EFC time.
C—Proceed immediately to the approach fix and hold until EFC.

If the clearance limit is not a fix from which an approach begins, leave the clearance limit at the expected-further-clearance time if one has been received. Then proceed to a fix from which an approach begins, and commence descent, or descent and approach, as close as possible to the estimated time of arrival as calculated from the filed, or amended (with ATC), estimated time en route. (J21) — 14 CFR §91.185(c)

Answer (A) is incorrect because the pilot is to leave for (not arrive at) the approach fix at the EFC time. Answer (C) is incorrect because the pilot should not leave the holding fix for the approach fix until the EFC time.

ALL
4464. You are in IMC and have two-way radio communications failure. If you do not exercise emergency authority, what procedure are you expected to follow?

A—Set transponder to code 7600, continue flight on assigned route and fly at the last assigned altitude or the MEA, whichever is higher.
B—Set transponder to code 7700 for 1 minute, then to 7600, and fly to an area with VFR weather conditions.
C—Set transponder to 7700 and fly to an area where you can let down in VFR conditions.

If an aircraft with a coded radar beacon transponder experiences a loss of two-way radio capability, the pilot should set transponder to code 7600. During a flight in IMC, the pilot should continue flight on the assigned route and fly at the last assigned altitude or the MEA, whichever is higher. (J24) — AIM ¶6-4-2(a)

Answers (B) and (C) are incorrect because code 7700 is not used in lost communication procedures, and if VFR conditions prevail, the flight should continue and land as soon as practicable.

ALL
4465. Which procedure should you follow if, during an IFR flight in VFR conditions, you have two-way radio communications failure?

A—Continue the flight under VFR and land as soon as practicable.
B—Continue the flight at assigned altitude and route, start approach at your ETA, or, if late, start approach upon arrival.
C—Land at the nearest airport that has VFR conditions.

Continued

Answers
4462 [B]   4463 [B]   4464 [A]   4465 [A]

Chapter 5 **Regulations and Procedures**

If the failure occurs in VFR conditions, or if VFR conditions are encountered after the failure, the pilot shall continue the flight under VFR and land as soon as practicable. (J24) — AIM ¶6-4-1

Answer (B) is incorrect because it only applies if in IMC. Answer (C) is incorrect because the pilot is may continue the flight and land as soon as practicable.

ALL
4466. What altitude and route should be used if you are flying in IMC and have two-way radio communications failure?

A—Continue on the route specified in your clearance, fly at an altitude that is the highest of last assigned altitude, altitude ATC has informed you to expect, or the MEA.
B—Fly direct to an area that has been forecast to have VFR conditions, fly at an altitude that is at least 1,000 feet above the highest obstacles along the route.
C—Descend to MEA and, if clear of clouds, proceed to the nearest appropriate airport. If not clear of clouds, maintain the highest of the MEA's along the clearance route.

If two-way radio communications are lost in IMC, the flight should be continued via the last assigned route by ATC, the route ATC advised or may be expected to advise, or by the flight plan. The altitude to be used is the highest of the altitude assigned in the last ATC clearance, the MEA, or the flight level ATC may be expected to assign. (J24) — AIM ¶6-4-1

Answer (B) is incorrect because the pilot should maintain the assigned route. Answer (C) is incorrect because the highest altitude assigned, expected to be assigned, or the MEA should be used.

ALL
4505. In the event of two way radio communications failure while operating on an IFR clearance in VFR conditions the pilot should continue

A—by the route assigned in the last ATC clearance received.
B—the flight under VFR and land as soon as practical.
C—the flight by the most direct route to the fix specified in the last clearance.

If a two-way radio communication failure occurs in VFR conditions, or if VFR conditions are encountered after the failure, each pilot shall continue the flight under VFR and land as soon as practicable. (B10) — 14 CFR §91.185

ALL
4500. (Refer to Figure 87.) While holding at the 10 DME fix east of LCH for an ILS approach to RWY 15 at Lake Charles Muni airport, ATC advises you to expect clearance for the approach at 1015. At 1000 you experience two-way radio communications failure. Which procedure should be followed?

A—Squawk 7600 and listen on the LOM frequency for instructions from ATC. If no instructions are received, start your approach at 1015.
B—Squawk 7700 for 1 minute, then 7600. After 1 minute, descend to the minimum final approach fix altitude. Start your approach at 1015.
C—Squawk 7600; plan to begin your approach at 1015.

If a pilot experiences a loss of two-way radio capability while on an IFR flight, the pilot should:

1. *Adjust the transponder to code 7600.*
2. *Leave the clearance limit to begin the approach as close as possible to the expect-further-clearance time.*

(J24) — AIM ¶6-4-1

Answer (A) is incorrect because there is no LOM for Lake Charles Muni Airport. Answer (B) is incorrect because 7700 should be squawked in an emergency situation, and altitude should be maintained until the expect-further-clearance time (the time at which you begin the approach).

ALL
4374. While flying on an IFR flight plan, you experience two-way communications radio failure while in VFR conditions. In this situation, you should continue your flight under

A—VFR and land as soon as practicable.
B—VFR and proceed to your flight plan destination.
C—IFR and maintain the last assigned route and altitude to your flight plan destination.

If the failure occurs in VFR conditions, or if VFR conditions are encountered after the failure, each pilot shall continue the flight under VFR and land as soon as practicable. (J24) — AIM ¶6-4-1

Answers

4466 [A]   4505 [B]   4500 [C]   4374 [A]

Chapter 6
Departure

Flight Plan Requirements   *6–3*

Clearance Requirements   *6–17*

ATC Clearance/Separations   *6–18*

Departure Procedures (DPs)   *6–21*

Services Available to Pilots   *6–25*

Pilot/Controller Roles and Responsibilities   *6–28*

VFR-On-Top   *6–30*

Chapter 6 **Departure**

Flight Plan Requirements

Prior to operating an aircraft in controlled airspace under IFR, the PIC must have filed an IFR flight plan and received an appropriate ATC clearance.

The most current en route and destination flight information for planning an instrument flight should be obtained from an FSS.

IFR flight plans should be filed with the FSS at least 30 minutes prior to the estimated time of departure in order to preclude any delay in receiving a clearance from ATC when ready to depart. If a pilot wishes to file an IFR flight plan while airborne, the pilot should contact the nearest FSS, submit the request, and remain VFR until a clearance is received.

The flight plan consists of:

Block 1

Indicate the type of flight planned. If part of the flight will be conducted under IFR and another portion under VFR, both blocks should be checked. This composite flight plan may specify VFR for either the first or last portion of the flight. If VFR is conducted for the first portion of the flight, the pilot should contact the nearest FSS to the point where he/she intends to change from VFR to IFR, close the VFR flight plan, request ATC clearance, and remain in VFR conditions until an IFR clearance is received. If the flight plan indicates that the pilot intends to fly IFR for the first portion of the flight and then to proceed VFR, the flight plan should indicate all points of transition from one airway to another, fixes defining direct route segments, and the clearance limit fix. Upon arrival at the clearance limit, the pilot should advise ATC to cancel the IFR flight plan, and then contact the nearest FSS to activate the VFR portion.

Block 2

Enter the complete aircraft identification.

Block 3

After entering the aircraft type, add a slant and the appropriate code indicating the maximum transponder or navigational capability of the aircraft. *See* FAA Legend 26.

Block 4

Enter the computed true airspeed in knots.

Block 5

Enter the identifier code of the departure airport. Review any applicable standard instrument departure procedures (DPs).

Block 6

Enter the proposed time of departure in Zulu time.

Block 7

Enter the requested initial enroute altitude.

Block 8

Enter the planned route of flight.

Continued

Chapter 6 **Departure**

Block 9

Enter the identifier code of the destination airport.

Block 10

Enter the estimated total time en route, based upon the actual distance to the airport.

Block 11

This remarks section is for the entry of information which the pilot feels is necessary to clarify any requests.

Block 12

The time entered in this block should represent the total usable fuel on board the aircraft. IFR flights in IFR weather conditions, require enough fuel to fly to the destination then to the alternate (if one is required), and then fly for 45 minutes at normal cruising speed (30 minutes for helicopters).

Block 13

If an alternate airport is required, it will be entered in this block. An alternate must be listed unless the ceiling at the destination, 1 hour before until 1 hour after the estimated time of arrival (ETA), is forecast to be at least 2,000 feet above the airport elevation and the visibility to be at least 3 miles. If an alternate airport is needed, its selection must consider certain weather requirements:

- To list an airport with a precision approach (ILS, PAR, MLS) as an alternate, the weather must be forecast to be at least a 600-foot ceiling, and visibility 2 statute miles at the estimated time of arrival.
- To list an airport with a nonprecision approach (VOR, NDB, ASR, etc.) as an alternate, the weather must be forecast to be at least an 800-foot ceiling, and visibility 2 statute miles at the estimated time of arrival.
- An airport without an authorized IAP may be listed as an alternate if the current weather forecast indicates the ceiling and visibility at the estimated time of arrival will allow for a descent from the minimum enroute altitude (MEA), approach, and landing under basic VFR.

Once a pilot elects to proceed to a selected alternate airport, that airport becomes the destination. The landing minimums for the new destination are the minimums that are published for the procedure to be flown.

ALL
4062. When is an IFR flight plan required?

A—When less than VFR conditions exist in either Class E or Class G airspace and in Class A airspace.
B—In all Class E airspace when conditions are below VFR, in Class A airspace, and in defense zone airspace.
C—In Class E airspace when IMC exists or in Class A airspace.

No person may operate an aircraft in controlled airspace under IFR unless the person has:

1. Filed an IFR flight plan; and
2. Received an appropriate ATC clearance.

No person may operate an aircraft within Class A airspace unless that aircraft is:

1. Operated under IFR at a specific flight level assigned by ATC;
2. Equipped with instruments and equipment required for IFR operations;
3. Flown by a pilot rated for instrument flight.

(B10) — 14 CFR §91.173, §91.135

Answer (A) is incorrect because an IFR flight plan is not required in Class G airspace. Answer (B) is incorrect because defense zone airspace does not require an IFR flight plan in VFR conditions.

Answers
4062 [C]

ALL
4063. Prior to which operation must an IFR flight plan be filed and an appropriate ATC clearance received?

A—Flying by reference to instruments in controlled airspace.
B—Entering controlled airspace when IMC exists.
C—Takeoff when IFR weather conditions exist.

No person may operate an aircraft in controlled airspace under IFR unless that person has:

1. *Filed an IFR flight plan; and*

2. *Received an appropriate ATC clearance.*

(B10) — 14 CFR §91.173

Answer (A) is incorrect because a pilot may fly with reference to instruments in controlled airspace in VFR conditions with a safety pilot without an IFR flight plan or clearance. Answer (C) is incorrect because an IFR flight plan is only required in controlled airspace.

ALL
4064. To operate under IFR below 18,000 feet, a pilot must file an IFR flight plan and receive an appropriate ATC clearance prior to

A—entering controlled airspace.
B—entering weather conditions below VFR minimums.
C—takeoff.

No person may operate an aircraft in controlled airspace under IFR unless that person has:

1. *Filed an IFR flight plan; and*

2. *Received an appropriate ATC clearance.*

(B10) — 14 CFR §91.173

Answers (B) and (C) are incorrect because an IFR flight plan and clearance is only required in controlled airspace.

ALL
4065. To operate an aircraft under IFR, a flight plan must have been filed and an ATC clearance received prior to

A—controlling the aircraft solely by use of instruments.
B—entering weather conditions in any airspace.
C—entering controlled airspace.

No person may operate an aircraft in controlled airspace under IFR unless that person has:

1. *Filed an IFR flight plan; and*

2. *Received an appropriate ATC clearance.*

(B10) — 14 CFR §91.173

Answer (A) is incorrect because a pilot may fly with reference to instruments in controlled airspace in VFR conditions with a safety pilot without an IFR flight plan or clearance. Answer (B) is incorrect because an IFR flight plan and clearance is only required in controlled airspace.

ALL
4066. When is an IFR clearance required during VFR weather conditions?

A—When operating in the Class E airspace.
B—When operating in a Class A airspace.
C—When operating in airspace above 14,500 feet.

No person may operate an aircraft within Class A airspace (regardless of the weather conditions) unless that aircraft is:

1. *Operated under IFR at a specific flight level assigned by ATC;*

2. *Equipped with instruments and equipment required for IFR operations;*

3. *Flown by a pilot rated for instrument flight.*

(B10) — 14 CFR §91.135

Answers (A) and (C) are incorrect because an IFR clearance is only required in Class E airspace when conditions are below VMC (Class E airspace includes that airspace from 14,000 feet up to but not including 18,000 feet MSL).

ALL
4067. Operation in which airspace requires filing an IFR flight plan?

A—Any airspace when the visibility is less than 1 mile.
B—Class E airspace with IMC and class A airspace.
C—Positive control area, Continental Control Area, and all other airspace, if the visibility is less than 1 mile.

No person may operate in controlled airspace under IFR unless that person has:

1. *Filed an IFR flight plan; and*

2. *Received an appropriate ATC clearance.*

Each person operating an aircraft within Class A airspace must conduct that operation under instrument flight rules, with an ATC clearance received prior to entering the airspace. (B10) — 14 CFR §91.173, §91.135

Answers (A) and (C) are incorrect because an IFR flight plan is only required in controlled airspace.

Answers

4063 [B]   4064 [A]   4065 [C]   4066 [B]   4067 [B]

Chapter 6 **Departure**

ALL
4068. When departing from an airport located outside controlled airspace during IMC, you must file an IFR flight plan and receive a clearance before

A—takeoff.
B—entering IFR conditions.
C—entering Class E airspace.

No person may operate an aircraft in controlled airspace under IFR unless that person has:

1. *Filed an IFR flight plan; and*
2. *Received an appropriate ATC clearance.*

(B10) — 14 CFR §91.173

Answers (A) and (B) are incorrect because an IFR flight plan and clearance is only required in controlled airspace.

ALL
4427. No person may operate an aircraft in controlled airspace under IFR unless he/she files a flight plan

A—and receives a clearance by telephone prior to takeoff.
B—prior to takeoff and requests the clearance upon arrival on an airway.
C—and receives a clearance prior to entering controlled airspace.

No person may operate an aircraft in controlled airspace under IFR unless that person has:

1. *Filed an IFR flight plan; and*
2. *Received an appropriate ATC clearance.*

(J08) — 14 CFR §91.173

Answer (A) is incorrect because it does not matter how the clearance is obtained. Answer (B) is incorrect because an IFR flight plan must have been filed before the pilot can operate in controlled airspace.

ALL
4005. During your preflight planning for an IFR flight, you determine that the first airport of intended landing has no instrument approach prescribed in 14 CFR part 97. The weather forecast for one hour before through one hour after your estimated time of arrival is 3000' scattered with 5 miles visibility. To meet the fuel requirements for this flight, you must be able to fly to the first airport of intended landing,

A—then to the alternate airport, and then for 30 minutes at normal cruising speed.
B—then to the alternate airport, and then for 45 minutes at normal cruising speed.
C—and then fly for 45 minutes at normal cruising speed.

No person may operate in IFR conditions unless the aircraft carries enough fuel (considering weather reports, forecasts, and conditions) to complete the flight to the first airport of intended landing, fly from that airport to the alternate airport and fly after that for 45 minutes at normal cruising speed. (B10) — 14 CFR §91.167(a)

ALL
4032. What are the minimum fuel requirements in IFR conditions, if the first airport of intended landing is forecast to have a 1,500-foot ceiling and 3 miles visibility at flight-planned ETA? Fuel to fly to the first airport of intended landing,

A—and fly thereafter for 45 minutes at normal cruising speed.
B—fly to the alternate, and fly thereafter for 45 minutes at normal cruising speed.
C—fly to the alternate, and fly thereafter for 30 minutes at normal cruising speed.

No person may operate a civil aircraft in IFR conditions unless it carries enough fuel (considering weather reports and forecasts, and weather conditions) to:

1. *Complete the flight to the first airport of intended landing;*
2. *Fly from that airport to the alternate airport; and*
3. *Fly after that for 45 minutes at normal cruising speed.*

The above does not apply if:

1. *There is a standard instrument approach procedure for the first airport of intended landing; and*
2. *For at least 1 hour before and 1 hour after the estimated time of arrival at the airport, the weather reports or forecasts or any combination of them, indicate:*
   a. *The ceiling will be at least 2,000 feet above the airport elevation; and*
   b. *Visibility will be at least 3 statute miles.*

An alternate, and fuel for flight to the alternate, plus 45 minutes reserve is required for this question because the forecast weather (1,500-foot ceiling) does not meet the 2,000-foot ceiling exemption. (B10) — 14 CFR §91.167 and §91.169

Answer (A) is incorrect because an alternate is required since the destination has a forecast ceiling of less than 2,000 feet AGL. Answer (C) is incorrect because the fuel reserve required after the alternate is 45 minutes (not 30 minutes).

Answers

4068 [C]　　　4427 [C]　　　4005 [B]　　　4032 [B]

ALL
4073. (Refer to Figure 1.) The time entered in block 12 for an IFR flight should be based on which fuel quantity?

A—Total fuel required for the flight.
B—Total useable fuel on board.
C—The amount of fuel required to fly to the destination airport, then to the alternate, plus a 45-minute reserve.

The pilot should fill in block 12 of the flight plan with the time computed from the departure point, using normal cruising speed and total usable fuel on board. (J15) — AIM ¶5-1-7(f)(12)

Answer (A) is incorrect because total fuel required for the flight is not an item listed on an IFR flight plan. Answer (C) is incorrect because it describes the IFR fuel requirements, which is not an item listed on an IFR flight plan.

ALL
4059. When may a pilot file a composite flight plan?

A—When requested or advised by ATC.
B—Any time a portion of the flight will be VFR.
C—Any time a landing is planned at an intermediate airport.

Composite flight plans can be filed whenever VFR operation will be used for one portion of a flight and IFR for another portion. (J15) — AIM ¶5-1-6

Answer (A) is incorrect because ATC can not request or advise on the type of flight plan to file, other than in certain weather conditions and in Class A airspace where IFR flight plans are required. Answer (C) is incorrect because the pilot is not required to convert a flight plan from IFR or VFR when landing at an intermediate airport.

ALL
4060. When filing a composite flight plan where the first portion of the flight is IFR, which fix(es) should be indicated on the flight plan form?

A—All points of transition from one airway to another, fixes defining direct route segments, and the clearance limit fix.
B—Only the fix where you plan to terminate the IFR portion of the flight.
C—Only those compulsory reporting points on the IFR route segment.

The flight plan should define the proposed route of flight by using airway designations, transition points between airways and fixes defining off-airway routes. A composite flight should also indicate the point at which the transition from IFR to VFR is proposed. (J15) — AIM ¶5-1-7(b)

Answer (B) is incorrect because the route must also be defined. Answer (C) is incorrect because compulsory reporting points are only necessary they define the route segments, point of transition, or the clearance limit fix.

ALL
4061. What is the recommended procedure for transitioning from VFR to IFR on a composite flight plan?

A—Prior to transitioning to IFR, contact the nearest FSS, close the VFR portion, and request ATC clearance.
B—Upon reaching the proposed point for change to IFR, contact the nearest FSS and cancel your VFR flight plan, then contact ARTCC and request an IFR clearance.
C—Prior to reaching the proposed point for change to IFR, contact ARTCC, request your IFR clearance, and instruct them to cancel the VFR flight plan.

When VFR flight is conducted for the first part of a composite flight, close the VFR portion and request ATC clearance from the FSS nearest the point at which the change from VFR to IFR is proposed. (J15) — AIM ¶5-1-6

Answer (B) is incorrect because the pilot must obtain an IFR clearance before the point where IFR operations begin. Answer (C) is incorrect because the VFR flight plan should be canceled with FSS, not ARTCC.

Answers

4073 [B]   4059 [B]   4060 [A]   4061 [A]

Chapter 6 Departure

ALL
4072. (Refer to Figure 1.) Which item(s) should be checked in block 1 for a composite flight plan?

A—VFR with an explanation in block 11.
B—IFR with an explanation in block 11.
C—VFR and IFR.

For a composite flight, the pilot should check both the VFR and IFR blocks in block 1 of the flight plan. (J15) — AIM ¶5-1-7

Answers (A) and (B) are incorrect because both VFR and IFR should be checked in block 1 of the flight plan for a composite flight.

RTC
4333. (Refer to Figure 62.) What aircraft equipment code should be entered in block 3 of the flight plan?

A—U.
B—A.
C—I.

Look at the Aircraft Equipment Status at the bottom of FAA Figure 62. The code for RNAV and transponder with altitude encoding capability is /I. The aircraft is also equipped with DME, but it is recommended that pilots file the maximum transponder or navigation capability of their aircraft in the equipment suffix. This will provide ATC with the necessary information to utilize all facets of navigational equipment and transponder capabilities available. See FAA Legend 26. (J15) — AIM ¶5-1-7

Answer (A) is incorrect because /U indicates transponder with altitude encoding capability, but no RNAV. Answer (B) is incorrect because /A indicates DME and transponder with altitude encoding capability, but no RNAV.

ALL
4075. (Refer to Figure 1.) Which equipment determines the code to be entered in block 3 as a suffix to aircraft type on the flight plan form?

A—DME, ADF, and airborne radar.
B—DME, transponder, and ADF.
C—DME, transponder, and RNAV.

When filing an IFR flight plan for flight in an aircraft equipped with a radar beacon transponder, DME equipment, TACAN-only equipment or a combination of both, identify equipment capability by adding a suffix to the aircraft type preceded by a slant. For example, /I means RNAV (which includes DME equipment) and altitude encoding transponder. (J15) — AIM ¶5-1-7

Answers (A) and (B) are incorrect because it is not necessary to report ADF or airborne radar.

ALL
4266. (Refer to Figure 27.) What aircraft equipment code should be entered in block 3 of the flight plan?

A—T.
B—U.
C—A.

Look at the Aircraft Equipment/Status at the bottom of FAA Figure 27. It indicates the aircraft has a transponder with Mode C and a DME. The code for DME and transponder with altitude encoding capability is /A. It is recommended that pilots file the maximum transponder or navigation capability of their aircraft in the equipment suffix. This will provide ATC with the necessary information to utilize all facets of navigational equipment and transponder capabilities available. See FAA Legend 26. (J15) — AIM ¶5-1-7

Answers (A) and (B) are incorrect because the T and U both indicate aircraft with no DME.

ALL
4277. (Refer to Figure 32.) What aircraft equipment code should be entered in block 3 of the flight plan?

A—A.
B—C.
C—I.

Look at the Aircraft Equipment/Status at the bottom of FAA Figure 32. It indicates the aircraft is equipped with transponder with Mode C and RNAV. The code for RNAV and transponder with altitude encoding capability is /I. It is recommended that pilots file the maximum transponder or navigation capability of their aircraft in the equipment suffix. This will provide ATC with the necessary information to utilize all facets of navigational equipment and transponder capabilities available. See FAA Legend 26. (J15) — AIM ¶5-1-7

Answer (A) is incorrect because /A indicates DME and transponder with Mode C. Answer (B) is incorrect because /C indicates RNAV and transponder, but with no Mode C.

ALL
4288. (Refer to Figure 38.) What aircraft equipment code should be entered in block 3 of the flight plan?

A—C.
B—I.
C—A.

Look at the Aircraft Equipment/Status of the bottom of FAA Figure 38. The aircraft is equipped with transponder with Mode C, RNAV, and DME. The code for RNAV and transponder with altitude encoding capability is /I.

Answers
4072 [C]   4333 [C]   4075 [C]   4266 [C]   4277 [C]   4288 [B]

It is recommended that pilots file the maximum transponder or navigation capability of their aircraft in the equipment suffix. This will provide ATC with the necessary information to utilize all facets of navigational equipment and transponder capabilities available. See *FAA Legend 26. (J15) — AIM ¶5-1-7*

Answer (A) is incorrect because /C indicates RNAV and transponder, but with no altitude encoding capability. Answer (C) is incorrect because /A indicates DME and transponder with altitude encoding capability, but no RNAV.

ALL
4300. (Refer to Figure 44.) What aircraft equipment code should be entered in block 3 of the flight plan?
A— A.
B— C.
C— I.

Look at the Aircraft/Equipment Status at the bottom of FAA Figure 44. The aircraft is equipped with transponder with Mode C, RNAV, and DME. The code for RNAV and transponder with altitude encoding capability is /I. It is recommended that pilots file the maximum transponder or navigation capability of their aircraft in the equipment suffix. This will provide ATC with the necessary information to utilize all facets of navigational equipment and transponder capabilities available. See *FAA Legend 26. (J15) — AIM ¶5-1-7*

Answer (A) is incorrect because /A indicates DME and transponder with altitude encoding capability, but no RNAV. Answer (B) is incorrect because /C indicates RNAV and transponder, but with no altitude encoding capability.

ALL
4312. (Refer to Figure 50.) What aircraft equipment code should be entered in block 3 of the flight plan?
A— I.
B— T.
C— U.

Look at the Aircraft/Equipment Status at the bottom of FAA Figure 50. The aircraft is equipped with a transponder with Mode C, RNAV, and DME. The code for RNAV and transponder with altitude encoding capability is /I. It is recommended that pilots file the maximum transponder or navigation capability of their aircraft and equipment suffix. This will provide ATC with the necessary information to utilize all facets of navigational equipment and transponder capabilities available. See *FAA Legend 26. (J15) — AIM ¶5-1-7*

Answer (B) is incorrect because /T indicates transponder with no altitude encoding capability and no RNAV. Answer (C) is incorrect because /U indicates transponder with altitude encoding capability, but no RNAV.

RTC
4322. (Refer to Figure 56.) What aircraft equipment code should be entered in block 3 of the flight plan?
A— U.
B— A.
C— I.

Look at the Aircraft/Equipment Status at the bottom of FAA Figure 56. The aircraft is equipped with transponder with Mode C, RNAV, and DME. The code for RNAV and transponder with altitude encoding capability is /I. It is recommended that pilots file the maximum transponder or navigation capability of their aircraft in the equipment suffix. See *FAA Legend 26. (J15) — AIM ¶5-1-7*

Answer (A) is incorrect because /U indicates transponder with altitude encoding capability, but no RNAV. Answer (B) is incorrect because /A indicates DME and transponder with altitude encoding capability, but no RNAV.

ALL
4344. (Refer to Figure 69.) What aircraft equipment code should be entered in block 3 of the flight plan?
A— A.
B— B.
C— U.

Look at the Aircraft Equipment/Status at the bottom of FAA Figure 69. The aircraft is equipped with transponder with Mode C, and DME. The code for DME and transponder with altitude encoding capability is /A. It is recommended that pilots file the maximum transponder or navigation capability of their aircraft in the equipment suffix. This will provide ATC with the necessary information to utilize all facets of navigational equipment and transponder capabilities available. See *FAA Legend 26. (J15) — AIM ¶5-1-7*

Answer (B) is incorrect because /B indicates DME and transponder, but no altitude encoding capability. Answer (C) is incorrect because /U indicates transponder with altitude encoding capability, but no DME.

Answers

| 4300 | [C] | 4312 | [A] | 4322 | [C] | 4344 | [A] |

Chapter 6 **Departure**

ALL
4358. (Refer to Figure 74.) What aircraft equipment code should be entered in block 3 of the flight plan?
A—T.
B—U.
C—A.

Look at the Aircraft Equipment/Status at the bottom of FAA Figure 74. The aircraft is equipped with transponder with Mode C, and DME. The code for DME and transponder with altitude encoding capability is /A. It is recommended that pilots file the maximum transponder or navigation capability of their aircraft in the equipment suffix. This will provide ATC with the necessary information to utilize all facets of navigational equipment and transponder capabilities available. See FAA Legend 26. (J15) — AIM ¶5-1-7

Answer (A) is incorrect because /T indicates transponder with no altitude encoding capability. Answer (B) is incorrect because /U indicates transponder with altitude encoding capability but no DME.

ALL
4074. (Refer to Figure 1.) What information should be entered in block 7 of an IFR flight plan if the flight has three legs, each at a different altitude?
A—Altitude for first leg.
B—Altitude for first leg and highest altitude.
C—Highest altitude.

Enter only the initial requested altitude in this block. When more than one IFR altitude or flight level is desired along the route of flight, it is best to make a subsequent request direct to the controller. (J15) — AIM ¶5-1-7

Answers (B) and (C) are incorrect because only the initial requested altitude is put into block 7 of the flight plan (not the highest altitude).

ALL
4081. What minimum weather conditions must be forecast for your ETA at an alternate airport, that has only a VOR approach with standard alternate minimums, for the airport to be listed as an alternate on the IFR flight plan?
A—800-foot ceiling and 1 statute mile visibility.
B—800-foot ceiling and 2 statute mile visibility.
C—1,000-foot ceiling and visibility to allow descent from minimum en route altitude (MEA), approach, and landing under basic VFR.

No person may include an airport as an alternate in an IFR flight plan unless current weather forecasts indicate that, at the estimated time of arrival at the alternate airport, the ceiling and visibility at that airport will be at or above the following alternate airport weather minimums:

1. *Precision approach procedure (ILS): Ceiling 600 feet and visibility 2 statute miles.*
2. *Nonprecision approach procedure (VOR, NDB, LOC): Ceiling 800 feet and visibility 2 statute miles.*

(If an instrument approach procedure has been published for that airport, use the alternate airport minimums specified in that procedure.) (B10) — 14 CFR §91.169(c)

Answer (A) is incorrect because the visibility requirements is 2 (not 1) SM. Answer (C) is incorrect because the ceiling and visibility minimums are those allowing descent from the MEA, approach, and landing under basic VFR (not a 1,000-foot ceiling).

AIR
4085-1. For aircraft other than helicopters, what forecast weather minimums are required to list an airport as an alternate on an IFR flight plan if the airport has VOR approach only?
A—Ceiling and visibility at ETA, 800 feet and 2 miles, respectively.
B—Ceiling and visibility from 2 hours before until 2 hours after ETA, 800 feet and 2 miles, respectively.
C—Ceiling and visibility at ETA, 600 feet and 2 miles, respectively.

An alternate airport with a VOR approach only (a nonprecision approach) can be used only if the current weather conditions indicate that at the ETA, the alternate airport ceiling is at least 800 feet and the visibility is at least 2 statute miles. (B10) — 14 CFR §91.169(c)

Answer (B) is incorrect because the alternate airport must have the specified minimums at the ETA (not 2 hours before until 2 hours after). Answer (C) is incorrect because 600 feet and 2 miles applies to alternate airports with precision approach procedures (ILS).

RTC
4085-2. For helicopters, what forecast weather minimums are required to list an airport as an alternate on an IFR flight plan if the airport has VOR approach only?
A—Ceiling 200 feet above the approach minimums and visibility 1 statute mile, but not less than the minimum visibility for the approach, at the alternate airport ETA.
B—Ceiling and visibility at ETA, 800 feet and 2 miles, respectively.
C—Ceiling 1,000 feet above airport elevation and 2 statute miles visibility for 1 hour after the alternate airport ETA.

Answers
4358 [C]   4074 [A]   4081 [B]   4085-1 [A]   4085-2 [A]

An alternate airport can be used only if the current weather conditions indicate that at the ETA, the alternate airport ceiling is 200 feet above the minimum for the approach to be flown, and visibility at least 1 statute mile but never less than the minimum visibility for the approach to be flown. (B10) — 14 CFR §91.169(c)

AIR
4087-1. For aircraft other than helicopters, what minimum weather conditions must be forecast for your ETA at an airport that has a precision approach procedure, with standard alternate minimums, in order to list it as an alternate for the IFR flight?

A— 600-foot ceiling and 2 SM visibility at your ETA.
B— 600-foot ceiling and 2 SM visibility from 2 hours before to 2 hours after your ETA.
C— 800-foot ceiling and 2 SM visibility at your ETA.

An alternate airport with a precision approach can be used only if the current weather conditions indicate that at the ETA, the alternate airport ceiling is at least 600 feet and the visibility is at least 2 statute miles. (B10) — 14 CFR §91.169(c)

Answer (B) is incorrect because the current weather conditions must specify the minimum requirements at the ETA (not 2 hours before to 2 hours after ETA). Answer (C) is incorrect because 800-foot ceilings and 2 SM visibility is the minimum for non-precision approaches.

RTC
4087-2. For helicopters, what minimum weather conditions must be forecast for your ETA at an alternate airport that has a precision approach procedure, with standard alternate minimums, in order to list it as an alternate for the IFR flight?

A— 600 foot ceiling and 2 SM visibility at your ETA.
B— 200 foot ceiling above the airport elevation and 1 SM visibility from 1 hour before to 1 hour after your ETA.
C— 200 foot ceiling above the approach minimums and 1 SM visibility, but not less than the visibility minimums for the approach, at your ETA.

An alternate airport can be used only if the current weather conditions indicate that at the ETA, the alternate airport ceiling is 200 feet above the minimum for the approach to be flown, and visibility at least 1 statute mile but never less than the minimum visibility for the approach to be flown. (B10) — 14 CFR §91.169(c)

ALL
4760-1. What are the alternate minimums that must be forecast at the ETA for an airport that has a precision approach procedure?

A— 400-foot ceiling and 2 miles visibility.
B— 600-foot ceiling and 2 miles visibility.
C— 800-foot ceiling and 2 miles visibility.

Unless otherwise authorized by the Administrator, no person may include an alternate airport in an IFR flight plan that has a precision (ILS) approach unless current weather forecasts indicate that, at the estimated time of arrival at the alternate airport, the ceiling will be at least 600 feet and visibility 2 statute miles. (B10) — 14 CFR §91.169(c)

Answer (A) is incorrect because 400 feet is not an alternative airport minimum. Answer (C) is incorrect because an 800-foot ceiling is the minimum for airports with non-precision approaches.

ALL
4760-2. When an alternate airport is required, what are the weather minimums that must be forecast at the ETA for an alternate airport that has a precision approach procedure?

A— Ceiling 200 feet above the approach minimums and at least 1 statute mile visibility, but not less than the minimum visibility for the approach.
B— 600 foot ceiling and 2 statute miles visibility.
C— Ceiling 200 feet above field elevation and visibility 1 statute mile, but not less than the minimum visibility for the approach.

Unless otherwise authorized by the Administrator, no person may include an alternate airport in an IFR flight plan that has a precision (ILS) approach unless current weather forecasts indicate that, at the estimated time of arrival at the alternate airport, the ceiling will be at least 600 feet and visibility 2 statute miles. (B10) — 14 CFR §91.169(c)

Answers

4087-1 [A] 4087-2 [C] 4760-1 [B] 4760-2 [B]

Chapter 6 **Departure**

RTC
4082-1. For helicopters, is an alternate airport required for an IFR flight to ATL (Atlanta Hartsfield) if the proposed ETA is 1930Z?

TAF KATL 121720Z 121818 20012KT 5SM HZ
   BKN030
 FM2000 3SM TSRA OVC025CB
 FM2200 33015G20KT P6SM BKN015 OVC040
   BECMG 0608
   02008KT BKN040 BECMG 1012 00000KT
     P6SM CLR=

A—Yes, because the ceiling could fall below 2,000 feet within 2 hours before to 2 hours after the ETC.
B—No, because the ceiling and visibility are forecast to remain at or above 1,000 feet and 3 miles, respectively.
C—No, because the ceiling and visibility are forecast to be at or above 1,000 feet above the airport elevation (and 400 feet above the approach minima) with 3 miles visibility at the ETA to 1 hour thereafter.

An alternate airport is required on a helicopter IFR flight plan unless, at the estimated time of arrival and for 1 hour thereafter weather reports and forecasts indicate that at the destination:

1. *The ceiling will be at least 1,000 feet above the airport elevation; or at least 400 feet above the approach minima.*
2. *The visibility will be at least 2 statute miles.*

In the forecast for Atlanta, the chance of reduced conditions occurs before and after the time frame. (B10) — 14 CFR §91.169(c)

AIR
4082-2. For aircraft other than helicopters, is an alternate airport required for an IFR flight to ATL (Atlanta Hartsfield) if the proposed ETA is 1930Z?

TAF KATL 121720Z 121818 20012KT 5SM HZ
   BKN030
 FM2000 3SM TSRA OVC025CB
 FM2200 33015G20KT P6SM BKN015 OVC040
     BECMG 0608 02008KT BKN040 BECMG 1012
 00000KT P6SM CLR=

A—No, because the ceiling and visibility are forecast to be at or above 2,000 feet and 3 miles within 1 hour before to 1 hour after the ETA.
B—No, because the ceiling and visibility are forecast to remain at or above 1,000 feet and 3 miles, respectively.
C—Yes, because the ceiling could fall below 2,000 feet within 2 hours before to 2 hours after the ETA.

An alternate airport is required on an IFR flight plan unless, from 1 hour before to 1 hour after the ETA at the destination airport, weather reports and forecasts indicate that at the destination:

1. *The ceiling will be at least 2,000 feet above the airport elevation; and*
2. *The visibility will be at least 3 statute miles.*

In the forecast for Atlanta, the chance of reduced conditions occurs before and after the 1 hour before to 1 hour after ETA. (B10) — 14 CFR §91.169(c)

Answers (B) and (C) are incorrect because the time frame for alternates is 1 hour before and after ETA.

AIR
4083-1. For aircraft other than helicopters, what minimum conditions must exist at the destination airport to avoid listing an alternate airport on an IFR flight plan when a standard IAP is available?

A—From 2 hours before to 2 hours after ETA, forecast ceiling 2,000, and visibility 2 and 1/2 miles.
B—From 2 hours before to 2 hours after ETA, forecast ceiling 3,000, and visibility 3 miles.
C—From 1 hour before to 1 hour after ETA, forecast ceiling 2,000, and visibility 3 miles.

An alternate airport is required on an IFR flight plan unless, from 1 hour before to 1 hour after the ETA at the destination airport, weather reports and forecasts indicate that at the destination:

1. *The ceiling will be at least 2,000 feet above the airport elevation; and*
2. *The visibility will be at least 3 statute miles.*

(B10) — 14 CFR §91.169(c)

RTC
4083-2. For helicopters, what minimum conditions must exist at the destination airport to avoid listing an alternate airport on an IFR flight plan when a standard IAP is available?

A—From 1 hour before to 1 hour after ETA, forecast ceiling 2,000, and visibility 3 miles.
B—From 1 hour before to 1 hour after ETA, reports and forecasts indicate a ceiling of 1,000 feet above the airport elevation and visibility 2 miles.
C—From the ETA to 1 hour after the ETA, reports and forecasts indicate a ceiling 1,000 feet above the airport elevation, or at least 400 feet above the lowest applicable approach minima, whichever is higher, and visibility 2 statute miles.

Answers

4082-1   [C]   4082-2   [A]   4083-1   [C]   4083-2   [C]

An alternate airport is required on helicopter IFR flight plans unless at the estimated time of arrival and for 1 hour after the ETA, weather reports and forecasts indicate that at the destination:

1. The ceiling will be at least 1,000 feet above the airport elevation, or at least 400 feet above the approach minima;
2. The visibility will be at least 2 statute miles.

(B10) — 14 CFR §91.169(c)

AIR
4083-3. For aircraft other than helicopters, under what conditions are you not required to list an alternate airport on an IFR flight plan if 14 CFR part 97 prescribes a standard IAP for the destination airport?

A—When the ceiling is forecast to be at least 1,000 feet above the lowest of the MEA, MOCA, or initial approach altitude and the visibility is 2 miles more than the minimum landing visibility within 2 hours of your ETA at the destination airport.
B—When the weather reports or forecasts indicate the ceiling and visibility will be at least 2,000 feet and 3 miles for 1 hour before to 1 hour after your ETA at the destination airport.
C—When the ceiling is forecast to be at least 1,000 feet above the lowest of the MEA, MOCA, or initial approach altitude within 2 hours of your ETA at the destination airport.

An alternate airport is required on an IFR flight plan unless, from 1 hour before to 1 hour after the ETA at the destination airport, weather reports and forecasts indicate that at the destination:

1. The ceiling will be at least 2,000 feet above the airport elevation; and
2. The visibility will be at least 3 statute miles

(B10) — 14 CFR §91.169(c)

ALL
4086. What are the minimum weather conditions that must be forecast to list an airport as an alternate when the airport has no approved IAP?

A—The ceiling and visibility at ETA, 2,000 feet and 3 miles, respectively.
B—The ceiling and visibility from 2 hours before until 2 hours after ETA, 2,000 feet and 3 miles, respectively.
C—The ceiling and visibility at ETA must allow descent from MEA, approach, and landing, under basic VFR.

If no instrument approach procedure has been published for that airport, the ceiling and visibility minimums are those allowing descent from the MEA, approach, and landing, under basic VFR. (B10) — 14 CFR §91.169(c)(2)

Answers (A) and (B) are incorrect because they describe the minimums when an alternate airport is required (inaccurately, because the alternate airport weather conditions apply 1 hour before and 1 hour after ETA).

ALL
4719. When a pilot elects to proceed to the selected alternate airport, which minimums apply for landing at the alternate?

A—600-1 if the airport has an ILS.
B—Ceiling 200 feet above the published minimum; visibility 2 miles.
C—The landing minimums for the approach to be used.

If the pilot elects to proceed to the selected alternate airport, the alternate ceiling and visibility minimums are disregarded, and the published landing minimum is applicable for the new destination, utilizing facilities as appropriate to the procedure. In other words, the alternate airport becomes a new destination, and the pilot uses the landing minimum appropriate to the type of procedure selected. (B10) — 14 CFR §91.169(c)

ALL
4630. If a pilot elects to proceed to the selected alternate, the landing minimums used at that airport should be the

A—minimums specified for the approach procedure selected.
B—alternate minimums shown on the approach chart.
C—minimums shown for that airport in a separate listing of "IFR Alternate Minimums."

If the pilot elects to proceed to the selected alternate airport, the alternate ceiling and visibility minimums are disregarded, and the published landing minimums are applicable for the new destination, utilizing facilities as appropriate to the procedure. In other words, the alternate airport becomes a new destination, and the pilot uses the landing minimums appropriate to the type of procedure selected. (B10) — 14 CFR §91.169(c)(1)

Answers (B) and (C) are incorrect because alternate minimums shown on the approach chart refer to the weather conditions required to list that airport as an alternate on your IFR flight plan (not to land there).

Answers

4083-3 [B] 4086 [C] 4719 [C] 4630 [A]

Chapter 6 **Departure**

ALL
4637. When making an instrument approach at the selected alternate airport, what landing minimums apply?

A—Standard alternate minimums (600-2 or 800-2).
B—The IFR alternate minimums listed for that airport.
C—The landing minimums published for the type of procedure selected.

If the pilot elects to proceed to the selected alternate airport, the alternate ceiling and visibility minimums are disregarded, and the published landing minimums are applicable for the new destination, utilizing facilities as appropriate to the procedure. In other words, the alternate airport becomes a new destination, and the pilot uses the landing minimums appropriate to the type of procedure selected. (B10) — 14 CFR §91.169(c)(1)

Answers (A) and (B) are incorrect because alternate minimums shown on the approach chart refer to the weather conditions required to list that airport as an alternate on your IFR flight plan (not to land there).

ALL
4769. An airport without an authorized IAP may be included on an IFR flight plan as an alternate, if the current weather forecast indicates that the ceiling and visibility at the ETA will

A—allow for descent from the IAF to landing under basic VFR conditions.
B—be at least 1,000 feet and 1 mile.
C—allow for a descent from the MEA, approach, and a landing under basic VFR.

If no instrument approach procedure has been published for that airport, the ceiling and visibility minimums are those allowing descent from the MEA, approach, and landing, under basic VFR. (B10) — 14 CFR §91.169(c)(2)

ALL
4070. Preferred IFR routes beginning with a fix indicate that departing aircraft will normally be routed to the fix by

A—the established airway(s) between the departure airport and the fix.
B—an instrument departure procedure (DP) or radar vectors.
C—direct route only.

Preferred IFR routes beginning/ending with a fix indicate that aircraft may be routed to/from these fixes via a Departure Procedure (DP) route, radar vectors (RV), or a Standard Terminal Arrival Route (STAR). (J34) — Airport/Facility Directory

Answer (A) is incorrect because established airway(s) may not even exist between the airport and the fix, or other routes may be more efficient. Answer (C) is incorrect because obstructions or traffic may prevent direct routing.

ALL
4275. (Refer to Figure 29.) What are the hours of operation (local standard time) of the control tower at Eugene/Mahlon Sweet Field?

A—0800 - 2300.
B—0600 - 0000.
C—0700 - 0100.

The hours of operation at Eugene/Mahlon Sweet Field are listed on FAA Figure 29 under Eugene Tower. Hours of operation are expressed in UTC (Coordinated Universal Time) and is shown as "Z" time or Zulu. The top line of the A/FD indicates the number of hours to be subtracted from UTC to obtain local standard time [GMT -8 (-7DT).]:

Published time of operation	1400 Z	to	0800Z
Time Zone correction	–800		–800
Local Standard Time			
hours of operations	0600	to	0000

(J34) — A/FD Legend

ALL
4305. (Refer to Figure 46.) What are the hours of operation (local time) of the ATIS for the Yakima Air Terminal when daylight savings time is in effect?

A—0500 to 2100 local.
B—0600 to 2200 local.
C—0700 to 2300 local.

The top line of the A/FD indicates the number of hours to be subtracted from UTC to obtain local standard time: GMT -8 (-7DT). The ATIS hours of operation can be found in the "Communications" line of the A/FD: ATIS 125.25 (1400-0600Z‡). The symbol (‡) indicates that during the periods of Daylight Savings Time, effective hours will be 1 hour earlier than shown. In this case, for daylight savings time:

1400 Z	to	0600 Z	
–0800		–0800	
0600	to	2200	Local daylight savings time

(J34) — Airport/Facility Directory

Answers

| 4637 | [C] | 4769 | [C] | 4070 | [B] | 4275 | [B] | 4305 | [B] |

ALL
4405. The most current en route and destination flight information for planning an instrument flight should be obtained from

A—the ATIS broadcast.
B—the FSS.
C—Notices to Airmen (Class II).

Every pilot is urged to receive a preflight briefing and to file a flight plan. This briefing should consist of the latest or most current weather, airport, and enroute NAVAID information. Briefing service may be obtained from an FSS either by telephone or interphone, by radio when airborne, or by a personal visit to the station. (J15) — AIM ¶5-1-1(a)

Answer (A) is incorrect because the ATIS broadcasts only information referring to landing and departing operations at one airport. Answer (C) is incorrect because Notices to Airmen (Class II) is a biweekly publication containing current NOTAMs and is only one part of a complete flight briefing.

ALL
4761. What point at the destination should be used to compute estimated time en route on an IFR flight plan?

A—The final approach fix on the expected instrument approach.
B—The initial approach fix on the expected instrument approach.
C—The point of first intended landing.

Unless otherwise authorized by ATC, each person filing an IFR or VFR flight plan shall include in it the point of first intended landing and the estimated elapsed time until over that point. (B10) — 14 CFR §91.169(a)

Answers (A) and (B) are incorrect because it is not possible to accurately tell which approach will be used.

RTC
4040. What are the fuel requirements for a night IFR flight in a helicopter when an alternate airport is not required?

A—Enough fuel to complete the flight to the first airport of intended landing and fly after that for 30 minutes.
B—Enough fuel to complete the flight to the first airport of intended landing, and fly thereafter for 45 minutes at cruising speed.
C—Enough fuel to complete the flight to the first airport of intended landing, make an approach, and fly thereafter for 45 minutes at cruising speed.

No person may operate a civil helicopter in IFR conditions unless it carries enough fuel (considering weather reports and forecasts, and weather conditions) to:

1. *Complete the flight to the first airport of intended landing;*
2. *Fly from that airport to the alternate airport; and*
3. *Fly after that for 30 minutes at normal cruising speed.*

(B10) — 14 CFR §91.167

Answer (B) is incorrect because a helicopter must be able to fly 30 minutes after the alternate (45 minutes is the requirement for airplanes). Answer (C) is incorrect because a helicopter must be able to fly to the destination, alternate, and fly after that for 30 minutes.

RTC
4041. When an alternate airport is required for helicopters on the flight plan, you must have sufficient fuel to complete the flight to the first airport of intended landing, fly to the alternate, and thereafter fly for at least

A—30 minutes at normal cruising speed.
B—45 minutes at holding speed.
C—45 minutes at normal cruising speed.

No person may operate a civil helicopter in IFR conditions unless it carries enough fuel (considering weather reports and forecasts, and weather conditions) to:

1. *Complete the flight to the first airport of intended landing;*
2. *Fly from that airport to the alternate airport; and*
3. *Fly after that for 30 minutes at normal cruising speed.*

(B10) — 14 CFR §91.167

Answers (B) and (C) are incorrect because the helicopter must be able to fly to the destination, alternate, and fly after that for 30 minutes at normal cruise.

Answers

4405 [B]   4761 [C]   4040 [A]   4041 [A]

Chapter 6 Departure

RTC
4084-1. For helicopters, what minimum weather conditions must be forecast for your ETA at an alternate airport that has only a VOR approach with standard alternate minimums, for the airport to be listed as an alternate on the IFR flight plan?

A—Ceiling 200 feet above the minimums for the approach to be flown and 1 statute mile visibility, but not less than the minimum visibility for the approach to be flown.
B—800 foot ceiling and 2 SM visibility.
C—800 foot ceiling and 1 statute mile (SM) visibility.

No person may include an airport as an alternate in a helicopter IFR flight plan unless current weather forecasts indicate that, at the estimated time of arrival at the alternate airport, the ceiling and visibility at that airport will be at or above the following alternate airport weather minimums: ceiling 200 feet above the minimum for the approach to be flown, and visibility at least 1 statute mile but never less than the minimum visibility for the approach to be flown. (B10) — 14 CFR §91.169(c)

RTC
4084-2. When 14 CFR part 97 prescribes a standard IAP for the destination airport, under what conditions are you not required to list an alternate airport on an IFR flight plan for an IFR flight in a helicopter?

A—When the weather reports or forecasts indicate the ceiling and visibility will be at least 2,000 feet and 3 miles for 1 hour before to 1 hour after your ETA at the destination airport.
B—When the ceiling is forecast to be at least 1,000 feet above the lowest of the MEA, MOCA, or initial approach altitude within 2 hours of your ETA at the destination airport.
C—At your ETA and for 1 hour after your ETA, the ceiling is forecast to be at least 1,000 feet above the field elevation, or at least 400 feet above the lowest applicable approach minima, whichever is higher, and visibility of at least 2 statute miles.

An alternate airport is required on a helicopter IFR flight plan unless at the estimated time of arrival and for 1 hour after the ETA, weather reports and forecasts indicate that at the destination:

1. The ceiling will be at least 1,000 feet above the airport elevation or at least 400 feet above the approach minima;

2. The visibility will be at least 2 statute miles.

(B10) — 14 CFR §91.169(c)

Answers
4084-1 [A]   4084-2 [C]

Clearance Requirements

If a PIC deviates from any ATC clearance or instruction, that pilot must notify ATC of the deviation as soon as possible.

If operating VFR, and compliance with an ATC clearance would cause a violation of a regulation, pilots should advise ATC and obtain a revised clearance.

ALL
4407. When may ATC request a detailed report of an emergency even though a rule has not been violated?

A—When priority has been given.
B—Any time an emergency occurs.
C—When the emergency occurs in controlled airspace.

Each pilot-in-command who, though not deviating from a rule, is given priority by ATC in an emergency, shall submit a detailed report of that emergency within 48 hours to the manager of that ATC facility, if requested by ATC. (J14) — 14 CFR §91.123(d)

Answer (B) is incorrect because a written report may be requested when priority has been granted, not any time an emergency occurs. Answer (C) is incorrect because a written report may be requested when priority has been granted in an emergency, regardless of where the emergency occurs.

ALL
4461. While on an IFR flight, a pilot has an emergency which causes a deviation from an ATC clearance. What action must be taken?

A—Notify ATC of the deviation as soon as possible.
B—Squawk 7700 for the duration of the emergency.
C—Submit a detailed report to the chief of the ATC facility within 48 hours.

Each pilot-in-command who, in an emergency, deviates from an ATC clearance or instruction shall notify ATC of that deviation as soon as possible. (J14) — 14 CFR §91.123(c)

Answer (B) is incorrect because the pilot must also notify ATC (not just squawk 7700) in an emergency causing deviation from a clearance. Answer (C) is incorrect because the requirement to submit a report within 48 hours only applies if given priority and/or ATC requests it.

ALL
4758. If during a VFR practice instrument approach, Radar Approach Control assigns an altitude or heading that will cause you to enter the clouds, what action should be taken?

A—Enter the clouds, since ATC authorization for practice approaches is considered an IFR clearance.
B—Avoid the clouds and inform ATC that altitude/heading will not permit VFR.
C—Abandon the approach.

If operating VFR and compliance with any radar vector or altitude would cause a violation of any regulation, pilots should advise ATC and obtain a revised clearance or instruction. (J11) — AIM ¶5-5-6(a)(3)

Answer (A) is incorrect because an IFR clearance is required to operate in the clouds. Answer (C) is incorrect because the approach need only be modified, not abandoned, to maintain VFR.

Answers

4407   [A]   4461   [A]   4758   [B]

ATC Clearance/Separations

All IFR clearances follow this same basic format:

1. Clearance limit
2. Departure procedure (as appropriate)
3. Route of flight
4. Initial altitude to be flown
5. Additional instructions as necessary.

If the route of flight filed in the flight plan can be approved with little or no revision, ATC will issue an abbreviated clearance. The abbreviated clearance will always contain the clearance limit, the instrument departure procedure (DP) name, number, and transition (if appropriate), the words "as filed," and the altitude to maintain.

There is no requirement for a pilot to read back an ATC clearance unless requested to do so. However, pilots of airborne aircraft should read back without request those parts of clearances containing altitude assignments or vectors.

The ATC IFR clearance may be received in many different ways, depending upon the point of departure. At locations not served by a control tower, the clearance may contain a void time. If the flight does not depart prior to the void time, ATC must be notified of the pilot's intentions as soon as possible, but no later than 30 minutes.

ATC may use the term "Cruise" to authorize flight at any altitude from the minimum IFR altitude up to and including the altitude specified. Climb, descent, and level-off within the block is at the pilot's discretion.

ALL
4395. What response is expected when ATC issues an IFR clearance to pilots of airborne aircraft?

A—Read back the entire clearance as required by regulation.
B—Read back those parts containing altitude assignments or vectors and any part requiring verification.
C—Read-back should be unsolicited and spontaneous to confirm that the pilot understands all instructions.

Pilots of airborne aircraft should read back those parts of ATC clearances and instructions containing altitude assignments or vectors as a means of mutual verification. The readback of the "numbers" serves as a double check between pilots and controllers and reduces the kinds of communications errors that occur when a number is either "misheard" or is incorrect. (J14) — AIM ¶4-4-6(b)

Answers (A) and (C) are incorrect because the readback is an expected, not required, procedure and only those parts containing assigned altitudes or vectors should be read back (not the entire clearance).

ALL
4396. Which clearance items are always given in an abbreviated IFR departure clearance? (Assume radar environment.)

A—Altitude, destination airport, and one or more fixes which identify the initial route of flight.
B—Destination airport, altitude, and DP Name-Number-Transition, if appropriate.
C—Clearance limit, and DP Name, Number, and/or Transition, if appropriate.

The clearance as issued will include the destination airport filed in the flight plan. An enroute altitude will be stated in the clearance or the pilot will be advised to expect an assigned or filed altitude within a given time frame or at a certain point after departure. ATC procedures now require the controller to state the DP name, the current number and the DP Transition name. (J14) — AIM ¶5-2-3

Answer (A) is incorrect because the fixes are included in the flight plan and will not be repeated in an abbreviated clearance. Answer (C) is incorrect because an enroute altitude is also given in an abbreviated clearance.

Answers

4395 [B]   4396 [B]

ALL
4398. On the runup pad, you receive the following clearance from ground control:

CLEARED TO THE DALLAS LOVE AIRPORT AS FILED—MAINTAIN SIX THOUSAND—SQUAWK ZERO SEVEN ZERO FOUR JUST BEFORE DEPARTURE—DEPARTURE CONTROL WILL BE ONE TWO FOUR POINT NINER.

An abbreviated clearance, such as this, will always contain the

A—departure control frequency.
B—destination airport and route.
C—requested enroute altitude.

An abbreviated clearance will also include an altitude phrase either directly stated or indirectly stated by assigning a DP. (J14) — AIM ¶5-2-3

ALL
4414. Which information is always given in an abbreviated departure clearance?

A—DP or transition name and altitude to maintain.
B—Name of destination airport or specific fix and altitude.
C—Altitude to maintain and code to squawk.

An abbreviated clearance always includes the name of the destination airport or clearance limit, altitude, and if a DP is to be flown, the DP name, current DP number, and the DP transition name. (J16) — AIM ¶5-2-3

Answer (A) is incorrect because the DP or transition name and altitude is only included when a DP is to be flown, and the name of destination or specific fix and altitude must be included. Answer (C) is incorrect because the abbreviated clearance must also include the name of destination airport or specific fix, and it does not have to include a code to squawk.

ALL
4486. An abbreviated departure clearance "...CLEARED AS FILED..." will always contain the name

A—and number of the STAR to be flown when filed in the flight plan.
B—of the destination airport filed in the flight plan.
C—of the first compulsory reporting point if not in a radar environment.

The abbreviated clearance will include the destination airport filed in the flight plan. (J16) — AIM ¶5-2-3

Answer (A) is incorrect because STARs are considered a part of the filed route of flight and will not normally be stated in an initial departure clearance. Answer (C) is incorrect because compulsory reporting points are not given in an abbreviated clearance.

ALL
4394. When departing from an airport not served by a control tower, the issuance of a clearance containing a void time indicates that

A—ATC will assume the pilot has not departed if no transmission is received before the void time.
B—the pilot must advise ATC as soon as possible, but no later than 30 minutes, of their intentions if not off by the void time.
C—ATC will protect the airspace only to the void time.

If operating from an airport not served by a control tower, the pilot may receive a clearance containing a provision that if the flight has not departed by a specific time, the clearance is void. In this situation, the pilot who does not depart prior to the void time must advise ATC as soon as possible, but no later than 30 minutes, of their intentions. (J16) — AIM ¶5-2-4(a)(1)

Answer (A) is incorrect because ATC will assume the pilot has departed unless they hear from the pilot. Answer (C) is incorrect because the airspace is protected until ATC hears from the pilot.

ALL
4392. What is the significance of an ATC clearance which reads "... CRUISE SIX THOUSAND ..."?

A—The pilot must maintain 6,000 feet until reaching the IAF serving the destination airport, then execute the published approach procedure.
B—Climbs may be made to, or descents made from, 6,000 feet at the pilot's discretion.
C—The pilot may utilize any altitude from the MEA/MOCA to 6,000 feet, but each change in altitude must be reported to ATC.

A cruise clearance is used in an ATC clearance to authorize a pilot to conduct flight at any altitude from the minimum IFR altitude up to, and including, the altitude specified in the clearance. The pilot may level-off at any intermediate altitude within this block of airspace. Climb/descent within the block is to be made at the discretion of the pilot. However, once the pilot starts descent, and verbally reports leaving an altitude in the block, he/she may not return to that altitude without additional ATC clearance. (J14) — AIM ¶4-4-3(d)(3)

Answer (A) is incorrect because the pilot does not have to maintain 6,000 feet. Answer (C) is incorrect because all airspace between the minimum IFR altitude and 6,000 feet is protected, and the pilot need not report altitude changes to ATC.

Answers

4398 [C]   4414 [B]   4486 [B]   4394 [B]   4392 [B]

ALL
4443. What is the significance of an ATC clearance which reads "...CRUISE SIX THOUSAND..."?

A—The pilot must maintain 6,000 until reaching the IAF serving the destination airport, then execute the published approach procedure.
B—It authorizes a pilot to conduct flight at any altitude from minimum IFR altitude up to and including 6,000.
C—The pilot is authorized to conduct flight at any altitude from minimum IFR altitude up to and including 6,000, but each change in altitude must be reported to ATC.

A cruise clearance is used in an ATC clearance to authorize a pilot to conduct flight at any altitude from the minimum IFR altitude up to and including the altitude specified in the clearance. The pilot may level-off at any intermediate altitude within this block of airspace. Climb/descent within the block is to be made at the discretion of the pilot. However, once the pilot starts descent and verbally reports leaving an altitude in the block, he/she may not return to that altitude without additional ATC clearance. (J14) — AIM ¶4-4-3

Answer (A) is incorrect because the pilot is not required to maintain 6,000 feet. Answer (C) is incorrect because the pilot may change altitude without reporting to ATC.

ALL
4458. A "CRUISE FOUR THOUSAND FEET" clearance would mean that the pilot is authorized to

A—vacate 4,000 feet without notifying ATC.
B—climb to, but not descend from 4,000 feet, without further ATC clearance.
C—use any altitude from minimum IFR to 4,000 feet, but must report leaving each altitude.

A cruise clearance is used in an ATC clearance to authorize a pilot to conduct flight at any altitude from the minimum IFR altitude up to and including the altitude specified in the clearance. The pilot may level-off at any intermediate altitude within this block of airspace. Climb/descent within the block is to be made at the discretion of the pilot. However, once the pilot starts descent and verbally reports leaving an altitude in the block, he/she may not return to that altitude without additional ATC clearance. (J14) — AIM ¶4-4-3

Answer (B) is incorrect because any airspace between the minimum IFR altitude and 4,000 feet MSL may be utilized without further ATC clearance. Answer (C) is incorrect because cruise clearances do not necessitate reporting a change in altitude to ATC.

RTC
4644. Does the ATC term "cleared to cruise" apply to helicopter IFR operations?

A—No, this term applies to airplane IFR operations only.
B—Yes, but the pilot must report leaving an altitude.
C—Yes, in part, it authorizes the pilot to commence the approach at the destination airport at pilot's discretion.

The term "cruise" may be used instead of "maintain" to assign a block of airspace, to a pilot, from the minimum IFR altitude up to and including the altitude specified in the cruise clearance. The pilot may level off at any intermediate altitude within this block of airspace. Climb/descent within the block is to be made at the discretion of the pilot. However, once the pilot starts descent and verbally reports leaving an altitude in the block, the pilot may not return to that altitude without additional ATC clearance. (J14) — AIM ¶4-4-3

ALL
4393. What is the recommended climb procedure when a nonradar departure control instructs a pilot to climb to the assigned altitude?

A—Maintain a continuous optimum climb until reaching assigned altitude and report passing each 1,000 foot level.
B—Climb at a maximum angle of climb to within 1,000 feet of the assigned altitude, then 500 feet per minute the last 1,000 feet.
C—Maintain an optimum climb on the centerline of the airway without intermediate level-offs until 1,000 feet below assigned altitude, then 500 to 1500 feet per minute.

When ATC has not used the phrase "at pilot's discretion" nor imposed any climb or descent restrictions, pilots should initiate climb or descent promptly and acknowledgment of the clearance. Descend or climb at an optimum rate consistent with the operating characteristics of the aircraft to 1,000 feet above or below the assigned altitude, and then attempt to descend or climb at a rate of between 500 and 1,500 fpm until the assigned altitude is reached. (J14) — AIM ¶4-4-9(d)

Answer (A) is incorrect because the pilot is not required to report passing each 1,000 feet of altitude. Answer (B) is incorrect because cruise climb (not maximum angle of climb) should be used.

Answers

4443 [B]   4458 [A]   4644 [C]   4393 [C]

ALL
4555. To comply with ATC instructions for altitude changes of more than 1,000 feet, what rate of climb or descent should be used?

A—As rapidly as practicable to 500 feet above/below the assigned altitude, and then at 500 feet per minute until the assigned altitude is reached.
B—1,000 feet per minute during climb and 500 feet per minute during descents until reaching the assigned altitude.
C—As rapidly as practicable to 1,000 feet above/below the assigned altitude, and then between 500 and 1,500 feet per minute until reaching the assigned altitude.

When ATC has not used the term "at pilot's discretion," nor imposed any climb or descent restrictions, pilots should initiate climb or descent promptly on acknowledgment of the clearance. Descend or climb at an optimum rate consistent with the operating characteristics of the aircraft to 1,000 feet above or below the assigned altitude, and then attempt to descend or climb at a rate of 500 to 1,500 feet per minute until the assigned altitude is reached. (J14) — AIM ¶4-4-9

ALL
4442. Which clearance procedures may be issued by ATC without prior pilot request?

A—DPs, STARs, and contact approaches.
B—Contact and visual approaches.
C—DPs, STARs, and visual approaches.

Contact approaches will only be authorized when requested by the pilot, and the reported ground visibility at the destination airport is at least 1 statute mile. ATC may initiate a visual approach when it will be operationally beneficial and certain conditions can be met. DPs and STARs are assigned as determined necessary by ATC. (J14) — AIM ¶5-2-3, ¶5-4-1, ¶5-4-20

Answers (A) and (B) are incorrect because contact approaches may not be issued unless the pilot requests one.

Departure Procedures (DPs)

To simplify ATC clearance delivery, coded ATC departure procedures, called Departure Procedures (DPs), have been established at certain airports. Pilots may be issued a DP whenever ATC deems it appropriate. If the PIC does not wish to use a DP, the pilot is expected to so advise ATC. If the pilot does elect to use a DP, he/she must possess at least the textual description. Figure 6-1 on the next page shows a DP for Detroit, Michigan.

On some DPs a climb gradient may be specified. The pilot can convert the "climb gradient per nautical mile" to "rate of climb in feet per minute" by using the rate of climb table in FAA Legend 16.

When ready for departure, the pilot will receive takeoff clearance from the tower. He/she should remain on the tower frequency until directed to contact departure control.

ALL
4270. (Refer to Figure 30.) Using an average ground speed of 120 knots, what minimum rate of climb must be maintained to meet the required climb rate (feet per NM) to 4,100 feet as specified on the instrument departure procedure?

A—400 feet per minute.
B—500 feet per minute.
C—800 feet per minute.

Using FAA Legend 16, enter the table in the required rate of climb column at the rate of climb required by the NOTE on the plan view of the DP (400 feet per NM).

Move directly to the right in that row until intercepting the 120-knot ground speed column and read 800 fpm. Or you can use the formula:

1. $\dfrac{\text{Ground speed} \times \text{Climb Rate Required}}{60} = \text{Rate of Climb Needed}$

2. $\dfrac{120 \times 400 \text{ feet/NM}}{60} = 800 \text{ feet per minute}$

(H342) — AC 61-23C, Chapter 8

Answers

4555 [C]   4442 [C]   4270 [C]

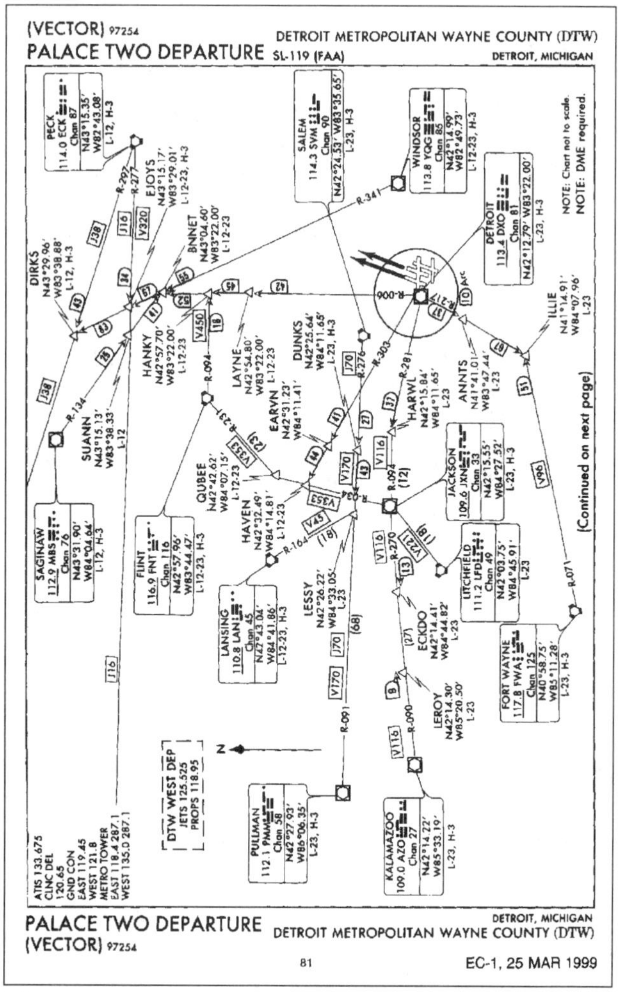

Chapter 6 **Departure**

see Legend (16) pg. 16

Figure 6-1. Departure Procedure, Detroit, Michigan

ALL
4272. (Refer to Figures 30 and 30A.) What is your position relative to GNATS intersection and the instrument departure routing?

A—On departure course and past GNATS.
B—Right of departure course and past GNATS.
C—Left of departure course and have not passed GNATS.

On the RMI in FAA Figure 30A, the thin needle is tuned to VIOLE LMM (356). The tail of the needle is on 280, which shows that the aircraft is on the 280° bearing from or 10° off course. This places the aircraft to the north (right) of the published route. The larger needle is set to indicate position from the MEDFORD (OED) VORTAC (113.6). The tail of the needle is on 224, which indicates that the aircraft is on the 225° radial of OED, therefore, west of and past GNATS INT. (J40) — DP Chart

ALL
4303. (Refer to Figure 46.) Using an average ground speed of 140 knots, what minimum indicated rate of climb must be maintained to meet the required climb rate (feet per NM) to 6,300 feet as specified on the instrument departure procedure?

A—350 feet per minute.
B—583 feet per minute.
C—816 feet per minute.

The note on the plan view of the DP states that a climb of 350 feet per NM is required to 6,300 feet. To convert this to a rate of climb, enter the rate-of-climb table (FAA Legend 16) at 350 feet per NM. Move directly to the right in that row until intercepting the 140-knot ground speed column and read 816 fpm. (J40) — U.S. Terminal Procedures, IAP Chart

130°

ALL
4304. (Refer to Figures 46 and 48.) What is your position relative to the 9 DME ARC and the 206° radial of the instrument departure procedure?

A—On the 9 DME arc and approaching R-206.
B—Outside the 9 DME arc and past R-206.
C—Inside the 9 DME arc and approaching R-206.

The DME readout shows the aircraft on the 9-mile DME arc. The HSI shows that if the aircraft turns to 206°, then the 206° radial will be to the pilot's left. The aircraft is heading 130°, so it is approaching R-206. (J40) — IAP Chart

ALL
4315. (Refer to Figures 52 and 54.) What is the aircraft's position relative to the HABUT intersection? (The VOR-2 is tuned to 116.5.)

A—South of the localizer and past the GVO R-163.
B—North of the localizer and approaching the GVO R-163.
C—South of the localizer and approaching the GVO R-163.

The standard indicator is tuned to the localizer (110.3). The aircraft is tracking "outbound" on the front course (the aircraft heading is 240), so the CDI will have reversed sensing. The pilot would have to fly left to get back on course (fly away from the needle), thus putting the aircraft north of the localizer course. The RMI indicator #2 (or large needle) is set to receive Gaviota (GVO) VORTAC. It is indicating that the aircraft is on the 130° radial of GVO; therefore, it's east of the HABUT INT. (J40) — IAP Chart

Answers

4272 [B]	4303 [C]	4304 [A]	4315 [B]

Chapter 6 **Departure**

see pg 16

ALL
4316. (Refer to Figure 52.) Using an average ground speed of 100 knots, what minimum rate of climb would meet the required minimum climb rate per NM as specified by the instrument departure procedure?

A—425 feet per minute.
B—580 feet per minute.
C—642 feet per minute.

The note on the plan view of the HABUT ONE DEPARTURE indicates this departure requires a minimum climb rate of 385 feet per NM to 6,000. The Rate-of-Climb Table (FAA Legend 16) is used to convert this to a rate of climb (fpm).

1. *Enter the Rate-of-Climb Table in the left-hand column at 350 feet per NM. Move directly to the right in that row until intercepting the 100-knot ground speed column and read 583 feet per minute rate of climb.*

2. *Enter the table in the left-hand column at 400 feet per NM. Move directly to the right in that row until intercepting the 100-knot ground speed column and read 667 fpm rate of climb.*

3. *Interpolate between 350 and 400 to find the rate of climb for 385: 642 fpm.*

(J40) — U.S. Terminal Procedures, IAP Chart

ALL
4361. (Refer to Figure 77.) At which point does the basic instrument departure procedure terminate?

A—When Helena Departure Control establishes radar contact.
B—At STAKK intersection.
C—Over the BOZEMAN VOR.

The heavy black lines with arrows portray the Departure Route, which, in this case, terminates at STAKK INT. Note also in the textual description all the transitions are indicated with the following statement, "...localizer course to STAKK. INT at or above 8,400 feet. Thence..." (J40) — U.S. Terminal Procedures, DP Chart Legend

Answer (A) is incorrect because Helena Departure Control will establish contact shortly after leaving the runway. Answer (C) is incorrect because the BOZEMAN VOR is the end of the BOZEMAN transition.

ALL
4363. (Refer to Figure 77.) At which minimum altitude should you cross the STAKK intersection?

A—6,500 feet MSL.
B—1,400 feet MSL.
C—10,200 feet MSL.

The departure route descriptions for the STAKK TWO DEPARTURE state "... cross STAKK INT at or above 10,200'..." for either runway. (J40) — U.S. Terminal Procedures, DP Chart Legend

ALL
4364. (Refer to Figure 77.) Using an average ground speed of 140 knots, what minimum rate of climb would meet the required minimum climb rate per NM as specified on the instrument departure procedure?

A—350 feet per minute.
B—475 feet per minute.
C—700 feet per minute.

Enter the Rate-of-Climb Table (FAA Legend 16) in the left-hand column at the required climb rate of 300 feet (see the Note on the left side of the plan view of FAA Figure 77). Move horizontally to the right to the 140-knot ground speed column and read the required rate of climb (700 fpm). (J40) — U.S. Terminal Procedures

ALL
4417. What action is recommended if a pilot does not wish to use an instrument departure procedure?

A—Advise clearance delivery or ground control before departure.
B—Advise departure control upon initial contact.
C—Enter "No DP" in the REMARKS section of the IFR flight plan.

If, for any reason, the pilot does not wish to use a DP, he/she is expected to advise ATC. Notification may be accomplished by writing "NO DP" in the remarks section of the flight plan, or by the less desirable method of verbally advising ATC. (J15) — AIM ¶5-2-6(a)(2)

Answers (A) and (B) are incorrect because entering "No DP" in the flight plan is preferred over a verbal request.

Answers

| 4316 [C] | 4361 [B] | 4363 [C] | 4364 [C] | 4417 [C] |

Chapter 6 **Departure**

ALL
4418. A particular instrument departure procedure requires a minimum climb rate of 210 feet per NM to 8,000 feet. If you climb with a ground speed of 140 knots, what is the rate of climb required in feet per minute?

A—210.
B—450.
C—490.

The question can be solved using the Rate-of-Climb Table (FAA Legend 16). At a ground speed of 140 knots, a climb rate of 210 feet per NM must be interpolated between 200 and 250: 583 − 467 = 116 x 20% = 23.2, added to 467 fpm equals 490.2 fpm. (J40) — AIM ¶5-2-6

ALL
4419. Which procedure applies to instrument departure procedures?

A—Instrument departure clearances will not be issued unless requested by the pilot.
B—The pilot in command must accept an instrument departure procedure when issued by ATC.
C—If an instrument departure procedure is accepted, the pilot must possess at least a textual description.

The use of a DP requires pilot possession of at least the textual description of the approved effective DP. (J16) — AIM ¶5-2-6

Answer (A) is incorrect because ATC will issue DP clearances when DP procedures exist. Answer (B) is incorrect because a DP clearance must be rejected if the pilot does not have the textual description of the DP or if it is otherwise unacceptable.

ALL
4488. (Refer to Figures 85 and 86.) Which combination of indications confirm that you are approaching WAGGE intersection slightly to the right of the LOC centerline on departure?

A—1 and 3.
B—1 and 4.
C—2 and 3.

When flying out on the back course of an ILS, you receive front course indications, so indicator #2 in FAA Figure 86 shows you are to the right of course. VOR indicators #3 and #4 show the 062 radial selected, which is the only way to determine WAGGE intersection. Visualize yourself flying out on Squaw Valley Radial 062. Indicator #3 shows you to the left of the 062 radial;

therefore, you are to the north of and approaching WAGGE intersection. (J40) — DP Chart

Answer (A) is incorrect because illustration 1 indicates the aircraft left (not right) of the localizer centerline. Answer (B) is incorrect because illustration 1 indicates the aircraft left (not right) of the localizer centerline, and illustration 4 indicates the aircraft south of R-062 which is beyond WAGGE intersection.

ALL
4489. (Refer to Figure 85.) What route should you take if cleared for the Washoe Two Departure and your assigned route is V6?

A—Climb on the LOC south course to WAGGE where you will be vectored to V6.
B—Climb on the LOC south course to cross WAGGE at 9,000, turn left and fly direct to FMG VORTAC and cross at or above 10,000, and proceed on FMG R-241.
C—Climb on the LOC south course to WAGGE, turn left and fly direct to FMG VORTAC. If at 10,000 turn left and proceed on FMG R-241; if not at 10,000 enter depicted holding pattern and climb to 10,000 before proceeding on FMG R-241.

The departure route description for the Washoe Two Departure is to climb via I-RNO localizer south course to WAGGE intersection then via radar vectors to assigned route. (J40) — AIM ¶5-2-6

Answers (B) and (C) are incorrect because FMG R-241 is V200-392 and V6 is RMG R-218.

ALL
4490. (Refer to Figure 85.) What procedure should be followed if communications are lost before reaching 9,000 feet?

A—At 9,000, turn left direct to FMG VORTAC, then via assigned route if at proper altitude; if not, climb in holding pattern until reaching the proper altitude.
B—Continue climb to WAGGE INT, turn left direct to FMG VORTAC, then if at or above MCA, proceed on assigned route; if not, continue climb in holding pattern until at the proper altitude.
C—Continue climb on LOC course to cross WAGGE INT at or above 9,000, turn left direct to FMG VORTAC to cross at 10,000 or above, and continue on assigned course.

The lost communication procedures state: if not in contact with departure control within one minute after takeoff, or if communications are lost before reaching 9,000 feet, continue climb via I-RNO localizer south course to WAGGE intersection. Turn left, proceed direct

Answers

| 4418 | [C] | 4419 | [C] | 4488 | [C] | 4489 | [A] | 4490 | [B] |

FMG VOR. Cross FMG VOR at or above MCA, thence via assigned route or climb in holding pattern northeast on FMG R-041. Left turns to cross FMG VOR at or above MCA for assigned route. (J40) — AIM ¶5-2-6

Answer (A) is incorrect because the procedure states to turn left to FMG VORTAC at WAGGE (regardless of altitude). Answer (C) is incorrect because there is not a requirement to cross WAGGE intersection at 9,000 feet.

ALL
4491. (Refer to Figure 85.) What is the minimum rate climb per NM to 9,000 feet required for the WASH2 WAGGE Departure?

A—400 feet.
B—750 feet.
C—875 feet.

See "note" on the plan view of the approach plate. It states: minimum climb rate at 400 feet per NM to 9,000 feet required. (J40) — AIM ¶5-2-6

ALL
4492. (Refer to Figure 85.) Of the following, which is the minimum acceptable rate of climb (feet per minute) to 9,000 feet required for the WASH2 WAGGE departure at a GS of 150 knots?

A—750 feet per minute.
B—825 feet per minute.
C—1,000 feet per minute.

See "note" on the plan view of the approach plate. It states the minimum climb rate of 400 feet per nautical mile to 9,000 feet required. Use the rate-of-climb table located in FAA Legend 16 to convert to feet per minute. Find the GS of 150 knots and the required climb rate of 400 feet. The required rate of 1,000 feet per minute can be read at the intersection of the two. (J40) — AIM ¶5-2-6

ALL
4638. Which is true regarding the use of an instrument departure procedure chart?

A—The use of instrument departure procedures is mandatory.
B—To use an instrument departure procedure, the pilot must possess at least the textual description of the approved standard departure.
C—To use an instrument departure procedure, the pilot must possess both the textual and graphic form of the approved procedure.

The use of a DP requires pilot possession of at least the textual description of the approved effective DP. (J16) — AIM ¶5-2-6

Answer (A) is incorrect because DP usage is not mandatory for IFR departures. Answer (C) is incorrect because the pilot must possess only a textual description of the approved effective DP.

Services Available to Pilots

At some locations within airspace defined as Terminal Radar Service Areas, ATC provides separation between all IFR aircraft and all participating VFR aircraft. Pilot participation is urged, but is not mandatory.

The FSS provides an Airport Advisory Service (AAS) on airports which do not have a control tower, or the tower is temporarily closed or operated on a part-time basis.

Controllers will issue traffic information with reference to a 12-hour clock. The traffic advisories are based on the radar position, and does not take into consideration any heading correction to account for a wind.

"Resume own navigation" means the pilot is to resume his/her own navigational responsibility. It is issued after a radar vector or when radar contact is lost.

"Radar contact" means ATC has identified the aircraft on the radar display and radar flight following will be provided until radar identification is terminated.

Answers

4491 [A]   4492 [C]   4638 [B]

Chapter 6 **Departure**

ALL
4409. What service is provided by departure control to an IFR flight when operating within the outer area of Class C airspace?

A—Separation from all aircraft.
B—Position and altitude of all traffic within 2 miles of the IFR pilot's line of flight and altitude.
C—Separation from all IFR aircraft and participating VFR aircraft.

Class C Service provides approved separation between IFR and VFR aircraft, and sequencing of VFR arrivals to the primary airport. (J08) — AIM ¶4-1-17(c)

Answer (A) is incorrect because separation from nonparticipating VFR aircraft cannot be provided. Answer (B) is incorrect because ATC will provide traffic advisories (to the extent possible in relation to higher priorities) in relation to your airplane's track, and may include altitude if known.

ALL
4390. When should your transponder be on Mode C while on an IFR flight?

A—Only when ATC requests Mode C.
B—At all times if the equipment has been calibrated, unless requested otherwise by ATC.
C—When passing 12,500 feet MSL.

Adjust transponder to reply on the Mode A/3 code specified by ATC and if equipped, to reply on Mode C with altitude reporting capability activated; that is, unless deactivation is directed by ATC, or unless the installed aircraft equipment has not been tested and calibrated as required by 14 CFR §91.413. (J11) — AIM ¶4-1-19

ALL
4415. If a control tower and an FSS are located on the same airport, which function is provided by the FSS during those periods when the tower is closed?

A—Automatic closing of the IFR flight plan.
B—Approach control services.
C—Airport Advisory Service.

Airport Advisory Service (AAS) is a service provided by an FSS physically located on an airport which does not have a control tower, or where the tower is temporarily closed or operated on a part-time basis. (J11) — AIM ¶4-1-9(d)

Answer (A) is incorrect because the pilot is responsible for closing the IFR flight plan if operating to an airport without a functioning control tower. Answer (B) is incorrect because the FSS can only issue advisories, not clearances.

ALL
4416. Which service is provided for IFR arrivals by a FSS located on an airport without a control tower?

A—Automatic closing of the IFR flight plan.
B—Airport advisories.
C—All functions of approach control.

Airport Advisory Service (AAS) is a service provided by an FSS physically located on an airport which does not have a control tower, or where the tower is temporarily closed or operated on a part-time basis. (J11) — AIM ¶4-1-9(d)

Answer (A) is incorrect because the pilot is responsible for closing the IFR flight plan if operating to an airport without a functioning control tower. Answer (C) is incorrect because the FSS can only issue advisories, not clearances.

ALL
4421. During a flight, the controller advises "traffic 2 o'clock 5 miles southbound." The pilot is holding 20 correction for a crosswind from the right. Where should the pilot look for the traffic?

A—40° to the right of the aircraft's nose.
B—20° to the right of the aircraft's nose.
C—Straight ahead.

Radar traffic information is given by the controller which includes: The azimuth from the aircraft's ground track (not the heading) in terms of a 12-hour clock, and each clock position is 30°. 2 o'clock is about 60° from the aircraft's course, but since the pilot is already holding a 20° right crosswind, the pilot will only have to look 40° to the right of the aircraft's nose. (J11) — AIM ¶4-1-14(c)

Answer (B) is incorrect because 20° to the right of the aircraft nose would be between 1 and 2 o'clock. Answer (C) is incorrect because straight ahead would be between 12 and 1 o'clock.

ALL
4422. What is meant when departure control instructs you to "resume own navigation" after you have been vectored to a Victor airway?

A—You should maintain the airway by use of your navigation equipment.
B—Radar service is terminated.
C—You are still in radar contact, but must make position reports.

"Resume own navigation" is used by ATC to advise a pilot to resume his/her own navigational responsibility. It is issued after completion of a radar vector or when

Answers

4409 [C]   4390 [B]   4415 [C]   4416 [B]   4421 [A]   4422 [A]

radar contact is lost while the aircraft is being radar vectored. (J33) — Pilot/Controller Glossary

Answer (B) is incorrect because "resume own navigation" means the radar controller will stop issuing vectors, but will not stop observing. Answer (C) is incorrect because position reports are not necessary when in radar contact.

ALL
4423. What does the ATC term "Radar Contact" signify?

A—Your aircraft has been identified and you will receive separation from all aircraft while in contact with this radar facility.
B—Your aircraft has been identified on the radar display and radar flight-following will be provided until radar identification is terminated.
C—You will be given traffic advisories until advised the service has been terminated or that radar contact has been lost.

Radar contact is used by ATC to inform an aircraft that it is identified on the radar display and radar flight following will be provided until radar identification is terminated. Radar service may also be provided within the limits of necessity and capability. When a pilot is informed of "radar contact," he/she automatically discontinues reporting over compulsory reporting points. (J33) — Pilot/Controller Glossary

Answer (A) is incorrect because separation from all aircraft only occurs in Class A, B, and C airspaces. Answer (C) is incorrect because traffic advisory service is only provided to the extent possible.

ALL
4424. Upon intercepting the assigned radial, the controller advises you that you are on the airway and to "RESUME OWN NAVIGATION." This phrase means that

A—you are still in radar contact, but must make position reports.
B—radar services are terminated and you will be responsible for position reports.
C—you are to assume responsibility for your own navigation.

"Resume own navigation" is used by ATC to advise a pilot to resume his/her own navigational responsibility. It is issued after completion of a radar vector or when radar contact is lost while the aircraft is being radar vectored. (J33) — Pilot/Controller Glossary

Answer (A) is incorrect because position reports are not necessary when in radar contact. Answer (B) is incorrect because "resume own navigation" means the radar controller will stop issuing vectors, but will not stop observing.

ALL
4736. When is radar service terminated during a visual approach?

A—Automatically when ATC instructs the pilot to contact the tower.
B—Immediately upon acceptance of the approach by the pilot.
C—When ATC advises, "Radar service terminated; resume own navigation."

On a visual approach, radar service is automatically terminated without advising the pilot, when the aircraft is instructed to contact the tower. (J19) — AIM ¶5-5-11(a)(6)

Answer (B) is incorrect because approach clearance is given long before radar service is terminated. Answer (C) is incorrect because "resume own navigation" is normally an enroute instruction.

ALL
4420. During a takeoff into IFR conditions with low ceilings, when should the pilot contact departure control?

A—Before penetrating the clouds.
B—When advised by the tower.
C—Upon completing the first turn after takeoff or upon establishing cruise climb on a straight-out departure.

The pilot should contact departure control on the assigned frequency upon release from the control tower. (J16) — AIM ¶5-2-5

Answers (A) and (C) are incorrect because tower control will advise the pilot when to contact departure control, which may or may not be before penetrating the clouds, upon completion of the first turn after takeoff, or upon establishing cruise climb on a straight-out departure.

Answers
4423 [B]   4424 [C]   4736 [A]   4420 [B]

Pilot/Controller Roles and Responsibilities

The PIC is responsible to see and avoid other traffic whenever meteorological conditions permit, regardless of the type of flight plan to be flown.

The PIC is responsible for complying with speed adjustments within ±10 knots or .02 Mach number of the specified speed.

ATC cannot issue a VFR clearance to a pilot on an IFR flight plan, unless that pilot requests it or to comply with noise abatement routes or altitudes.

ALL
4471. What responsibility does the pilot in command of an IFR flight assume upon entering VFR conditions?

A—Report VFR conditions to ARTCC so that an amended clearance may be issued.
B—Use VFR operating procedures.
C—To see and avoid other traffic.

When meteorological conditions permit, regardless of type of flight plan or whether or not under control of a radar facility, the pilot is responsible to see and avoid other traffic, terrain, or obstacles. (J19) — AIM ¶5-5-8

Answer (A) is incorrect because it is not required to report VFR conditions unless requested to do so by ATC. Answer (B) is incorrect because the pilot must follow IFR procedures when on an IFR flight.

ALL
4373. When is a pilot on an IFR flight plan responsible for avoiding other aircraft?

A—At all times when not in radar contact with ATC.
B—When weather conditions permit, regardless of whether operating under IFR or VFR.
C—Only when advised by ATC.

When weather conditions permit, regardless of whether an operation is conducted under instrument flight rules or visual flight rules, vigilance shall be maintained by each person operating an aircraft so as to see and avoid other aircraft in compliance with this section. (J02) — 14 CFR §91.113(b)

Answers (A) and (C) are incorrect because if weather conditions permit, each pilot is responsible for seeing and avoiding other aircraft.

ALL
4538. When should pilots state their position on the airport when calling the tower for takeoff?

A—When visibility is less than 1 mile.
B—When parallel runways are in use.
C—When departing from a runway intersection.

Pilots should state their position on the airport when calling the tower for takeoff from a runway intersection. (J13) — AIM ¶4-3-10(c)

ALL
4633. Under which of the following circumstances will ATC issue a VFR restriction to an IFR flight?

A—Whenever the pilot reports the loss of any navigational aid.
B—When it is necessary to provide separation between IFR and special VFR traffic.
C—When the pilot requests it.

ATC may not authorize VFR-On-Top/VFR conditions operations unless the pilot requests the VFR operation, or a clearance to operate in VFR conditions will result in noise abatement benefits where part of the IFR departure route does not conform to an FAA-approved noise abatement route or altitude. (J19) — AIM ¶4-4-7(d)

Answer (A) is incorrect because a pilot would only report a malfunction of navigation equipment to ATC. Answer (B) is incorrect because Special VFR traffic would be found in Class D airspace that is currently experiencing IMC, so ATC would not issue a VFR restriction to an IFR flight.

Answers

4471 [C]	4373 [B]	4538 [C]	4633 [C]

ALL
4725. What is the pilot in command's responsibility when flying a propeller aircraft within 20 miles of the airport of intended landing and ATC requests the pilot to reduce speed to 160? (Pilot complies with speed adjustment.)

A—Reduce TAS to 160 knots and maintain until advised by ATC.
B—Reduce IAS to 160 MPH and maintain until advised by ATC.
C—Reduce IAS to 160 knots and maintain that speed within 10 knots.

When complying with speed adjustments assignment, maintain an indicated airspeed within ±10 knots or .02 Mach number of the specified speed. (J19) — AIM ¶5-5-9(a)(3)

Answer (A) is incorrect because the speed restrictions issued by ATC are based on indicated (not true) airspeed. Answer (B) is incorrect because MPH are not used in ATC instructions.

ALL
4534. (Refer to Figure 94.) Mandatory airport instruction signs are designated by having—

A—Yellow lettering with a black background.
B—White lettering with a red background.
C—Black lettering with a yellow background.

Mandatory instruction signs have a red background with a white inscription and are used to denote:

1. *An entrance to a runway or critical area; and*
2. *Areas where an aircraft is prohibited from entering.*

(J05) — AIM ¶2-3-8

ALL
4535. (Refer to Figure 94.) What sign is designated by illustration 7?

A—Location sign.
B—Mandatory instruction sign.
C—Direction sign.

The holding position sign is a mandatory instruction sign, used to hold an aircraft on a taxiway located in the approach or departure area for a runway so the aircraft does not interfere with operations on that runway. (J05) — AIM ¶2-3-8

ALL
4536. (Refer to Figure 94.) What color are runway holding position signs?

A—White with a red background.
B—Red with a white background.
C—Yellow with a black background.

Holding position signs are mandatory instruction signs, and have a red background with a white inscription. (J05) — AIM ¶2-3-8

ALL
4537. (Refer to Figure 94.) Hold line markings at the intersection of taxiways and runways consist of four lines that extend across the width of the taxiway. These lines are—

A—white and the dashed lines are nearest the runway.
B—yellow and the dashed lines are nearest the runway.
C—yellow and the solid lines are nearest the runway.

Holding position markings indicate where an aircraft is supposed to stop. They consist of four yellow lines: two solid and two dashed, spaced six inches apart and extending across the width of the taxiway or runway. The solid lines are always on the side where the aircraft is to hold. (J05) — AIM ¶2-3-5

Answers

| 4725 | [C] | 4534 | [B] | 4535 | [B] | 4536 | [A] | 4537 | [B] |

VFR-On-Top

A pilot operating in Visual Meteorological Conditions (VMC) while on an IFR flight plan may wish to select an altitude of his/her choice. The pilot may request VFR-On-Top, and if the request is approved, select an appropriate VFR altitude based on the magnetic course being flown, which is at or above the minimum IFR altitude (0° through 179°, odd thousands plus 500 feet; 180° through 359°, even thousands plus 500 feet).

ATC may provide traffic information on other IFR or VFR aircraft to pilots operating VFR-On-Top, but when in VMC, it is the pilot's responsibility to see and avoid other aircraft.

Pilots should advise ATC prior to any altitude change to ensure the exchange of accurate traffic information.

VFR-On-Top will not be authorized at or above 18,000 feet MSL.

ALL
4430. What altitude may a pilot select upon receiving a VFR-On-Top clearance?

A— Any altitude at least 1,000 feet above the meteorological condition.
B— Any appropriate VFR altitude at or above the MEA in VFR weather conditions.
C— Any VFR altitude appropriate for the direction of flight at least 1,000 feet above the meteorological condition.

The VFR-On-Top clearance is the ATC authorization for an IFR aircraft to operate in VFR conditions at any appropriate VFR altitude (as specified in 14 CFR and as restricted by ATC). A pilot receiving this authorization must comply with VFR visibility, distance from cloud criteria, and the minimum IFR altitudes. (J19) — 14 CFR §91.159

Answers (A) and (C) are incorrect because with a VFR-On-Top clearance, the pilot is not restricted to 1,000 feet above the meteorological condition.

ALL
4431. When must a pilot fly at a cardinal altitude plus 500 feet on an IFR flight plan?

A— When flying above 18,000 feet in VFR conditions.
B— When flying in VFR conditions above clouds.
C— When assigned a VFR-On-Top clearance.

A pilot operating on an IFR flight plan, in VFR conditions, with an ATC authorization to "Maintain VFR-On-Top, maintain VFR conditions," must fly at appropriate VFR altitudes which are VFR cardinal altitudes plus 500 feet. (J19) — 14 CFR §91.179

Answer (A) is incorrect because VFR-On-Top is not permitted in Class A airspace, which is from 18,000 feet MSL up to and including FL600. Answer (B) is incorrect because a pilot on a IFR flight plan can only utilize VFR altitudes when assigned a VFR-On-Top clearance.

ALL
4433. You have filed an IFR flight plan with a VFR-On-Top clearance in lieu of an assigned altitude. If you receive this clearance and fly a course of 180°, at what altitude should you fly? (Assume VFR conditions.)

A— Any IFR altitude which will enable you to remain in VFR conditions.
B— An odd thousand-foot MSL altitude plus 500 feet.
C— An even thousand-foot MSL altitude plus 500 feet.

A pilot operating on an IFR flight plan, in VFR conditions, with a VFR-On-Top clearance, must fly at appropriate VFR altitudes, which are VFR cardinal altitudes plus 500 feet. On a magnetic course of 180° through 359°, any even thousand-foot MSL altitude plus 500 feet must be flown. (J06) — 14 CFR §91.179

Answer (A) is incorrect because when operating under a VFR-On-Top clearance, the pilot must use VFR (not IFR) altitudes. Answer (B) is incorrect because odd thousand-foot altitudes plus 500 feet must be flown on a magnetic course of 0° through 179°.

ALL
4447. Where are VFR-On-Top operations prohibited?

A— In Class A airspace.
B— During off-airways direct flights.
C— When flying through Class B airspace.

VFR-On-Top is not permitted in Class A airspace. Consequently, IFR flights operating VFR-On-Top are not allowed to operate above FL180. (J08) — AIM ¶4-4-7(h)

Answers (B) and (C) are incorrect because VFR-On-Top operations are allowed (not prohibited) while flying during off-airways direct flights or when flying through Class B airspace.

Answers

4430 [B]   4431 [C]   4433 [C]   4447 [A]

Chapter 6 Departure

ALL
4449. Which rules apply to the pilot in command when operating on a VFR-On-Top clearance?

A—VFR only.
B—VFR and IFR.
C—VFR when "in the clear" and IFR when "in the clouds."

When operating in VFR conditions with an ATC authorization to "Maintain VFR-On-Top/Maintain VFR Conditions," pilots on IFR flight plans must:

1. *Fly at the appropriate VFR altitude as prescribed in 14 CFR §91.159.*
2. *Comply with the VFR visibility and distance from cloud criteria in 14 CFR §91.155 (Basic VFR Weather Minimums).*
3. *Comply with instrument flight rules that are applicable to this flight (i.e., minimum IFR altitudes, position reporting, radio communications, course to be flown, adherence to ATC clearance, etc.).*

(J14) — AIM ¶4-4-7

Answer (A) is incorrect because a VFR-On-Top clearance does not cancel the IFR flight plan. Answer (C) is incorrect because a VFR-On-Top clearance does not permit a pilot to fly in IMC.

ALL
4450. When can a VFR-On-Top clearance be assigned by ATC?

A—Only upon request of the pilot when conditions are indicated to be suitable.
B—Any time suitable conditions exist and ATC wishes to expedite traffic flow.
C—When VFR conditions exist, but there is a layer of clouds below the MEA.

ATC may not authorize VFR-On-Top operations unless the pilot requests the VFR operation while operating in VMC weather. (J14) — AIM ¶4-4-7

Answers (B) and (C) are incorrect because ATC can only issue a VFR-On-Top clearance upon a pilot's request.

ALL
4451. Which ATC clearance should instrument-rated pilots request in order to climb through a cloud layer or an area of reduced visibility and then continue the flight VFR?

A—To VFR-On-Top.
B—Special VFR to VFR Over-the-Top.
C—VFR Over-the-Top.

Pilots desiring to climb through a cloud, haze, smoke, or other meteorological formation, and then either cancel their IFR flight plan or operate VFR-On-Top, may request a climb to VFR-On-Top. (J14) — AIM ¶4-4-7

Answer (B) is incorrect because a special VFR clearance is issued in Class D airspace, and the pilot must remain clear of clouds (not climb through them). Answer (C) is incorrect because the pilot must request a VFR-On-Top (not VFR Over-the-Top) clearance.

ALL
4452. When on a VFR-On-Top clearance, the cruising altitude is based on

A—true course.
B—magnetic course.
C—magnetic heading.

When operating on a VFR-On-Top clearance, the pilot must fly VFR cruising altitudes which are based on Magnetic Course. (J14) — AIM ¶4-4-7

ALL
4453. In which airspace is VFR-On-Top operation prohibited?

A—Class B airspace.
B—Class E airspace.
C—Class A airspace.

ATC will not authorize VFR or VFR-On-Top operations in Class A airspace. (J14) — AIM ¶4-4-7

Answers (A) and (B) are incorrect because VFR-On-Top is allowed (not prohibited) in Class B and E airspace.

ALL
4454. What cruising altitude is appropriate for VFR on Top on a westbound flight below 18,000 feet?

A—Even thousand-foot levels.
B—Even thousand-foot levels plus 500 feet, but not below MEA.
C—Odd thousand-foot levels plus 500 feet, but not below MEA.

When operating on a VFR-On-Top clearance, the pilot must fly appropriate VFR cruising altitudes. On a magnetic course of 180° through 359°, the pilot must fly any even thousand-foot MSL altitude plus 500 feet. (J14) — AIM ¶4-4-7

Answer (A) is incorrect because an even thousand-foot level is an IFR cruising altitude and VFR-On-Top requires VFR altitudes. Answer (C) is incorrect because odd thousand-foot levels plus 500 feet are for eastbound flights.

Answers

| 4449 | [B] | 4450 | [A] | 4451 | [A] | 4452 | [B] | 4453 | [C] | 4454 | [B] |

Chapter 6 **Departure**

ALL
4510. (Refer to Figure 91.) What are the two limiting cruising altitudes useable on V343 for a VFR-On-Top flight from DBS VORTAC to RANEY intersection?

A—14,500 and 16,500 feet.
B—15,000 and 17,000 feet.
C—15,500 and 17,500 feet.

The eastbound course on V343 from DBS VORTAC to RANEY intersection requires VFR-On-Top altitudes of odd plus 500 feet above the MEA (15,000 feet MSL), and below Class A (18,000 feet MSL). (J35) — Enroute Low Altitude Chart Legend

ALL
4455. What reports are required of a flight operating on an IFR clearance specifying VFR on Top in a nonradar environment?

A—The same reports that are required for any IFR flight.
B—All normal IFR reports except vacating altitudes.
C—Only the reporting of any unforecast weather.

When operating in VFR conditions with an ATC authorization to "Maintain VFR-On-Top" pilots on IFR flight plans must comply with instrument flight rules that are applicable to this flight (i.e., minimum IFR altitudes, position reporting, radio communications, course to be flown, adherence to ATC clearance, etc.). (J14) — AIM ¶4-4-7

Answer (B) is incorrect because all normal IFR reports are required while flying on a VFR-On-Top clearance, including any altitude change. Answer (C) is incorrect because all IFR reports, not only unforecast weather, must be made while flying on a VFR-On-Top clearance.

ALL
4457. What minimums must be considered in selecting an altitude when operating with a VFR-On-Top clearance?

A—At least 500 feet above the lowest MEA, or appropriate MOCA, and at least 1,000 feet above the existing meteorological condition.
B—At least 1,000 feet above the lowest MEA, appropriate MOCA, or existing meteorological condition.
C—Minimum IFR altitude, minimum distance from clouds, and visibility appropriate to altitude selected.

VFR-On-Top flight must be conducted at or above the minimum IFR altitude and in basic VFR weather conditions. (J14) — AIM ¶4-4-7

Answers (A) and (B) are incorrect because the pilot must fly at or above (not a specified distance from) the minimum IFR altitude and the pilot may fly above, below, or between layers of the existing weather conditions.

ALL
4643. When operating under IFR with a VFR-On-Top clearance, what altitude should be maintained?

A—The last IFR altitude assigned by ATC.
B—An IFR cruising altitude appropriate to the magnetic course being flown.
C—A VFR cruising altitude appropriate to the magnetic course being flown and as restricted by ATC.

If the ATC clearance assigns "VFR conditions on top" the pilot shall maintain an appropriate VFR altitude or flight level as prescribed by 14 CFR §91.159 (which describes the VFR hemispheric rule). (J14) — AIM ¶4-4-7

Chapter 7
En Route

Instrument Altitudes   *7–3*

Enroute Low Altitude Chart   *7–5*

Holding   *7–12*

Estimated Time Enroute (ETE)   *7–22*

Fuel Consumption   *7–28*

Chapter 7 **En Route**

Instrument Altitudes

Minimum Enroute Altitude (MEA) is the lowest published altitude between radio fixes which ensures acceptable navigational signal coverage and meets obstacle clearance requirements between those fixes.

Minimum Reception Altitude (MRA) is the lowest altitude at which an intersection can be determined.

Minimum Crossing Altitude (MCA) is the lowest altitude at certain fixes at which an aircraft must cross when proceeding in the direction of a higher minimum enroute IFR altitude.

Minimum Obstruction Clearance Altitude (MOCA) is the lowest published altitude in effect between radio fixes on VOR airways, off-airway routes, or route segments which meets obstacle clearance requirements for the entire route segment, and which ensures acceptable navigational signal coverage within 25 statute (22 nautical) miles of a VOR.

Minimum Sector Altitude (MSA) is the lowest altitude which may be used under emergency conditions which will provide a minimum clearance of 1,000 feet above all obstacles located in an area contained within a 25 NM sector centered on a radio aid to navigation.

ALL
4429. What is the definition of MEA?

A—The lowest published altitude which meets obstacle clearance requirements and assures acceptable navigational signal coverage.
B—The lowest published altitude which meets obstacle requirements, assures acceptable navigational signal coverage, two-way radio communications, and provides adequate radar coverage.
C—An altitude which meets obstacle clearance requirements, assures acceptable navigation signal coverage, two-way radio communications, adequate radar coverage, and accurate DME mileage.

Minimum Enroute IFR Altitude (MEA) is the lowest published altitude between radio fixes which assures acceptable navigational signal coverage and meets obstacle clearance requirements between those fixes. The MEA prescribed for a Federal airway or segment thereof, area navigation low or high route or other direct route applies to the entire width of the airway, segment or route between the radio fixes defining the airway, segment or route. (J33) — Pilot/Controller Glossary

Answers (B) and (C) are incorrect because the MEA does not guarantee that two-way radio communications are adequate radar coverage.

ALL
4432. The altitude that provides acceptable navigational signal coverage for the route, and meets obstacle clearance requirements, is the minimum:

A—enroute altitude.
B—reception altitude.
C—obstacle clearance altitude.

The minimum enroute altitude (MEA) is the lowest published altitude between radio fixes that ensures acceptable navigational signal coverage and meets obstacle clearance requirements between those fixes. (J33) — Pilot/Controller Glossary

ALL
4435. Reception of signals from an off-airway radio facility may be inadequate to identify the fix at the designated MEA. In this case, which altitude is designated for the fix?

A—MRA.
B—MCA.
C—MOCA.

Minimum Reception Altitude (MRA) is the lowest altitude required to receive adequate signals to determine specific fixes. Reception of signals from a radio facility located off the airway being flown may be inadequate at the designated MEA, in which case, a MRA is designated for the fix. (J33) — Pilot/Controller Glossary

Answer (B) is incorrect because MCA (minimum crossing altitude) is the lowest altitude at a fix at which an aircraft must cross when flying in the direction of a higher MEA. Answer (C) is incorrect because the MOCA (minimum obstruction clearance altitude) is the lowest published altitude which meets the obstacle clearance requirements of the entire route segment and also ensures acceptable navigation signal coverage only within 22 NM of a VOR.

Answers
4429 [A]   4432 [A]   4435 [A]

Chapter 7 **En Route**

ALL
4436. Which condition is guaranteed for all of the following altitude limits: MAA, MCA, MRA, MOCA, and MEA? (Non-mountainous area.)

A—Adequate navigation signals.
B—Adequate communications.
C—1,000-foot obstacle clearance.

Obstruction clearance is normally at least 1,000 feet in non-mountainous areas above the highest terrain or obstruction 4 nautical miles either side of centerline of the airway or route. (J06) — 14 CFR §91.177

Answer (A) is incorrect because the MOCA is the only altitude which assures acceptable navigational signals within 22 NM of a VOR. Answer (B) is incorrect because minimum IFR altitudes do not assure adequate communications coverage.

ALL
4437. If no MCA is specified, what is the lowest altitude for crossing a radio fix, beyond which a higher minimum applies?

A—The MEA at which the fix is approached.
B—The MRA at which the fix is approached.
C—The MOCA for the route segment beyond the fix.

If a normal climb, commenced immediately after passing a fix beyond which a higher MEA applies, would not assure adequate obstruction clearance, then a MCA is specified. Otherwise, cross the radio fix at the MEA at which it is approached, and then climb to the next higher MEA. (J33) — Pilot/Controller Glossary

ALL
4542. MEA is an altitude which assures

A—obstacle clearance, accurate navigational signals from more than one VORTAC, and accurate DME mileage.
B—a 1,000-foot obstacle clearance within 2 miles of an airway and assures accurate DME mileage.
C—acceptable navigational signal coverage and meets obstruction clearance requirements.

Minimum Enroute IFR Altitude (MEA) is the lowest published altitude which assures acceptable navigational signal coverage and meets obstacle clearance requirements between those fixes. (J33) — Pilot/Controller Glossary

ALL
4544. Reception of signals from a radio facility, located off the airway being flown, may be inadequate at the designated MEA to identify the fix. In this case, which altitude is designated for the fix?

A—MOCA.
B—MRA.
C—MCA.

MRA (Minimum Reception Altitude) is the lowest altitude required to receive adequate signals to determine specific fixes. Reception of signals from a radio facility located off the airway being flown may be inadequate at the designated MEA, in which case, an MRA is designated for the fix. (J33) — Pilot/Controller Glossary

Answer (A) is incorrect because MOCA (Minimum Obstruction Clearance Altitude) is the lowest altitude which meets obstacle clearance requirements for the entire route segment and which assures acceptable navigation coverage only within 22 NM of a VOR. Answer (C) is incorrect because MCA (Minimum Crossing Altitude) is the lowest altitude at a fix which an aircraft must cross when proceeding in the direction of a higher MEA.

ALL
4545. ATC may assign the MOCA when certain special conditions exist, and when within

A—22 NM of a VOR.
B—25 NM of a VOR.
C—30 NM of a VOR.

Minimum Obstruction Clearance Altitude (MOCA) is the lowest published altitude in effect between radio fixes on VOR airways, off-airway routes, or route segments which meets obstacle clearance requirements for the entire route segment and which assures acceptable navigational signal coverage only within 25 statute (22 nautical) miles of a VOR. (J33) — Pilot/Controller Glossary

Answers (B) and (C) are incorrect because ATC may assign the MOCA as an altitude only with 22 NM of a VOR.

Answers

4436 [C]   4437 [A]   4542 [C]   4544 [B]   4545 [A]

ALL
4547. Acceptable navigational signal coverage at the MOCA is assured for a distance from the VOR of only

A—12 NM.
B—22 NM.
C—25 NM.

Minimum Obstruction Clearance Altitude (MOCA) is the lowest published altitude in effect between radio fixes on VOR airways, off-airway routes, or route segments which meets obstacle clearance requirements for the entire route segment and which assures acceptable navigational signal coverage only within 25 statute (22 nautical) miles of a VOR. (J33) — Pilot/Controller Glossary

Answers (A) and (C) are incorrect because the MOCA is only acceptable when within 25 statute (22 nautical) miles of a VOR.

ALL
4540. What obstacle clearance and navigation signal coverage is a pilot assured with the Minimum Sector Altitudes depicted on the IAP charts?

A—1,000 feet and acceptable navigation signal coverage within a 25 NM radius of the navigation facility.
B—1,000 feet within a 25 NM radius of the navigation facility but not acceptable navigation signal coverage.
C—500 feet and acceptable navigation signal coverage within a 10 NM radius of the navigation facility.

Minimum Sector Altitudes provide at least 1,000 feet of clearance above the highest obstacle in the defined sector to a distance of 25 NM from the facility. Navigational course guidance is not assured at the MSA. (J18) — AIM ¶5-4-5

Answer (A) is incorrect because MSAs do not provide acceptable navigation signal coverage. Answer (C) is incorrect because MSAs provide at least 1,000 feet (not 500) of obstacle clearance, up to a distance of 25 (not 10) NM from the facility, and they do not provide acceptable navigation signal coverage.

Enroute Low Altitude Chart

The Enroute Low Altitude Charts are valid for use up to, but not including, 18,000 feet MSL. You should become familiar with FAA Legends 23, 24, and 25, which depict the legends for the charts. These will be available to you when you take the FAA Knowledge Test.

ALL
4546. Which aeronautical chart depicts Military Training Routes (MTR) above 1,500 feet?

A—IFR Planning Chart.
B—IFR Low Altitude En Route Chart.
C—IFR High Altitude En Route Chart.

Military Training Routes (MTRs) are charted on IFR Low Altitude Enroute charts, VFR Sectional charts, and Area Planning charts. (J10) — AIM ¶3-5-2

Answer (A) is incorrect because MTRs are charted on Area (not IFR) Planning charts. Answer (C) is incorrect because MTRs are charted on IFR Low (not High) Altitude Enroute charts.

ALL
4327. (Refer to Figures 59 and 60.) What are the operating hours (local standard time) of the Houston EFAS?

A—0600 to 2200.
B—0700 to 2300.
C—1800 to 1000.

Look at the bottom of FAA Figure 60 to find the Houston EFAS is operated by Montgomery Co. FSS. The hours of operations are given as 1200-0400Z. To convert this to local time, subtract 6 hours (see the top line of the A/FD: UTC–6). Therefore, the local hours of operation are 0600-2200Z. (J34) — A/FD

Answers

4547 [B]   4540 [B]   4546 [B]   4327 [A]

Chapter 7 **En Route**

ALL
4365. (Refer to Figures 76 and 77.) Which en route low altitude navigation chart would cover the proposed routing at the BOZEMAN VORTAC?

A—L-2.
B—L-7.
C—L-9.

The name and letter numbers in the top right corner of an airport/facility directory are the charts on which the airport and surrounding area are found. FAA Figure 76 indicates L-9 is the chart to cover the BOZEMAN area. (J34) — U.S. Terminal Procedures, DP Chart Legend

ALL
4339. (Refer to Figures 65 and 67.) What is the significance of the symbol at GRICE intersection?

A—It signifies a localizer-only approach is available at Harry P. Williams Memorial.
B—The localizer has an additional navigation function.
C—GRICE intersection also serves as the FAF for the ILS approach procedure to Harry P. Williams Memorial.

The localizer symbol at GRICE INT indicates that the localizer has a navigation function in addition to course guidance. See FAA Legend 23. (J35) — Enroute Low Altitude Chart Legend

ALL
4506. (Refer to Figure 89.) What is the ARTCC discrete frequency at the COP on V208 southwest bound from HVE to PGA VOR/DME?

A—122.1.
B—122.4.
C—133.6.

Discrete frequencies are normally designated for each control sector in enroute terminal facilities. Since the COP for southwest-bound traffic between Hanksville and Page on V208 is still in Salt Lake City ARTCC's boundary, the frequency should be the one contained in the Hanksville Remote Site scalloped box by the Hanksville VORTAC: 133.6. (J35) — Enroute Low Altitude Chart Legend

ALL
4645. (Refer to Figure 47.) En route on V112 from BTG VORTAC to LTJ VORTAC, the minimum altitude crossing Gymme intersection is

A—6,400 feet.
B—6,500 feet.
C—7,000 feet.

In the absence of a minimum crossing altitude, you would cross at the MEA. Looking at GYMME intersection, the minimum enroute altitude eastbound is 7,000 feet and westbound is 6,500 feet. Since the question asks the altitude while traveling west to east, 7,000 is the minimum altitude to be flown crossing GYMME. (J35) — Enroute Low Altitude Chart Legend

Answer (A) is incorrect because 6,400 is the MOCA along V112. Answer (B) is incorrect because 6,500 is the MEA for a westbound (not eastbound) flight on V112.

ALL
4646. (Refer to Figure 47.) When en route on V448 from YKM VORTAC to BTG VORTAC, what minimum navigation equipment is required to identify ANGOO intersection?

A—One VOR receiver.
B—One VOR receiver and DME.
C—Two VOR receivers.

Magnetic bearings are read directly from the VOR indicator. You can easily identify ANGOO intersection while on the radial of YKM with one VOR. Identify the intersecting radial of DLS simply by alternately tuning the single VOR to each station. (J35) — Enroute Low Altitude Chart Legend

ALL
4647. (Refer to Figure 47.) En route on V468 from BTG VORTAC to YKM VORTAC, the minimum altitude at TROTS intersection is

A—7,100 feet.
B—10,000 feet.
C—11,500 feet.

TROTS intersection has a flag with an "x" in it, which means that there is a Minimum Crossing Altitude (MCA). Under the intersection name the MCA is shown as 11,500 on V468 when proceeding to the northeast. (J35) — Enroute Low Altitude Chart Legend

Answer (A) is incorrect because 7,100 feet is the MOCA along V468. Answer (B) is incorrect because 10,000 feet is the MEA west of TROTS intersection.

Answers

4365 [C]   4339 [B]   4506 [C]   4645 [C]   4646 [A]   4647 [C]

Chapter 7 **En Route**

ALL
4077. Which types of airspace are depicted on the En Route Low Altitude Chart?

A—Class D, Class C, Class B, Class E and special use airspace.
B—Class A, special use airspace, Class D and Class E.
C—Special use airspace, Class E, Class D, Class A, Class B and Class C.

All Class B, C, D and E airspace, victor airways, and special use airspace, all up to 18,000 feet MSL, are shown on low altitude enroute charts. (J15) — AIM ¶9-1-4(c)

Answers (B) and (C) are incorrect because Class A airspace is not shown on low altitude enroute charts.

ALL
4287. (Refer to Figure 34.) For planning purposes, what is the highest useable altitude for an IFR flight on V573 from the HOT VORTAC to the TXK VORTAC?

A—16,000 feet MSL.
B—14,500 feet MSL.
C—13,999 feet MSL.

The Enroute Low Altitude U.S. IFR Chart is for use up to, but not including, 18,000 feet. In controlled airspace, each person operating an aircraft under IFR in level cruising flight shall maintain the altitude assigned by ATC or the regulatory IFR altitudes. Flying on V573 from HOT VORTAC to the TXK VORTAC is a westerly course which requires an even-thousands altitude. (J35) — Enroute Low Altitude Chart

Answer (B) is incorrect because 14,500 feet MSL is a VFR cruising altitude. Answer (C) is incorrect because IFR flight requires rounded-off odd or even thousands of feet (13,999 is neither).

ALL
4501. (Refer to Figure 89.) When flying from Milford Municipal to Bryce Canyon via V235 and V293, what minimum altitude should you be at when crossing Cedar City VOR?

A—11,400 feet.
B—12,000 feet.
C—13,000 feet.

The Minimum Crossing Altitude (MCA) is the lowest altitude at which an aircraft must cross when proceeding in the direction of a higher Minimum Enroute Altitude (MEA). The MCA southbound on V235 is 11,400 feet and 12,000 feet eastbound on V293. The 12,000 feet eastbound must be reached by the Cedar City VOR. (J35) — Enroute Low Altitude Chart Legend

ALL
4291. (Refer to Figure 40.) For planning purposes, what is the highest useable altitude for an IFR flight on V16 from BGS VORTAC to ABI VORTAC?

A—17,000 feet MSL.
B—18,000 feet MSL.
C—6,500 feet MSL.

The Enroute Low Altitude U.S. IFR Chart is for use up to, but not including, 18,000 feet. (J35) — Enroute Low Altitude Chart

Answer (B) is incorrect because the airways go up to 17,999 feet, not including 18,000 feet. Answer (C) is incorrect because the 6,500 feet is the MRA at LORAN intersection.

ALL
4317. (Refer to Figure 53.) Where is the VOR COP on V27 between the GVO and MQO VORTACs?

A—20 DME from GVO VORTAC.
B—20 DME from MQO VORTAC.
C—30 DME from SBA VORTAC.

The change-over point (COP) is located 20 NM from GVO or 34 NM from MQO. (J35) — Enroute Low Altitude Chart Legend

ALL
4263. (Refer to Figure 24.) At what point should a VOR changeover be made from JNC VOR to MANCA intersection southbound on V187?

A—36 NM south of JNC.
B—52 NM south of JNC.
C—74 NM south of JNC.

The changeover should be made at the changeover symbol if depicted, where there is a change in the direction of the airway, or in the absence of these, at the halfway point between the VORs. In this case, the changeover point (COP) is indicated with a symbol, and should be made 52 miles from JNC and 90 miles from the next VOR. (J35) — Enroute Low Altitude Chart Legend

Answer (A) is incorrect because 36 NM south of JNC is just south of HERRM intersection. Answer (C) is incorrect because 74 NM south of JNC is a little over halfway between the VORs, which is incorrect since there is a specifically designated changeover point.

Answers

4077 [A]	4287 [A]	4501 [B]	4291 [A]	4317 [A]	4263 [B]

Chapter 7 **En Route**

ALL
4318. (Refer to Figure 53.) What service is indicated by the inverse "H" symbol in the radio aids to navigation box for PRB VORTAC?

A—VOR with TACAN compatible DME.
B—Availability of HIWAS.
C—En Route Flight Advisory Service available.

The black circle with a reversed-out "H" in the upper-right corner of the PRB VORTAC identifier box indicates availability of Hazardous Inflight Weather Advisory Service (HIWAS). (J35) — Enroute Low Altitude Chart Legend

Answer (A) is incorrect because the VORTAC symbol indicates TACAN with DME. Answer (C) is incorrect because 122.0 is the nationwide frequency for Enroute Flight Advisory Service.

ALL
4348. (Refer to Figures 70 and 71.) Which VORTAC along the proposed route of flight could provide HIWAS information?

A—SPARTA VORTAC.
B—HUGUENOT VORTAC.
C—KINGSTON VORTAC.

Kingston VORTAC has a black circle with a reversed-out "H" in the upper-right corner of the information box. This indicates HIWAS availability. (J35) — Enroute Low Altitude Chart Legend

Answers (A) and (B) are incorrect because the Sparta and Huguenot VORTACs do not have the solid square in the upper-right corner of their information boxes.

ALL
4336. (Refer to Figure 65.) Which point would be the appropriate VOR COP on V552 from the LFT to the TBD VORTACs?

A—CLYNT intersection.
B—HATCH intersection.
C—34 DME from the LFT VORTAC.

The total distance between LFT and TBD is 68 NM (shown in the box between NAVAIDs). The changeover point (COP) is half of 68, or 34 NM from LFT. VOR changeover points are where the pilot should change navigation receiver frequency from the station behind the aircraft to the station ahead and are:

1. *Halfway between the VORs, if no other designation is given;*
2. *At a course change; or*
3. *At the COP symbol.*

(J35) — Enroute Low Altitude Chart Legend

Answers (A) and (B) are incorrect because they are not the halfway points, and there is no symbol designating these locations as the COP.

ALL
4516. (Refer to Figure 91.) Where should you change VOR frequencies when en route from DBS VORTAC to JAC VOR/DME on V520?

A—35 NM from DBS VORTAC.
B—60 NM from DBS VORTAC.
C—60 NM from JAC VOR/DME.

The changeover point (COP) symbol indicates the point where the pilot should change navigation receiver frequency from the station behind the aircraft to the station ahead. The distance from each VORTAC can be read above and below the symbol, pointing to the VORTAC which it is referring to. The COP from DBS to JAC on V520 is 60 NM from DBS VORTAC. (J35) — Enroute Low Altitude Chart Legend

ALL
4366. (Refer to Figure 78.) What is the maximum altitude that you may flight plan an IFR flight on V-86 EASTBOUND between BOZEMAN and BILLINGS VORTACs?

A—14,500 feet MSL.
B—17,000 feet MSL.
C—18,000 feet MSL.

The Enroute Low Altitude U.S. IFR Chart is for use up to, but not including, 18,000 feet. (J35) — Enroute Low Altitude Chart Legend

Answer (A) is incorrect because 14,500 feet is a VFR (not IFR) cruising altitude. Answer (C) is incorrect because Victor airways extend up to, but not including FL180.

Answers

4318 [B]   4348 [C]   4336 [C]   4516 [B]   4366 [B]

ALL
4370-1. (Refer to Figure 78.) What is the minimum crossing altitude over the BOZEMAN VORTAC for a flight southeast bound on V86?

A— 8,500 feet MSL.
B— 9,300 feet MSL.
C— 9,700 feet MSL.

Bozeman (BZN) has a flag with an X in it, which indicates it has a minimum crossing altitude (MCA). Above the Bozeman communications box is the list of MCAs. For a flight southeast bound on V86, the MCA is 9,300 feet MSL. (J35) — Enroute Low Altitude Chart Legend

Answer (A) is incorrect because 8,500 feet MSL is the MEA on V86 prior to BZN. Answer (C) is incorrect because 9,700 feet MSL is the MEA on V365.

ALL
4370-2. (Refer to Figure 78.) When eastbound on V86 between Whitehall and Livingston, the minimum altitude that you should cross BZN is

A— 9,300 feet.
B— 10,400 feet.
C— 8,500 feet.

Bozeman (BAN) has a flag with an X in it, indicating a minimum crossing altitude (MCA). Above the Bozeman communications box is the list of MCAs. For a flight eastbound on V86, the MCA is 9,300 feet MSL. (H842) — FAA-H-8083-15, Chapter 8

ALL
4280. (Refer to Figure 34.) At which altitude and location on V573 would you expect the navigational signal of the HOT VOR/DME to be unreliable?

A— 3,000 feet at APINE intersection.
B— 2,600 feet at MARKI intersection.
C— 4,000 feet at ELMMO intersection.

The navigational signal at APINE intersection would be unreliable below 3,500 feet because the MEA and MOCA are 3,500 and 2,500 feet, respectively. Since APINE is further than 22 NM from the HOT VOR, navigational signal is only assured with the MEA. (J35) — Enroute Low Altitude Chart Legend

Answer (B) is incorrect because MARKI should have a reliable VOR signal down to the MOCA (which is 2,500 feet), since it is within 22 NM from HOT VOR. Answer (C) is incorrect because the signal at ELMMO would be reliable at the MEA, which is 3,500 feet.

ALL
4493. (Refer to Figure 87.) Where is the VOR COP when flying east on V306 from Daisetta to Lake Charles?

A— 50 NM east of DAS.
B— 40 NM east of DAS.
C— 30 NM east of DAS.

VOR changeover points (COP) are the points at which the pilot should change navigation receiver frequency from the station behind the aircraft to the station ahead and are:

1. *halfway between stations, if no other designation is given;*
2. *at a course change in the route;*
3. *at the COP symbol.*

In this case, the COP symbol indicates the change should be made 30 NM from Daisetta (DAS). (J35) — AIM ¶5-3-6(b)

ALL
4496. (Refer to Figure 87.) What is indicated by the localizer course symbol at Jefferson County Airport?

A— A published LDA localizer course.
B— A published SDF localizer course.
C— A published ILS localizer course, which has an additional navigation function.

The localizer course symbol at Jefferson County Field indicates an ILC Localizer Course with ATC function is available. (J35) — Enroute Low Altitude Chart Legend

Answer (A) is incorrect because although it could be used for an LDA approach, this symbol is not representative of a strictly LDA approach. Answer (B) is incorrect because SDF sites do not have back courses, and this symbol is accompanied by the words "back course."

ALL
4497. (Refer to Figure 87.) Which VHF frequencies, other than 121.5, can be used to receive De Ridder FSS in the Lake Charles area?

A— 122.1, 126.4.
B— 123.6, 122.65.
C— 122.2, 122.3.

Frequencies 122.2 and 121.5 are normally available at all FSS's. In addition, 122.3 is listed on the top of the Lake Charles VOR identifier box, which indicates there is a transceiver at that site remoted to the DE RIDDER flight service station. (J35) — Enroute Low Altitude Chart Legend.

Answers

| 4370-1 | [B] | 4370-2 | [A] | 4280 | [A] | 4493 | [C] | 4496 | [C] | 4497 | [C] |

ALL
4498. (Refer to Figure 87.) Why is the localizer back course at Jefferson County Airport depicted?

A—The back course is not aligned with a runway.
B—The back course has a glide slope.
C—The back course has an additional navigation function.

An ILS Localizer Course at Jefferson County Airport with an additional navigation function is available. The feathered side indicates Blue Sector. The "back course" indicates that it is a back course ILS Localizer Course with additional navigation function. (J35) — Enroute Low Altitude Chart Legend

ALL
4499. (Refer to Figure 87.) Where is the VOR changeover point on V20 between Beaumont and Hobby?

A—Halfway point.
B—MOCKS intersection.
C—Anahuac Beacon.

The changeover point (COP) is located midway between the navigation facilities for straight route segments. In the case of dogleg route segments, the COP is located at the intersection of radials or courses forming the dogleg. When the COP is not located at the midway point, the location will be identified with a COP symbol, giving the mileage to the radio aids.
There is no COP symbol on this route segment; therefore, the COP is at the halfway point. (J35) — Enroute Low Altitude Chart Legend

ALL
4502. (Refer to Figure 89.) What VHF frequencies are available for communications with Cedar City FSS?

A—123.6, 121.5, 108.6, and 112.8.
B—122.2, 121.5, 122.6, and 112.1.
C—122.2, 121.5, 122.0, and 123.6.

There is an FSS located on the field at Cedar City (area D) that is in operation part-time. The frequencies available at Cedar City are 122.6 (published), 121.5 and 122.2 (the box is bold). The pilot can also use 122.1 to transmit to the FSS, and they will respond through the MLF VORTAC on 112.1 (area C). (J35) — Enroute Low Altitude Chart Legend

Answers (A) and (C) are incorrect because 123.6 and 122.0 are not listed above any communication boxes controlled by Cedar City FSS, and 112.8 (BCE VORTAC) is underlined, which indicates there is no voice on that frequency.

ALL
4261. (Refer to Figure 24.) Proceeding southbound on V187, (vicinity of Cortez VOR) contact is lost with Denver Center. You should attempt to reestablish contact with Denver Center on:

A—133.425 MHz.
B—122.1 MHz and receive on 108.4 MHz.
C—122.35 MHz.

Air Route Traffic Control Center names and frequencies are shown on low altitude charts enclosed in a box of scalloped lines. Just northwest of the Cortez VOR symbol is the box showing Denver Center available at Cortez, using frequency 133.425. (J17) — Enroute Low Altitude Chart Legend

ALL
4517. (Refer to Figure 91.) What is the minimum crossing altitude at SABAT intersection when eastbound from DBS VORTAC on V298?

A—8,300 feet.
B—11,100 feet.
C—13,000 feet.

SABAT Intersection has a flag with an "X" in it, indicating a Minimum Crossing Altitude (MCA). The MCA at SABAT when eastbound from DBS VORTAC on V298 is 11,100 feet. (J35) — Enroute Low Altitude Chart Legend

ALL
4509. (Refer to Figure 91.) What is the minimum crossing altitude at DBS VORTAC for a northbound IFR flight on V257?

A—7,500 feet.
B—8,600 feet.
C—11,100 feet.

DBS VORTAC can be found in the bottom left-hand quadrant of the FAA figure. The minimum crossing altitude (MCA) is identified in the middle of the DBS VORTAC compass rose, beneath the flag with the "X" inside. The minimum crossing altitude at DBS VORTAC for an IFR flight northbound on V257 is 8,600 feet. (J35) — Enroute Low Altitude Chart Legend

Answer (A) is incorrect because 7,500 feet is the MEA on V257. Answer (C) is incorrect because 11,100 feet is the MOCA on V210257.

Answers

| 4498 | [C] | 4499 | [A] | 4502 | [B] | 4261 | [A] | 4517 | [B] | 4509 | [B] |

ALL
4512. (Refer to Figure 91.) What lighting is indicated on the chart for Jackson Hole Airport?

A—Lights on prior request.
B—No lighting available.
C—Pilot controlled lighting.

The Jackson Hole Airport is located in the bottom right quadrant of FAA Figure 91. The "L" in the circle indicates pilot controlled lighting. (J03) — Enroute Low Altitude Chart Legend

ALL
4325. (Refer to Figure 58.) On which frequencies could you communicate with the Montgomery County FSS while on the ground at College Station?

A—122.65, 122.2, 122.1, 113.3.
B—122.65, 122.2.
C—118.5, 122.65, 122.2.

Look at the Communications section of the excerpt from the A/FD. College Station RCO gives the frequencies 122.65 and 122.2. (J34) — A/FD Legend

ALL
4515. (Refer to Figure 91.) What is the function of the Great Falls RCO (Yellowstone vicinity)?

A—Long range communications outlet for Great Falls Center.
B—Remote communications outlet for Great Falls FSS.
C—Satellite remote controlled by Salt Lake Center with limited service.

A Remote Communications Outlet (RCO) is an unmanned communications facilities remotely controlled by air traffic personnel. RCOs serve FSS's. An RCO may be UHF or VHF and will extend the communications range of the air traffic facility. (J35) — Enroute Low Altitude Chart Legend

ALL
4262. (Refer to Figures 22 and 24.) For planning purposes, what would be the highest MEA on V187 between Grand Junction, Walker Airport, and Durango, La Plata Co. Airport?

A—12,000 feet.
B—15,000 feet.
C—16,000 feet.

The highest published MEA along the route is 15,000 feet MSL, on V187 between HERRM and MANCA intersections. (J35) — Enroute Low Altitude Chart Legend

Answer (A) is incorrect because 12,000 feet is the MEA between JNC and HERRM intersection, but it is not the highest MEA on the route. Answer (C) is incorrect because 16,000 feet is not an MEA along any part of this route.

ALL
4264. (Refer to Figure 24.) What is the MOCA between JNC and MANCA intersection on V187?

A—10,900 feet MSL.
B—12,000 feet MSL.
C—13,700 feet MSL.

*If an altitude is written under the MEA with an asterisk [*], this altitude represents the MOCA (minimum obstacle clearance altitude). In this case, it is 13,700 feet MSL is the MOCA on V187. (J35) — Enroute Low Altitude Chart Legend*

ALL
4639. For IFR operations off of established airways below 18,000 feet, VOR navigational aids used to describe the "route of flight" should be

A—40 NM apart.
B—70 NM apart.
C—80 NM apart.

For operation off established airways below 18,000 feet MSL, pilots should use aids not more than 80 NM apart. These aids are depicted on Enroute Low Altitude Charts. (J15) — AIM ¶5-1-7(c)(3)(c)

Answers

| 4512 | [C] | 4325 | [B] | 4515 | [B] | 4262 | [B] | 4264 | [C] | 4639 | [C] |

Holding

Holding may be necessary when ATC is unable to clear a flight to its destination. VORs, non-directional beacons, airway intersections, and DME fixes may all be used as holding points. Flying a holding pattern involves two turns and two straight-and-level legs as shown in Figure 7-1.

At and below 14,000 feet MSL (no wind), the aircraft flies the specified course inbound to the fix, turns to the right 180°, flies a parallel course outbound for 1 minute, again turns 180° to the right, and flies 1 minute inbound to the fix. Above 14,000 feet MSL, the inbound leg length is 1-1/2 minutes. If a nonstandard pattern is to be flown, ATC will specify left turns.

When 3 minutes or less from the holding fix, the pilot is expected to start a speed reduction so as to cross the fix at or below the maximum holding airspeed. For all aircraft between MHA (minimum holding altitude) and 6,000 feet MSL, holding speed is 200 KIAS. For all aircraft between 6,001 and 14,000 feet MSL, holding speed is 230 KIAS. For all aircraft 14,000 feet MSL and above, holding speed is 265 KIAS. Exceptions to these speeds will be indicated by an icon.

The aircraft is holding as of the initial time of arrival over the fix, and that time should be reported to ATC. The initial outbound leg is flown for 1 minute at or below 14,000 feet MSL. Subsequently, timing of the outbound leg should be adjusted as necessary to arrive at the proper inbound leg length. Timing of the outbound leg begins over or abeam the fix, whichever occurs later. If the abeam position cannot be determined, start timing when the turn to outbound is completed. The same entry and holding procedures apply to DME holding, except distances in nautical miles are used to establish leg length.

The FAA has three recommended methods for entering a holding pattern, as shown in Figure 7-2. An aircraft approaching from within sector (A) would fly a parallel entry by turning left to parallel the outbound course, making another left turn to remain in protected airspace, and returning to the holding fix. Aircraft approaching from sector (B) would fly a teardrop entry, by flying outbound on a track of 30° or less to the holding course, and then making a right turn to intercept the holding course inbound to the fix. Those approaching from within sector (C) would fly a direct entry by turning right to fly the pattern.

If the holding pattern is charted, the controller may omit all holding instructions, except the holding direction and the statement "as published." Pilots are expected to hold in the pattern depicted even if it means crossing the clearance limit. If the holding pattern to be used is not depicted on charts, ATC will issue general holding instructions. The holding clearance will include the following information: direction of holding from the fix in terms of the eight cardinal compass points; holding fix; radial, course, bearing, airway, or route on which the aircraft is to hold; leg length in miles if DME or RNAV is to be used; direction of turn if left turns are to be made; time to expect further clearance and any pertinent additional delay information.

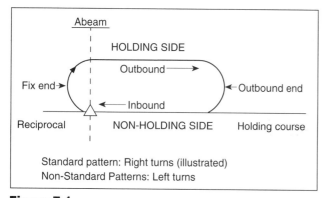

Figure 7-1

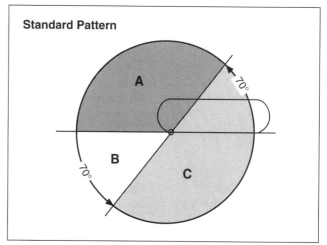

Figure 7-2

Chapter 7 **En Route**

ALL
4494. (Refer to Figure 87.) At STRUT intersection headed eastbound, ATC instructs you to hold west on the 10 DME fix west of LCH on V306, standard turns, what entry procedure is recommended?

A—Direct.
B—Teardrop.
C—Parallel.

Determine the holding pattern by placing your pencil on the holding fix and dragging it on the holding radial given by ATC, then returning back to the fix. Then draw the pattern from the fix with turns in the direction specified. In this question, ATC has specified standard turns (right turns).
   The entry procedure is based on the aircraft's heading. To determine which entry procedure to use, draw a line at a 70 angle from the holding fix, and cutting the outbound leg at about one-third its length. With the aircraft at STRUT intersection heading eastbound, we are in the largest piece of the pie, so a direct entry would be used. See the figure below. (J17) — AIM ¶5-3-7

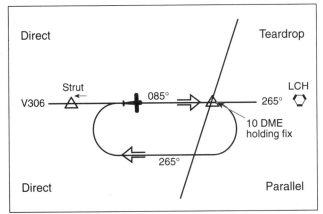

Question 4494

ALL
4609. (Refer to Figure 112.) You arrive at the 15 DME fix on a heading of 350°. Which holding pattern correctly complies with the ATC clearance below, and what is the recommended entry procedure?

"...HOLD WEST OF THE ONE FIVE DME FIX ON THE ZERO EIGHT SIX RADIAL OF THE ABC VORTAC, FIVE MILE LEGS, LEFT TURNS..."

A—1; teardrop entry.
B—1; direct entry.
C—2; direct entry.

Determine the holding pattern by placing your pencil on the holding fix and dragging it on the holding radial given by ATC, then returning back to the fix. Then draw the pattern from the fix with turns in the direction specified. In this question, ATC has specified left-hand turns, which corresponds to holding pattern 1.
   The entry procedure is based on the aircraft's heading. To determine which entry procedure to use, draw a line at a 70° angle from the holding fix, and cutting the outbound leg at about one-third its length. With a heading of 350°, we are in the largest piece of the pie, so a direct entry would be used. See the figure below. (J17) — AIM ¶5-3-7

Answer (A) is incorrect because a heading of 350° requires a direct entry. Answer (C) is incorrect because the second holding pattern does not have the holding fix at the end of the inbound leg.

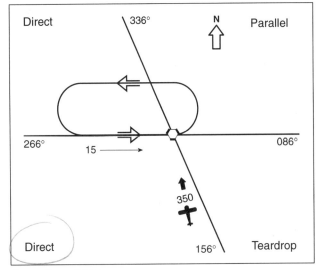

Question 4609

Answers

4494   [A]   4609   [B]

ALL
4610. (Refer to Figure 113.) You receive this ATC clearance:

"...HOLD EAST OF THE ABC VORTAC ON THE ZERO NINER ZERO RADIAL, LEFT TURNS..."

What is the recommended procedure to enter the holding pattern?

A—Parallel only.
B—Direct only.
C—Teardrop only.

Determine the holding pattern by placing your pencil on the holding fix and dragging it on the holding radial given by ATC, then returning back to the fix. Then draw the pattern from the fix with turns in the direction specified. Holding east on the 090° radial with left turns means you will be south of R-090.

The entry procedure is based on the aircraft's heading. To determine which entry procedure to use, draw a line at a 70° angle from the holding fix, and cutting the outbound leg at about one-third its length. With a heading of 055°, we are in the second-largest piece of the pie, so a parallel entry would be used. See the figure below. (J17) — AIM ¶5-3-7

Answer (B) is incorrect because a direct entry would only be appropriate between the 340° and 160° radials. Answer (C) is incorrect because a teardrop entry would only be appropriate between the 270° and 340° radials.

ALL
4611. (Refer to Figure 113.) You receive this ATC clearance:

"...CLEARED TO THE ABC VORTAC. HOLD SOUTH ON THE ONE EIGHT ZERO RADIAL..."

What is the recommended procedure to enter the holding pattern?

A—Teardrop only.
B—Direct only.
C—Parallel only.

Determine the holding pattern by placing your pencil on the holding fix and dragging it on the holding radial given by ATC, then returning back to the fix. Then draw the pattern from the fix with turns in the direction specified. Holding south on the 180° radial with right turns means you will be east of R-180.

The entry procedure is based on the aircraft's heading. To determine which entry procedure to use, draw a line at a 70° angle from the holding fix, and cutting the outbound leg at about one-third its length. With a heading of 055°, we are in the largest piece of the pie, so a direct entry would be used. See the figure below. (J17) — AIM ¶5-3-7

Answer (A) is incorrect because a teardrop entry would only be appropriate between the 290° and 360° radials. Answer (C) is incorrect because a parallel entry would only be appropriate between the 360° and 110° radials.

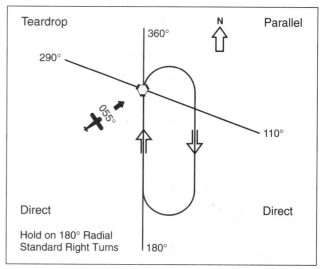

Question 4610

Question 4611

Answers

4610 [A]   4611 [B]

ALL
4612. (Refer to Figure 113.) You receive this ATC clearance:

"...CLEARED TO THE XYZ VORTAC. HOLD NORTH ON THE THREE SIX ZERO RADIAL, LEFT TURNS..."

What is the recommended procedure to enter the holding pattern.

A—Parallel only.
B—Direct only.
C—Teardrop only.

Determine the holding pattern by placing your pencil on the holding fix and dragging it on the holding radial given by ATC, then returning back to the fix. Then draw the pattern from the fix with turns in the direction specified. Holding north on the 360° radial with left turns means you will be east of R-360.
  The entry procedure is based on the aircraft's heading. To determine which entry procedure to use, draw a line at a 70° angle from the holding fix, and cutting the outbound leg at about one-third its length. With a heading of 055°, we are in the smallest piece of the pie, so a teardrop entry would be used. See the figure below. (J17) — AIM ¶5-3-7

Answer (A) is incorrect because a parallel entry would only be appropriate between the 070° and 180° radials. Answer (B) is incorrect because a direct entry would only be appropriate between the 070° and 250° radials.

ALL
4613. (Refer to Figure 113.) You receive this ATC clearance:

"...CLEARED TO THE ABC VORTAC. HOLD WEST ON THE TWO SEVEN ZERO RADIAL..."

What is the recommended procedure to enter the holding pattern?

A—Parallel only.
B—Direct only.
C—Teardrop only.

Determine the holding pattern by placing your pencil on the holding fix and dragging it on the holding radial given by ATC, then returning back to the fix. Then draw the pattern from the fix with turns in the direction specified. Holding west on the 270° radial with right turns means you will be south of R-270.
  The entry procedure is based on the aircraft's heading. To determine which entry procedure to use, draw a line at a 70° angle from the holding fix, and cutting the outbound leg at about one-third its length. With a heading of 055°, we are in the largest piece of the pie, so a direct entry would be used. See the figure below. (J17) — AIM ¶5-3-7

Answer (A) is incorrect because a parallel entry would only be appropriate between the 090° and 200° radials. Answer (C) is incorrect because a teardrop entry would only be appropriate between the 020° and 090° radials.

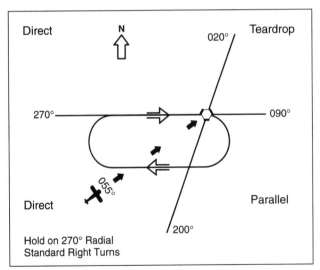

Question 4612

Question 4613

Answers
4612 [C]   4613 [B]

ALL
4614. (Refer to Figure 114.) A pilot receives this ATC clearance:

"...CLEARED TO THE ABC VORTAC. HOLD WEST ON THE TWO SEVEN ZERO RADIAL..."

What is the recommended procedure to enter the holding pattern?

A—Parallel or teardrop.
B—Parallel only.
C—Direct only.

Determine the holding pattern by placing your pencil on the holding fix and dragging it on the holding radial given by ATC, then returning back to the fix. Then draw the pattern from the fix with turns in the direction specified. Holding west on the 270° radial with right turns means you will be south of R-270.

The entry procedure is based on the aircraft's heading. To determine which entry procedure to use, draw a line at a 70° angle from the holding fix, and cutting the outbound leg at about one-third its length. With a heading of 055°, we are in the largest piece of the pie, so a direct entry would be used. See the figure below. (J17) — AIM ¶5-3-7

Answer (A) is incorrect because either the parallel or teardrop entry is only appropriate when on the 090° radial. Answer (B) is incorrect because a parallel entry would only be appropriate between the 090° and 200° radials.

AIR
4615. (Refer to Figure 114.) A pilot receives this ATC clearance:

"...CLEARED TO THE XYZ VORTAC. HOLD NORTH ON THE THREE SIX ZERO RADIAL, LEFT TURNS..."

What is the recommended procedure to enter the holding pattern?

A—Teardrop only.
B—Parallel only.
C—Direct only.

Determine the holding pattern by placing your pencil on the holding fix and dragging it on the holding radial given by ATC, then returning back to the fix. Then draw the pattern from the fix with turns in the direction specified. Holding north on the 360° radial with left turn means you will be east of R-360.

The entry procedure is based on the aircraft's heading. To determine which entry procedure to use, draw a line at a 70° angle from the holding fix, and cutting the outbound leg at about one-third its length. With a heading of 055°, we are in the largest piece of the pie, so a direct entry would be used. See the figure below. (J17) — AIM ¶5-3-7

Answer (A) is incorrect because a teardrop entry would only be appropriate between the 180° and 250° radials. Answer (B) is incorrect because a parallel entry would only be appropriate between the 360° and 250° radials.

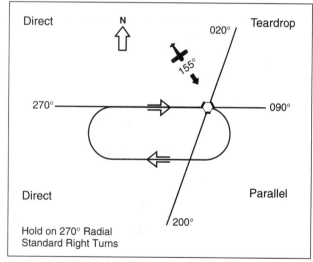

Question 4614

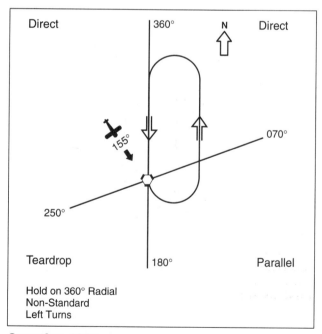

Question 4615

Answers

4614 [C]   4615 [C]

ALL
4616. (Refer to Figure 114.) A pilot receives this ATC clearance:

"...CLEARED TO THE ABC VORTAC. HOLD SOUTH ON THE ONE EIGHT ZERO RADIAL..."

What is the recommended procedure to enter the holding pattern?

A—Teardrop only.
B—Parallel only.
C—Direct only.

Determine the holding pattern by placing your pencil on the holding fix and dragging it on the holding radial given by ATC, then returning back to the fix. Then draw the pattern from the fix with turns in the direction specified. Holding south on the 180° radial with right turn means you will be east of R-180.

The entry procedure is based on the aircraft's heading. To determine which entry procedure to use, draw a line at a 70° angle from the holding fix, and cutting the outbound leg at about one-third its length. With a heading of 055°, we are in the smallest piece of the pie, so a teardrop entry would be used. See the figure below. (J17) — AIM ¶5-3-7

Answer (B) is incorrect because a parallel entry would only be appropriate between the 360° and 110° radials. Answer (C) is incorrect because a direct entry would only be appropriate between the 110° and 290° radials.

ALL
4618. (Refer to Figure 115.) You receive this ATC clearance:

"...HOLD WEST OF THE ONE FIVE DME FIX ON THE ZERO EIGHT SIX RADIAL OF ABC VORTAC, FIVE MILE LEGS, LEFT TURNS..."

You arrive at the 15 DME fix on a heading of 350°. Which holding pattern correctly complies with these instructions, and what is the recommended entry procedure?

A—1; teardrop.
B—2; direct.
C—1; direct.

Determine the holding pattern by placing your pencil on the holding fix and dragging it on the holding radial given by ATC, then returning back to the fix. Then draw the pattern from the fix with turns in the direction specified. In this question, ATC has specified left-hand turns, which corresponds to holding pattern 1.

The entry procedure is based on the aircraft's heading. To determine which entry procedure to use, draw a line at a 70° angle from the holding fix, and cutting the outbound leg at about one-third its length. With a heading of 350°, we are in the largest piece of the pie, so a direct entry would be used. See the figure below. (J17) — AIM ¶5-3-7

Answer (A) is incorrect because a heading of 350° requires a direct entry. Answer (B) is incorrect because the second holding pattern does not have the holding fix at the end of the inbound leg.

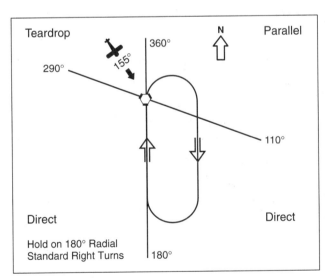

Question 4616

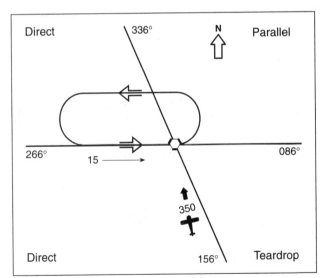

Question 4618

Answers

4616 [A]   4618 [C]

ALL
4619. (Refer to Figure 116.) You arrive over the 15 DME fix on a heading of 350°. Which holding pattern correctly complies with the ATC clearance below, and what is the recommended entry procedure?

"...HOLD WEST OF THE ONE FIVE DME FIX ON THE TWO SIX EIGHT RADIAL OF THE ABC VORTAC, FIVE MILE LEGS, LEFT TURNS..."

A—1; teardrop entry.
B—2; direct entry.
C—1; direct entry.

Determine the holding pattern by placing your pencil on the holding fix and dragging it on the holding radial given by ATC, then returning back to the fix. Then draw the pattern from the fix with turns in the direction specified. In this question, ATC has specified left-hand turns, which corresponds to holding pattern 2.

The entry procedure is based on the aircraft's heading. To determine which entry procedure to use, draw a line at a 70° angle from the holding fix, and cutting the outbound leg at about one-third its length. With a heading of 350°, we are in the largest piece of the pie, so a direct entry would be used. See the figure below. (J17) — AIM ¶5-3-7

Answer (A) is incorrect because a heading of 350° requires a direct entry, and the first holding pattern does not have the holding fix at the end of the inbound leg. Answer (C) is incorrect because the first holding pattern does not have the holding fix at the end of the inbound leg.

ALL
4621. (Refer to Figure 117.) You receive this ATC clearance:

"...CLEARED TO THE ABC NDB. HOLD SOUTHEAST ON THE ONE FOUR ZERO DEGREE BEARING FROM THE NDB. LEFT TURNS..."

At station passage you note the indications in Figure 117. What is the recommended procedure to enter the holding pattern?

A—Direct only.
B—Teardrop only.
C—Parallel only

Determine the holding pattern by placing your pencil on the holding fix and dragging it on the holding radial given by ATC, then returning back to the fix. Then draw the pattern from the fix with turns in the direction specified. In this question, ATC has specified left-hand turns.

The entry procedure is based on the aircraft's heading. To determine which entry procedure to use, draw a line at a 70° angle from the holding fix, and cutting the outbound leg at about one-third its length. With a heading of 055°, we are in the second-largest piece of the pie, so a parallel entry would be used. See the figure below. (J17) — AIM ¶5-3-7

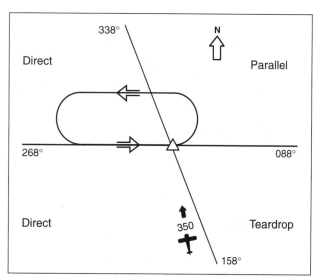

Question 4619

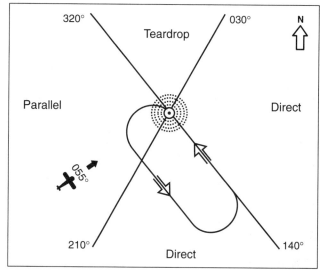

Question 4621

Answers

4619 [B]   4621 [C]

ALL
4622. (Refer to Figure 117.) You receive this ATC clearance:

"...CLEARED TO THE XYZ NDB. HOLD NORTHEAST ON THE ZERO FOUR ZERO DEGREE BEARING FROM THE NDB. LEFT TURNS..."

At station passage you note the indications in Figure 117. What is the recommended procedure to enter the holding pattern?

A—Direct only.
B—Teardrop only.
C—Parallel only.

Determine the holding pattern by placing your pencil on the holding fix and dragging it on the holding radial given by ATC, then returning back to the fix. Then draw the pattern from the fix with turns in the direction specified. In this question, ATC has specified left-hand turns.

The entry procedure is based on the aircraft's heading. To determine which entry procedure to use, draw a line at a 70° angle from the holding fix, and cutting the outbound leg at about one-third its length. With a heading of 055°, we are in the smallest piece of the pie, so a teardrop entry would be used. See the figure below. (J17) — AIM ¶5-3-7

ALL
4623. (Refer to Figure 117.) You receive this ATC clearance:

"...CLEARED TO THE ABC NDB. HOLD SOUTHWEST ON THE TWO THREE ZERO DEGREE BEARING FROM THE NDB..."

At station passage you note the indications in Figure 117. What is the recommended procedure to enter the holding pattern?

A—Direct only.
B—Teardrop only.
C—Parallel only.

Determine the holding pattern by placing your pencil on the holding fix and dragging it on the holding radial given by ATC, then returning back to the fix. Then draw the pattern from the fix with turns in the direction specified. In this question, ATC has not specified which way to turn, so the standard direction is right turns.

The entry procedure is based on the aircraft's heading. To determine which entry procedure to use, draw a line at a 70° angle from the holding fix, and cutting the outbound leg at about one-third its length. With a heading of 055°, we are in the largest piece of the pie, so a direct entry would be used. See the figure below. (J17) — AIM ¶5-3-7

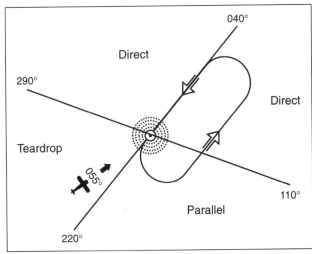

Question 4622

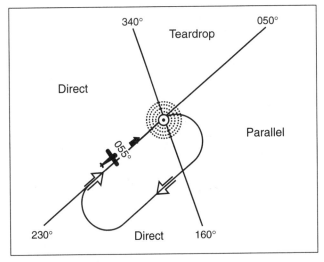

Question 4623

Answers

4622 [B]   4623 [A]

Chapter 7 En Route

ALL
4624. What timing procedure should be used when performing a holding pattern at a VOR?

A—Timing for the outbound leg begins over or abeam the VOR, whichever occurs later.
B—Timing for the inbound leg begins when initiating the turn inbound.
C—Adjustments in timing of each pattern should be made on the inbound leg.

Outbound leg timing begins over/abeam the fix, whichever occurs later. (J17) — AIM ¶5-3-7

Answers (B) and (C) are incorrect because timing procedures coordinate with the outbound (not inbound) leg.

ALL
4620. At what point should the timing begin for the first leg outbound in a nonstandard holding pattern?

A—Abeam the holding fix, or wings level, whichever occurs last.
B—When the wings are level at the completion of the 180° turn outbound.
C—When over or abeam the holding fix, whichever occurs later.

Outbound leg timing begins over/abeam the fix, whichever occurs later. If the abeam position cannot be determined, start timing when the turn to the outbound is completed. (J17) — AIM ¶5-3-7

Answer (A) is incorrect because timing begins on the outbound leg when the wings are level only if the abeam point cannot be determined. Answer (B) is incorrect because you may roll wings level before abeam the fix.

ALL
4625. When holding at an NDB, at what point should the timing begin for the second leg outbound?

A—When the wings are level and the wind drift correction angle is established after completing the turn to the outbound heading.
B—When the wings are level after completing the turn to the outbound heading, or abeam the fix, whichever occurs first.
C—When abeam the holding fix.

Outbound leg timing begins over/abeam the fix, whichever occurs later. (J17) — AIM ¶5-3-7

Answer (A) is incorrect because the timing should start when the turn is complete only when the position abeam the fix cannot be determined. Answer (B) is incorrect because timing of the outbound leg begins when abeam the fix (not wings level).

ALL
4617. To ensure proper airspace protection while in a holding pattern, what is the recommended maximum airspeed above 14,000 feet for civil turbojet aircraft?

A—230 knots.
B—265 knots.
C—200 knots.

The maximum airspeed above 14,000 feet is 265 KIAS. (J17) — AIM ¶5-3-7

ALL
4626. To ensure proper airspace protection while holding at 5,000 feet in a civil aircraft, what is the recommended maximum indicated airspeed a pilot should use?

A—230 knots.
B—200 knots.
C—210 knots.

The maximum holding airspeed between MHA and 6,000 feet MSL is 200 KIAS. (J17) — AIM ¶5-3-7

ALL
4766. To ensure proper airspace protection while in a holding pattern, what is the recommended maximum indicated airspeed above 14,000 feet?

A—220 knots.
B—265 knots.
C—200 knots.

While in a holding pattern, the maximum indicated airspeed above 14,000 feet is 265 KIAS. (J17) — AIM ¶5-3-7

Answers

| 4624 [A] | 4620 [C] | 4625 [C] | 4617 [B] | 4626 [B] | 4766 [B] |

Chapter 7 En Route

ALL
4668. When more than one circuit of the holding pattern is needed to lose altitude or become better established on course, the additional circuits can be made

A—at pilot's discretion.
B—only in an emergency.
C—only if pilot advises ATC and ATC approves.

If timed approaches are in use, aircraft are expected to proceed inbound on final approach when clearance is received. Pilots are responsible for advising ATC if they elect to make additional circuits to lose altitude or become established on course. There is no AIM language regarding ATC approval, but it is implied. (J18) — AIM ¶5-4-9

Answer (A) is incorrect because although additional circuits may be made at your discretion, you must advise ATC. Answer (B) is incorrect because it is not necessary to declare an emergency.

ALL
4681. (Refer to Figure 129.) What type of entry is recommended to the missed approach holding pattern if the inbound heading is 050°?

A—Direct.
B—Parallel.
C—Teardrop.

The missed approach procedure is to fly direct to Bendy WPT. The holding pattern depicted is left-hand turns, so the pilot should execute a teardrop turn to enter the holding pattern. (J17) — AIM ¶5-3-7

Answer (A) is incorrect because a direct entry would be used if the pilot were coming from the north. Answer (B) is incorrect because a parallel entry would be used if the pilot were coming from the southeast.

ALL
4698. (Refer to Figure 133.) What type of entry is recommended for the missed approach holding pattern at Riverside Municipal?

A—Direct.
B—Parallel.
C—Teardrop.

The pilot will use a direct entry for the holding pattern at Riverside Municipal. The missed approach procedures route the pilot directly to R-078 where they can join the depicted holding pattern. (J17) — AIM ¶5-3-7

Answer (B) is incorrect because a parallel entry would be used if the pilot were entering from the northwest. Answer (C) is incorrect because a teardrop entry would be used if the pilot were entering from the southwest.

ALL
4675. (Refer to Figure 128.) What type entry is recommended for the missed approach holding pattern depicted on the VOR RWY 36 approach chart for Price/Carbon County Airport?

A—Direct only.
B—Teardrop only.
C—Parallel only.

The missed approach takes you outbound on the PUC 127° radial then right turn direct to PUC on a no-wind course of 310 or more, which will be a direct entry. (J17) — AIM ¶5-3-7

Answers

4668 [C]   4681 [C]   4698 [A]   4675 [A]

Chapter 7 **En Route**

Estimated Time Enroute (ETE)

In order to answer the questions looking for the Estimated Time Enroute (ETE), you must complete the flight log provided with each question. Each question will go through the following steps:

1. Calculate the magnetic winds using the winds aloft and the variation.
2. Find the distance of the first leg of the flight.
3. Calculate the ground speed for the first leg of the flight.
4. Compute the time en route for the first leg of the flight.
5. Repeat the first four steps to complete the flight log.

ALL
4268. (Refer to the FD excerpt below, and use the wind entry closest to the flight planned altitude.) Determine the time to be entered in block 10 of the flight plan.

Route of flight .... Figures 27, 28, 29, 30 and 31
Flight log & MAG VAR .... Figure 28
GNATS ONE DEPARTURE
   and Excerpt from AFD .... Figure 30

FT	3000	6000	9000
OTH	0507	2006+03	2215-05

A—1 hour 10 minutes.
B—1 hour 15 minutes.
C—1 hour 20 minutes.

To answer this question, complete the flight log in FAA Figure 28, using the information given in the question:

1. Change the winds aloft for 9,000 feet at OTH (North Bend, OR) from True to Magnetic, using the variation from the flight log (FAA Figure 28):

   Winds at OTH at 9,000 are 2215-05 =
   220 true north / 15 knots / -05°C

   $\begin{array}{rl} 220° & \text{true} \\ -20° & \text{East variation} \\ \hline 200° & \text{magnetic} \end{array}$

2. Find the distances for the leg from MERLI INT to MOURN INT using the following DME formula.

   a. Miles Flown = $\dfrac{\text{DME Distance} \times \text{Number of Degrees}}{60}$

   b. Miles Flown = $\dfrac{15 \times (R\text{-}333° - R\text{-}251°)}{60}$

   c. $\dfrac{15 \times 82}{60} = 20.5$ miles

   d. $\begin{array}{rl} 20.5 & \text{miles from the 251° radial to the 333° radial of the arc} \\ +16.0 & \text{miles from the arc to MOURN INT} \\ \hline 36.5 & \text{NM} \end{array}$

3. Find time en route for the MERLI INT to MOURN INT leg using a flight computer:

   Distance 36.5 NM (calculated in step 2)
   Ground speed 135 Knots (given in flight log FAA Figure 28)
   Time en route is 00:16:13

4. Find the distance for the leg from MOURN INT to the Roseburg VOR on V121. (Remember the distance value that is not boxed is the mileage between reporting points, radio aids, and/or mileage breakdown points.) This distance is 19 miles.

5. Compute the ground speed for the leg from MOURN INT to the Roseburg VOR on V121:

   Wind direction 200 (found in Step 1)
   Wind speed 15 knots (found in Step 1)
   Course 287 (found in FAA Figure 31)
   TAS 155 (found in FAA Figure 28)

   Therefore, ground speed is 153.5 Knots

6. Compute the time en route for the leg from MOURN INT to the Roseburg VOR on V121:

   Distance 19 miles (found in Step 4)
   Ground speed 153.5 (found in Step 5)
   Time en route is 00:07:26

7. Repeat the first six steps to fill in the flight log:

FROM	TO	CRS	NM	GS	ETE
MFR	MERLI	—	—	—	:11:00 (given)
MERLI	MOURN	—	37	135	:16:13
MOURN	RBG	287	19	153	:07:26
RBG	OTH	271	38	149	:15:15
OTH	EUG	024	59	170	:20:50
EUG	App/Ldg	—	—	—	:10:00 (given)
					1:20:44

(H342) — AC 61-23C, Chapter 8

Answers
4268 [C]

ALL
4290. (Refer to the FD excerpt below, and use the wind entry closest to the flight planned altitude.) Determine the time to be entered in block 10 of the flight plan.

Route of flight .... Figures 38, 39, and 40
Flight log & MAG VAR .... Figure 39
ACTON TWO ARRIVAL .... Figure 41

FT	3000	6000	9000	12000
ABI		2033+13	2141+13	2142+05

A—1 hour 24 minutes.
B—1 hour 26 minutes.
C—1 hour 31 minutes.

To answer this question, complete the flight log in FAA Figure 39, using the information given in the problem:

1. *Change the winds aloft for 12,000 feet at ABI (Abilene) from True to Magnetic, using the variation from the flight log (FAA Figure 39):*

   *Winds at ABI at 12,000 are 2142+05 =
   210 true north / 42 knots / +5°C*

   $$\begin{array}{rl} 210° & true \\ -11° & East\ variation \\ \hline 199° & magnetic \end{array}$$

2. *Find the distances for the leg from Big Spring VORTAC to Loran INT on V16. The distance that is not boxed is the mileage between reporting points, radio aids, and/or mileage breakdown points. The distance is 42 NM (22 + 20).*

3. *Find the ground speed for the leg from Big Spring VORTAC to Loran INT on V16 using a flight computer:*

   *Wind direction 199 (calculated in step 1)
   Wind speed 42 (found in step 1)
   Course 075 (V16 is the 075 radial from BGS)
   TAS 156 (FAA Figure 38, box 4)*

   Therefore, ground speed is 175.6 knots.

4. *Compute the time en route for the leg from Big Spring VORTAC to Loran INT on V16:*

   *Distance 42 miles (found in Step 2)
   Ground speed 175.6 (found in Step 3)*

   Therefore, time en route is 00:14:21.

5. *Repeat the first four steps to fill in the flight log:*

FROM	TO	CRS	NM	GS	ETE
21XS	BGS	—	—	—	:06:00 (given)
BGS	Loran	075	42	176	:14:21
Loran	ABI	076	40	175	:13:44
ABI	COTTN	087	63	167	:22:40
COTTN	AQN	075	50	176	:17:05
AQN	Creek	040	32	195	:09:52
Creek	App/Ldg	—	—	—	:08:00 (given)
					1:31:42

(H342) — AC 61-23C, Chapter 8

ALL
4279. (Refer to the FD excerpt below, and use the wind entry closest to the flight planned altitude.) Determine the time to be entered in block 10 of the flight plan.

Route of flight .... Figures 32,33,34,35,35A, & 36
Flight log & MAG VAR .... Figure 33
RNAV RWY 33 & Excerpt from AFD .... Figure 36

FT	3000	6000	9000	12000
DAL	2027	2239+13	2240+08	2248+05

A—1 hour 35 minutes.
B—1 hour 41 minutes.
C—1 hour 46 minutes.

To answer this question, the flight log in FAA Figure 33 must be completed. The related BUJ.BUJ3 STAR (FAA Figure 35) and the enroute chart (FAA Figure 34) provide some alterations to the flight log. On V573 the bend in the airway 10 NM southwest of MARKI intersection requires an additional leg. The check points in the STAR must be accounted for, as well.

1. *Interpolate winds aloft, convert to magnetic:*

6000	8000	9000
220°	220° True (216° Mag)	220°
39K	39.66K	40K

2. *Using a flight computer, calculate the ground speeds and ETE to complete the flight log:*

FROM	TO	CRS	DIST	GS	ETE
HOT	Marki	—	—	—	00:12:00 (given)
Marki	VOR COP	221	10	140.5	00:04:16
COP	TXK	210	45	140.5	00:19:13
TXK	Conny	272	61	154.8	00:23:39
Conny	BUJ3	239	59	142.8	00:24:47
BUJ3	D/A	—	—	—	00:10:00 (given)
					1:33:55

(H342) — AC 61-23C, Chapter 8

Answers

4290 [C]   4279 [A]

Chapter 7 **En Route**

ALL
4302. Determine the time to be entered in block 10 of the flight plan. (Refer to the FD excerpt below, and use the wind entry closest to the flight planned altitude.)

Route of flight .... Figures 44, 45, 46, and 47
Flight log & MAG VAR .... Figure 45
GROMO TWO DEPARTURE
    and Excerpt from AFD .... Figure 46

FT	3000	6000	9000	12000
YKM	1615	1926+12	2032+08	2035+05

A—54 minutes.
B—1 hour 02 minutes.
C—1 hour 07 minutes.

To answer this question, complete the flight log in FAA Figure 45, using the information given in the problem:

1. Change the winds aloft for 12,000 feet at the VOR COP from True to Magnetic, using the variation from the flight log (FAA Figure 45):

    Winds at YKM at 12,000 are 2035+05 =
    200 true north / 35 knots / 5°C

    200°   true
    −20°   East variation
    ─────
    180°   magnetic

2. Find the distances for the leg from Hitch to COP; 20 + 17 = 37 NM. The distance value that is not boxed is the mileage between reporting points, radio aids, and/or mileage breakdown points.

3. Compute the ground speed for the leg from Hitch to COP:

    Wind direction 180 (found in Step 1)
    Wind speed 35 knots (found in Step 1)
    Course 206 (found in FAA Figure 45)
    TAS 180 (found in FAA Figure 44, block 4)

    Therefore, ground speed is 147.9 Knots.

4. Compute the time en route for the leg from Hitch to COP:

    Distance 37 miles (found in Step 2)
    Ground speed 147.9 (found in Step 3)

    Therefore, time en route is 00:15:01.

5. Repeat the first six steps to fill in the flight log:

FROM	TO	CRS	NM	GS	ETE
YKM	Hitch	—	—	—	:10:00 (given)
Hitch	COP	206	37	147.9	:15:00
COP	BTG	234	53	157.2	:20:00
BTG	PDX	160	10	146.7	:04:00
PDX	App/Ldg	—	—	—	:13:00 (given)
					1:02:00

(H342) — AC 61-23C, Chapter 8

ALL
4314. Determine the time to be entered in block 10 of the flight plan. (Refer to the FD excerpt below, and use the wind entry closest to the flight planned altitude.)

Route of flight .... Figures 50, 51, 52, and 53
Flight log and MAG VAR .... Figure 51
HABUT ONE DEPARTURE
    and Excerpt from AFD .... Figure 52

FT	3000	6000	9000
SBA	0610	2115+05	2525+00

A—43 minutes.
B—46 minutes.
C—51 minutes.

To answer this question, complete the flight log in FAA Figure 51, using the information given in the problem:

1. Change the winds aloft for 9,000 feet at SBA (San Luis Obispo) from True to Magnetic, using the variation from the flight log (FAA Figure 51):

    Winds at SBA at 9,000 are 2525+00 =
    250 true north / 25 knots / 0°C

    250°   true
    −16°   East variation
    ─────
    234°   magnetic

2. Find the distance for the leg from HABUT INT to Gaviota VORTAC (GVO) on the HABUT one departure (FAA Figure 52). This distance is 6.4 NM.

3. Compute the ground speed for the leg from HABUT INT to Gaviota VORTAC (GVO):

    Wind direction 234 (found in Step 1)
    Wind speed 25 knots (found in Step 1)
    Course 343 (found in FAA Figure 52)
    TAS 158 (found in FAA Figure 50, block 4)

    Therefore, ground speed is 164.4 Knots.

4. Compute the time en route for the leg from HABUT INT to Gaviota VORTAC (GVO):

    Distance 6.4 miles (found in Step 2)
    Ground speed 164.4 (found in Step 3)

    Therefore, time en route is 00:02:20.

5. Repeat the first four steps to fill in the flight log:

FROM	TO	CRS	NM	GS	ETE
SBA	HABUT 1	—	—	—	:08:00 (given)
HABUT 1	GVO	343	6.4	164.4	:02:20
GVO	MQO	306	54	148.7	:22:07
MQO	PRB	358	26	170.6	:09:09
PRB	App/Ldg	—	—	—	:10:00 (given)
					00:51:16

(H342) — AC 61-23C, Chapter 8

Answers

4302 [B]   4314 [C]

Chapter 7 **En Route**

RTC
4324. Determine the time to be entered in block 10 of the flight plan. (Refer to the FD excerpt below, and use the wind entry closest to the flight planned altitude.)

Route of flight .... Figures 56, 57, 58, and 59
Flight log & MAG VAR .... Figure 57
Excerpt from AFD (CLL) .... Figure 58

FT	3000	6000	9000
HOU	0507	1015+05	2225+00

A—1 hour 06 minutes.
B—1 hour 10 minutes.
C—1 hour 14 minutes.

To answer this question, complete the flight log in FAA Figure 57, using the information given in the problem:

1. Change the winds aloft for 6,000 feet at HOU (Houston, TX) from True to Magnetic, using the variation from the flight log (FAA Figure 57):

    Winds at HOU at 6,000 are 1015+05 =
    100 true north / 15 knots / 5°C

    100°$$true
    $\underline{-6°\text{East variation}}$
    094°$$magnetic

2. Find the distance for the leg from CLL to TNV. (Remember, the distance value that is boxed is the mileage between total mileage between compulsory reporting points and/or radio aids.) This distance is 27 NM.

3. Compute the ground speed for the leg from CLL to TNV:

    Wind direction 094 (found in Step 1)
    Wind speed 15 knots (found in Step 1)
    Course 127 (found in FAA Figure 57)
    TAS 110 (found in FAA Figure 56)

    Therefore, ground speed is 97.1 Knots.

4. Compute the time en route for the leg from CLL to TNV:

    Distance 27 miles (found in Step 2)
    Ground speed 97.1 (found in Step 3)

    Therefore, time en route is 00:16:41.

5. Repeat the first four steps to fill in the flight log:

FROM	TO	CRS	NM	GS	ETE
E.W.	CLL	—	—	—	:05:00 (given)
CLL	TNV	127	27	97.1	:16:41
TNV	IAH	110	42	95.5	:26:23
IAH	HUB	161	18	103.3	:10:27
HUB	App/Ldg	—	—	—	:15:00 (given)
					01:13:31

(H342) — AC 61-23C, Chapter 8

ALL
4360. Determine the time to be entered in block 10 of the flight plan. (Refer to the FD excerpt below, and use the wind entry closest to the flight planned altitude.)

Route of flight .... Figures 74, 75, 76, 77, and 78
Flight log & MAG VAR .... Figure 75
VOR indications
and Excerpts from AFD .... Figure 76

FT	6000	9000	12000	18000
BIL	2414	2422+11	2324+05	2126-11

A—1 hour 15 minutes.
B—1 hour 20 minutes.
C—1 hour 25 minutes.

To answer this question, complete the flight log in FAA Figure 75, using the information given in the problem:

1. Change the winds aloft for 12,000 feet at BIL (Logan Intl Airport) from True to Magnetic, using the variation from the flight log (FAA Figure 75):

    Winds at BIL at 12,000 are 2324+05 =
    230 true north / 24 knots / 5°C

    230°$$true
    $\underline{-18°\text{East variation}}$
    212°$$magnetic

2. Find the distance for the leg from Vests to BZN using FAA Figure 77. The distance is 44 NM.

3. Compute the ground speed for the leg from Vests to BZN:

    Wind direction 212 (found in Step 1)
    Wind speed 24 knots (found in Step 1)
    Course 140 (found in FAA Figure 75)
    TAS 160 (found in FAA Figure 74)

    Therefore, ground speed is 150.9 Knots.

4. Compute the time en route for the leg from Vests to BZN:

    Distance 44 miles (found in Step 2)
    Ground speed 150.9 (found in Step 3)

    Therefore, time en route is 00:17:29.

5. Repeat the first four steps to fill in the flight log:

FROM	TO	CRS	NM	GS	ETE
HLN	Vests	—	—	—	:15:00 (given)
Vests	BZN	140	44	150.9	:17:29
BZN	"x"	110	13	163.3	:04:47
"x"	LVM	063	20	180.1	:06:40
LVM	Reepo	067	39	179.1	:13:04
Reepo	BIL	069	38	178.5	:12:46
BIL	App/Ldg	—	—	—	:15:00 (given)
					01:24:46

(H342) — AC 61-23C, Chapter 8

Answers

4324 [C]   4360 [C]

Chapter 7 **En Route**

ALL
4346. Determine the time to be entered in block 10 of the flight plan. (Refer to the FD excerpt below, and use the wind entry closest to the flight planned altitude.)

Route of flight .... Figures 69, 70, and 71
Flight log and MAG VAR .... Figure 70
JUDDS TWO ARRIVAL
 and Excerpt from AFD .... Figure 72

FT	3000	6000	9000
BDL	3320	3425+05	3430+00

A—1 hour 14 minutes.
B—58 minutes.
C—50 minutes.

To answer this question, complete the flight log in FAA Figure 70, using the information given in the problem:

1. *Change the winds aloft for 3,000 feet from True to Magnetic, using the variation from the flight log (FAA Figure 70) and use the winds at 6,000 feet for the planned altitude of 5,000 feet:*

    Winds at 3,000 feet are 3320 =
     330 true north/20 knots
    Winds at 5,000 feet will be around =
     340 true north/25 knots

3,000 feet:	5,000 feet:
330	340
+14 W	+14 W
344	354

2. *Using a flight computer, calculate the ground speeds and ETE to complete the flight log:*

FROM	TO	Alt/Climb	CRS	DIST	GS	ETE	
GrLake	Shaff	Climb	—	—	—	00:08:00	(given)
Shaff	Helon	5,000	029	24	107	00:13:27	
Helon	IGN	5,000	102	21	133	00:09:28	
IGN	JD2X	3,000	112	15	139	00:06:28	
JD2X	Judds	3,000	100	17	135	00:07:33	
Judds	Briss	3,000	057	6	121	00:02:59	
Briss	D/A	—	—	—	—	00:12:00	(given)
						00:59:55	

(H342) — AC 61-23C, Chapter 8

RTC
4335. Determine the time to be entered in block 10 of the flight plan. (Refer to the FD excerpt below, and use the wind entry closest to the flight planned altitude.)

Route of flight .... Figures 62, 63, 64, and 65
Flight log & MAG VAR................................ Figure 63
Excerpt from AFD (LFT) .... Figure 64

FT	3000	6000	9000
MSY	1422	1626+15	1630+10

A—56 minutes.
B—1 hour 02 minutes.
C—1 hour 07 minutes.

To answer this question, complete the flight log in FAA Figure 63, using the information given in the problem:

1. *Change the winds aloft for 6,000 feet at MSY from True to Magnetic, using the variation from the flight log (FAA Figure 63):*

    Winds at MSY at 6,000 are 1626+15 =
     160 true north / 26 knots / 15°C

160°	true
− 6°	East variation
154°	magnetic

2. *Find the distance for the leg from Lafayette VORTAC to the Hatch INT. This distance is 29 miles (found in the D with the arrow).*

3. *Compute the ground speed for the leg from Lafayette VORTAC to the Hatch INT:*

    Wind direction 154 (found in Step 1)
    Wind speed 26 knots (found in Step 1)
    Course 109 (found in FAA Figure 65, V552 is
     R-109 from LKM)
    TAS 105 (found in FAA Figure 62)

    Therefore, ground speed is 85.0 Knots.

4. *Compute the time en route for the leg from Lafayette VORTAC to the Hatch INT:*

    Distance 29 miles (found in Step 2)
    Ground speed 85.0 (found in Step 3)
    Therefore, time en route is 00:19:46.

5. *Repeat the first four steps to fill in the flight log:*

FROM	TO	CRS	NM	GS	ETE	
LFT A/P	LFT VOR	—	—	—	:05:00	(given)
LFT VOR	HATCH	109	29	85	:20:28	
HATCH	GRICE	110	20	85	:14:10	
GRICE	TBD	110	19	85	:13:27	
TBD	App/Ldg	117	—	—	:10:00	(given)
					1:03:05	

(H342) — AC 61-23C, Chapter 8

Answers

4346 [B]   4335 [B]

ALL
4260. (Refer to FD excerpt below, and use the wind entry closest to the flight planned altitude.) Determine the time to be entered in block 10 of the flight from GJT to DRO.

Route of flight .... Figure 21
Flight log & MAG VAR .... Figure 22
En route chart .... Figure 24

FT	12,000	18,000
FNM	2408-05	2208-21

A—1 hour 08 minutes.
B—1 hour 03 minutes.
C—58 minutes.

To answer this question, the flight log (FAA Figure 22) must be completed. Interpolate the 15,000-foot winds aloft as 230° at 8 knots. Convert wind to magnetic: 230° – 14°E = 216°. Perform standard cross-country calculations:

FROM	TO	CRS	NM	GS	ETE
GJT	Herrm	—	(given)	—	24:00
Herrm	Manca	152°	75	171.345	26:16
Manca	DRO	(given)	—	—	18:30
					1:08:46

(H342) — AC 61-23C, Chapter 8

ALL
4259. (Refer to Figures 21 and 21A, 22 and 22A, 23, 24, 25, and 26.) After departing GJT and arriving at Durango Co., La Plata Co. Airport, you are unable to land because of weather. How long can you hold over DRO before departing for return flight to the alternate, Grand Junction Co., Walker Field Airport?

Total useable fuel on board, 68 gallons.
Average fuel consumption 15 GPH.
Wind and velocity at 16,000, 2308-16°.

A—1 hour 33 minutes.
B—1 hour 37 minutes.
C—1 hour 42 minutes.

To answer this question, the times en route and fuel consumption for both trips (GJT-DRO and DRO-GJT) must be calculated. There must be enough fuel for the round trip, plus 45 minutes. Subtract this amount from the total fuel on board, 68 gallons. Any extra is available for holding at DRO.

Convert wind to magnetic: 230° – 14°E = 216°. Perform standard cross-country calculations, and complete the flight logs for each trip (FAA Figures 22 and 22A):

FROM	TO	CRS	NM	GS	ETE	Fuel
GJT	Herrm	—	(given)	—	24:00	6.0
Herrm	Manca	151°	75	171.5	26:14	6.6
Manca	DRO	—	(given)	—	18:30	4.6
					1:08:44	17.2
DRO	Manca	—	(given)	—	14:30	3.6
Manca	Herrm	333°	75	177.5	25:21	6.3
Herrm	JNC	331°	35	177.2	11:51	3.0
JNC	GJT	—	(given)	—	12:00	3.0
					1:03:42	15.9

68 gal @ 15 GPH = 4:32:00
4:32:00 – (2:12:26 + 0:45:00) = 1 hr 34 minutes

(H342) — AC 61-23C, Chapter 8

ALL
4511. (Refer to Figure 91.) What should be the approximate elapsed time from BZN VOR to DBS VORTAC, if the wind is 24 knots from 260° and your intended TAS is 185 knots? (VAR 17° E.)

A—33 minutes.
B—37 minutes.
C—39 minutes.

1. Convert wind to magnetic using the 17° variation:
   260° – 17°E = 243°

2. Calculate the ground speed using your E6-B or CX-2:

Wind direction	243°
Wind speed	24 knots
Course	186°
TAS	185 knots
GS	170.8 knots

3. Calculate your ETE (leg time) using your E6-B or CX-2:

Distance	111 NM
GS	170.8 knots
Time	00:38:59

(H342) — AC 61-23C, Chapter 8

Answer (A) is incorrect because 33 minutes would be the result of flying from DBS to BZN. Answer (B) is incorrect because 37 minutes would be the result if 260° was used for the wind, instead 243°.

Answers

4260 [A]   4259 [A]   4511 [C]

Chapter 7 **En Route**

ALL
4514. (Refer to Figure 91.) Southbound on V257, at what time should you arrive at DBS VORTAC if you crossed over CPN VORTAC at 0850 and over DIVID intersection at 0854?

A—0939.
B—0943.
C—0947.

1. Calculate your ground speed between CPN and DIVID intersection (upper left quadrant of FAA Figure 91) using your E6-B or CX-2:

   Distance   9 NM
   Time   4 minutes
   GS   135.0 knots

2. Calculate your leg time between DIVID and DBS VORTAC:

   Distance   110 NM
   GS   135 knots
   Time   00:48:53 or 49 minutes

3. Calculate your ETA at DBS VORTAC:

   0854 + 49 minutes = 0943

(H342) — AC 61-23C, Chapter 8

Fuel Consumption

In order to calculate the fuel consumption, you must complete the flight log provided with each question. Each question will go through the following steps:

1. Calculate the ground speed.
2. Find the time en route for the flight.
3. Calculate the total fuel burn.

ALL
4265. (Refer to Figures 21, 22, and 24.) What fuel would be consumed on the flight between Grand Junction, Co. and Durango, Co. if the average fuel consumption is 15 GPH.

A—17 gallons.
B—20 gallons.
C—25 gallons.

Use the following steps to complete the flight log and find the fuel consumption at 15 GPH:

1. Find the ground speed. The winds are given 230° (True) at 8 knots. The true course is 165° (151° magnetic + 14° variation), and the true airspeed is 175. Using a flight computer you have an estimated ground speed of 171 knots.

2. Find the time en route for the flight. Looking at FAA Figure 24, the Enroute Low Altitude Chart shows the leg from HERRM to MANCA to be 75 miles. Using a flight computer calculate the time to the second leg to be 0:26:20. Adding the time it takes for each leg (the first leg is given as 00:24:00, and the last leg is given as 00:18:30), the total time en route is 01:08:50.

3. Find the total fuel burn. Using a flight computer, set fuel burn at 15 GPH and find the total fuel burn for 01:09:00 as 17.2 gallons.

(H342) — AC 61-23C, Chapter 8

Answers

4514   [B]   4265   [A]

Chapter 8
Arrival and Approach

STARs *8-3*

Communications During Arrival *8-5*

Instrument Approach Terms and Abbreviations *8-8*

Instrument Approach Procedures Chart (IAP) *8-12*

Radar Approaches *8-29*

Visual and Contact Approaches *8-30*

Timed Approaches from Holding *8-32*

Missed Approach *8-34*

RVR *8-34*

Inoperative Components *8-36*

Hazards on Approach *8-38*

Airport Lighting and Marking Aids *8-40*

Closing the Flight Plan *8-45*

Chapter 8 **Arrival and Approach**

STARs

Standard Terminal Arrivals (STARs) are used in much the same way as DPs — to relieve frequency congestion and to expedite the arrival of aircraft into the terminal area. Like DPs, STARs may be issued by ATC whenever it is deemed appropriate. As with DPs, the PIC may either accept or decline a STAR, but if the STAR is accepted, the pilot must possess at least the textual description of the procedure. Should the pilot not wish to use a STAR, a notation to that effect in the remarks section of the flight plan will make ATC aware of the PIC's decision.

The STAR is depicted by a procedural track line. For example, the ACTON EIGHT arrival shown in Figure 8-1 begins at the ACTON VORTAC and ends when the aircraft has departed Creek Intersection heading 350 and is given radar vectors to final approach course. Familiarize yourself with FAA Legends 12 and 17, as they have information on STARs which will be available for you during the FAA Knowledge Test.

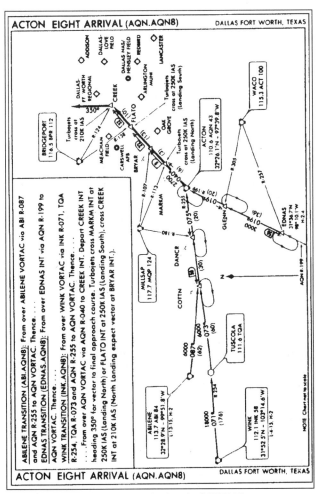

Figure 8-1. Standard terminal arrival (STAR)

ALL
4640. Which is true regarding STARs?

A—STARs are used to separate IFR and VFR traffic.
B—STARs are established to simplify clearance delivery procedures.
C—STARs are used at certain airports to decrease traffic congestion.

Standard Terminal Arrival Routes have been established to simplify clearance delivery procedures for arriving aircraft at certain areas having high density traffic. (J18) — AIM ¶5-4-1

ALL
4286. (Refer to Figures 35 and 35A.) At which point does the BUJ.BUJ3 arrival begin?

A—At the TXK VORTAC.
B—At BOGAR intersection.
C—At the BUJ VORTAC.

The heavy black lines with arrows portray the Arrival Route, which, in this case, begins at the Blue Ridge VOR (BUJ). Note also, in the textual description, all the transitions end with the following statement, "To BUJ VORTAC thence..." (J41) — STAR Chart

Answers

4640 [B]   4286 [C]

Chapter 8 Arrival and Approach

ALL

4292. (Refer to Figures 41 and 41A.) At which point does the AQN.AQN2 arrival begin?

A—ABI VORTAC.
B—ACTON VORTAC.
C—CREEK intersection.

Heavy lines ending in arrows are arrival routes. Also, arrival descriptions in FAA Figure 41A state that transitions end and arrivals begin ("Thence...") at AQN VORTAC. (J41) — U.S. Terminal Procedures, STAR Legend

Answer (A) is incorrect because the beginning of the Abilene transition is ABI VORTAC. Answer (C) is incorrect because CREEK intersection is only a part of the AQN.AQN8 arrival.

ALL

4294. (Refer to Figures 41 and 41A.) On which heading should you plan to depart CREEK intersection?

A—010°.
B—040°.
C—350°.

Turbojets landing south at the major area airports will follow the short, thick 350° arrow from CREEK and receive vectors to appropriate final approach courses. (See arrival descriptions in FAA Figure 41A.) (J41) — U.S. Terminal Procedures, STAR Legend

ALL

4349. (Refer to Figure 72.) At which location or condition does the IGN.JUDDS2 arrival begin?

A—JUDDS intersection.
B—IGN VORTAC.
C—BRISS intersection.

The heavy black lines with arrows portray the Arrival Route, which, in this case, begins at the Kingston VORTAC (IGN). Note also, the textual description of the transition: "From over IGN VORTAC..." (J41) — U.S. Terminal Procedures, STAR Chart Legend

ALL

4751. Under which condition does ATC issue a STAR?

A—To all pilots wherever STAR's are available.
B—Only if the pilot requests a STAR in the "Remarks" section of the flight plan.
C—When ATC deems it appropriate, unless the pilot requests "No STAR."

Pilots of IFR civil aircraft destined to locations for which STARs have been published may be issued a clearance containing a STAR whenever ATC deems it appropriate. Use of STARs requires pilot possession of at least the approved textual description. As with any ATC clearance or portion thereof, it is the responsibility of each pilot to accept or refuse an issued STAR. A pilot should notify ATC if he/she does not wish to use a STAR by placing "NO STAR" in the remarks section of the flight plan, or by the less desirable method of verbally stating the same to ATC. (J18) — AIM ¶5-4-1(c)

Answer (A) is incorrect because STARs are not mandatory to pilots or ATC. Answer (B) is incorrect because a request is only necessary when the pilot does not wish to use a STAR.

ALL

4281. (Refer to Figures 35 and 37.) What is your position relative to the CONNY intersection on the BUJ.BUJ3 transition?

A—Left of the TXK R-272 and approaching the BUJ R-059°.
B—Left of the TXK R-266 and past the BUJ R-065.
C—Right of the TXK R-270 and approaching the BUJ R-245.

FAA Figure 37 has an RMI tuned to 116.3 (TXK VORTAC), with indications that show the airplane on the 270° radial (thick needle tail pointing to 270°) which is left of the 272°. FAA Figure 37 also has an HSI tuned to 114.9 (BUJ VORTAC) and set to 239°. With the bar deflected to the northwest, the airplane is southeast of the BUJ R-059 and approaching it. Thus, the airplane is left of the TXK R-272 and approaching the BUJ R-059. (J41) — STAR Chart

Answer (B) is incorrect because the airplane is on R-270 which is right (not left) of R-266, and the airplane is on (not past) the BUJ R-065. Answer (C) is incorrect because the airplane is on R-270 (not right of) and approaching R-244 of LIT (not BUJ).

Answers

4292 [B]   4294 [C]   4349 [B]   4751 [C]   4281 [A]

Chapter 8 **Arrival and Approach**

Communications During Arrival

Should the pilot elect to cancel the IFR flight plan and proceed VFR to the destination, the pilot is expected to monitor the ATIS (if available) for non-control information regarding runway in use, weather, etc. The ATIS broadcast will be updated upon receipt of any official weather. Should the visibility exceed 5 statute miles, or the ceiling be above 5,000 feet AGL, those items may be omitted. If inbound to an airport with an operating air traffic control tower, a pilot who has canceled his/her IFR flight plan must contact the tower at least 5 miles from the airport to comply with Class D airspace requirements. If Class B airspace or Class C airspace exists at the destination, remaining VFR requirements for those must also be met.

As the IFR flight enters the terminal area, it is common to receive radar vectors to the final approach course, or be constantly reassigned altitudes. Climbs and descents should be made at an optimum rate of between 500 and 1,500 feet per minute. If at any time, the aircraft is unable to climb or descend at 500 feet, ATC must be advised. During a radar vector to the final approach course, it is good operating practice to read back the portions of clearances which contain heading and altitude assignments.

In order to avoid excessive vectoring in the terminal environment, ATC may request pilots to adjust their speed. ATC will express all speeds in terms of knots indicated airspeed (KIAS) and use only 10-knot increments. Pilots complying with such a request are expected to maintain that indicated speed within ±10 knots. If unable to comply with the request, the pilot is expected to advise ATC immediately of the speed that will be used.

If radar vectored and about to pass through the final approach course, the pilot may not turn inbound unless an approach clearance has been received. Instead, the pilot should maintain the course and query ATC. ATC occasionally needs to vector an aircraft through the approach course, and in such cases the controllers should state their intentions to do so. Controllers are required to advise you if they will be vectoring you across the localizer.

If ATC uses the phrase "cleared for approach," the pilot may use any authorized instrument approach for the airport as long as the entire procedure is followed. If cleared for a specific procedure, such as "cleared for the ILS runway 31R approach," the pilot must execute only that ILS approach.

If the pilot is being radar vectored or operating on an unpublished route when approach clearance is received, the pilot must maintain the last assigned altitude until established on a segment of a published route or instrument approach procedure.

ALL
4403. When are ATIS broadcasts updated?

A—Every 30 minutes if weather conditions are below basic VFR; otherwise, hourly.
B—Upon receipt of any official weather, regardless of content change or reported values.
C—Only when the ceiling and/or visibility changes by a reportable value.

ATIS broadcasts shall be updated upon the receipt of any official weather, regardless of content change and reported values. A new recording will also be made when there is a change in other pertinent data such as runway change, instrument approach in use, etc. (J11) — AIM ¶4-1-13(b)

Answer (A) is incorrect because the ATIS updates are not different under VFR or IFR conditions. Answer (C) is incorrect because the recording will be updated when official weather is received, even if there is no change.

Answers
4403 [B]

Chapter 8 **Arrival and Approach**

ALL
4404. Absence of the sky condition and visibility on an ATIS broadcast specifically implies that

A—the ceiling is more than 5,000 feet and visibility is 5 miles or more.
B—the sky condition is clear and visibility is unrestricted.
C—the ceiling is at least 3,000 feet and visibility is 5 miles or more.

If the weather is above a ceiling/sky condition of 5,000 feet and the visibility is 5 miles or more, inclusion of the ceiling/sky condition, visibility, and obstructions to vision in the ATIS message is optional. (J11) — AIM ¶4-1-13(b)

Answers (B) and (C) are incorrect because the absence of the sky conditions and visibility on an ATIS broadcast means the ceilings are more than 5,000 feet and visibility 5 SM or better.

ALL
4293. (Refer to Figures 41 and 41A.) Which frequency would you anticipate using to contact Regional Approach Control? (ACTON TWO ARRIVAL).

A—119.05.
B—124.15.
C—125.8.

Frequencies are in the upper left corner of FAA Figure 41 (with the procedure turned to read correctly). Approaching from the southwest, anticipate using 125.8. (J41) — U.S. Terminal Procedures, STAR Legend

ALL
4295. (Refer to Figures 41, 42, and 42A.) Approaching DFW from Abilene, which frequencies should you expect to use for Regional Approach Control, control tower, and ground control respectively?

A—119.05; 126.55; 121.65.
B—119.05; 124.15; 121.8.
C—125.8; 124.15; 121.8.

Looking at the A/FD excerpt in FAA Figure 42, the Communications section, aircraft approaching from the west would anticipate using 125.8 or 132.1 for approach control, 124.15 for tower control, and 121.65 or 121.8 for ground control. (J18) — U.S. Terminal Procedures

ALL
4469. When are you required to establish communications with the tower, (Class D airspace) if you cancel your IFR flight plan 10 miles from the destination?

A—Immediately after canceling the flight plan.
B—When advised by ARTCC.
C—Before entering Class D airspace.

When flying in Class D airspace, communications must be established with the tower. Class D airspace has a radius of at least 4.4 nautical miles, which equals 5 statute miles. (J08) — 14 CFR §91.129

Answer (A) is incorrect because it not necessary to contact tower 10 miles out. Answer (B) is incorrect because the pilot is only required to contact tower prior to entering the Class D airspace.

ALL
4726. You are being vectored to the ILS approach course, but have not been cleared for the approach. It becomes evident that you will pass through the localizer course. What action should be taken?

A—Turn outbound and make a procedure turn.
B—Continue on the assigned heading and query ATC.
C—Start a turn to the inbound heading and inquire if you are cleared for the approach.

Radar vectors and altitude/flight levels will be issued as required for spacing and separating aircraft; therefore, you must not deviate from the headings as issued by approach control. You will normally be informed when it becomes necessary to vector you across the final approach course for spacing or other reasons. If you determine that approach course crossing is imminent and you have not been informed that you will be vectored across it, you should question the controller. You should not turn inbound on the final approach course unless you have received an approach clearance. (J18) — AIM ¶5-4-7

Answers

| 4404 | [A] | 4293 | [C] | 4295 | [C] | 4469 | [C] | 4726 | [B] |

Chapter 8 **Arrival and Approach**

ALL
4734. When being radar vectored for an ILS approach, at what point may you start a descent from your last assigned altitude to a lower minimum altitude if cleared for the approach?

A—When established on a segment of a published route or IAP.
B—You may descend immediately to published glide slope interception altitude.
C—Only after you are established on the final approach unless informed otherwise by ATC.

When operating on an unpublished route or while being radar vectored, the pilot, when an approach clearance is received, shall maintain the last altitude assigned to that pilot until the aircraft is established on a segment of a published route or instrument approach procedure, unless a different altitude is assigned by ATC. (J18) — AIM ¶5-4-7

Answer (B) is incorrect because the pilot must be on part of the published approach in order to descend. Answer (C) is incorrect because the pilot should follow the prescribed altitude whenever on a segment of a published route or IAP.

ALL
4757. While being vectored, if crossing the ILS final approach course becomes imminent and an approach clearance has not been issued, what action should be taken by the pilot?

A—Turn outbound on the final approach course, execute a procedure turn, and inform ATC.
B—Turn inbound and execute the missed approach procedure at the outer marker if approach clearance has not been received.
C—Maintain the last assigned heading and query ATC.

Radar vectors and altitude/flight levels will be issued as required for spacing and separating aircraft; therefore, pilots must not deviate from the headings as issued by approach control. Aircraft will normally be informed when it becomes necessary to vector across the final approach course for spacing or other reasons. If approach course crossing is imminent and the pilot has not been informed that the aircraft will be vectored across the final approach course, the pilot should query the controller. (J18) — AIM ¶5-4-3(b)

Answers (A) and (B) are incorrect because the pilot should not deviate from headings or altitudes issued by approach control until an amended clearance has been received.

ALL
4641. While being radar vectored, an approach clearance is received. The last assigned altitude should be maintained until

A—reaching the FAF.
B—advised to begin descent.
C—established on a segment of a published route or IAP.

Upon receipt of an approach clearance while on an unpublished route or being radar vectored, the pilots must maintain the last assigned altitude until established on a segment of a published route or IAP, at which time published altitudes apply. (J18) — AIM ¶5-5-4(a)(3)(b)

Answer (A) is incorrect because published altitudes may be used as soon as the aircraft is on a published route, which should be before the final approach fix (FAF). Answer (B) is incorrect because once ATC issues an approach clearance, descent to published altitudes is left to the discretion of the pilot.

Answers

4734 [A]   4757 [C]   4641 [C]

Chapter 8 **Arrival and Approach**

Instrument Approach Terms and Abbreviations

The **Decision Height (DH)** is the altitude at which, on a precision approach, a decision must be made to either continue the approach or execute a missed approach.

The **Minimum Descent Altitude (MDA)** is the lowest altitude, expressed in feet above MSL, to which descent is authorized on final approach or during circle-to-land maneuvering in execution of a standard instrument approach procedure where no electronic glide slope is provided.

The **Final Approach Fix (FAF)** identifies the beginning of the final approach segment of an instrument approach procedure. The FAF for nonprecision approaches is designated by a Maltese Cross. On precision approaches, a lightning bolt symbol indicates the FAF. If ATC directs a glide slope intercept altitude which is lower than that published, the actual point of glide slope intercept becomes the FAF.

The **Final Approach Point (FAP)** applies only to nonprecision approaches with no designated FAF, such as an on-airport VOR or NDB. It is the point at which an aircraft has completed the procedure turn, is established inbound on the final approach course, and may start the final descent. The FAP serves as the FAF and identifies the beginning of the final approach segment.

The **Glide Slope (GS)** or glide path provides vertical guidance for aircraft during approach and landing. Applying the glide slope angle and the ground speed to the rate of descent table gives a recommended vertical speed.

An **Instrument Approach Procedure (IAP)** is a series of predetermined maneuvers for the orderly transfer of an aircraft under instrument flight conditions, from the beginning of the initial approach to a landing or to a point from which a landing may be made visually.

A **Nonprecision Approach** is a standard instrument approach procedure in which no electronic glide slope is provided; for example, NDB, VOR, TACAN, ASR, LDA, or SDF approaches.

A **Precision Approach** is a standard instrument approach procedure in which an electronic glide slope/glide path is provided; for example, ILS, MLS, or PAR approaches.

A **Procedure Turn (PT)** is the maneuver prescribed when it is necessary to reverse direction in order to establish an aircraft on the intermediate approach segment or on the final approach course. A procedure turn (and the initial approach segment) begins by overheading a facility or fix. The maximum speed for a PT is 200 KIAS.

In the case of a radar vector to a final approach fix, or position or a timed approach from a holding fix, or where the procedure is specified "NoPT," no pilot may make a procedure turn unless, upon receiving the final approach clearance, the pilot so advises ATC and a clearance is received.

Chapter 8 **Arrival and Approach**

A barb indicates the direction or side of the outbound course on which the procedure turn is made. If no barb is depicted then a procedure turn is not authorized. Headings are provided for course reversal using the 45°-type procedure turn. However, the point at which the turn may be commenced and the type and rate of turn is left to the discretion of the pilot. When a teardrop procedure turn is depicted and a course reversal is required, this type of turn must be executed. When a one-minute holding pattern replaces the procedure turn, the standard entry and the holding pattern must be followed, except when radar vectoring is provided or when "NoPT" is shown on the approach course. The holding maneuver must be executed within the 1-minute time limitation or within the published leg length.

A **Straight-in Approach** is one in which the final approach segment is begun without first having executed a procedure turn (such as a holding pattern, teardrop, or 45°-type). Radar vectors to final are straight-in approaches. The straight-in approach may be followed by either a straight-in landing or a circle-to-land maneuver.

Straight-in landing minimums are published when the runway alignment and the IAP final approach course are within 30° of each other, and will be identified by the letter "S" followed by designation of the runway to which they apply; for example, "S-ILS 34."

The same IAP may furnish circling minimums for use if it is necessary to land on a runway other than the one aligned with the final approach course. These will be identified by the word "circling" and no runway will be designated.

In some cases, the IAP final approach course may not be within 30 of alignment with a runway. Then, only circling minimums will be published.

The **Initial Approach Fix (IAF)** is the point at which the procedure turn begins, or any fix labeled "IAF" that identifies the beginning of the initial approach segment of an instrument approach procedure. Follow the published feeder route unless instructed otherwise.

The **Initial Approach Segment** is the segment of an IAP between IAF and the intermediate fix, or the point where the aircraft is established on the intermediate or final approach course. The intermediate approach segment is the segment between the intermediate fix or point and the final approach fix. The final approach segment is the segment between the final approach fix or point and the runway or missed approach point. The missed approach segment is the segment between the missed approach point or the point of arrival at decision height and the missed approach fix at the prescribed altitude.

The **Visual Descent Point (VDP)** is a point on a final approach course of a nonprecision approach from which a normal descent from the MDA to the runway can be commenced, provided the pilot has the runway, lights, etc., identified.

The **Sidestep Maneuver** is one that ATC may authorize to a runway that is within 1,200 feet of an adjacent parallel runway. Pilots are expected to commence the sidestep maneuvers as soon as possible after the runway, or runway environment is in sight.

Chapter 8 Arrival and Approach

ALL
4740. When cleared to execute a published sidestep maneuver for a specific approach and landing on the parallel runway, at what point is the pilot expected to commence this maneuver?

A—At the published minimum altitude for a circling approach.
B—As soon as possible after the runway or runway environment is in sight.
C—At the localizer MDA minimum and when the runway is in sight.

Aircraft that execute a side-step maneuver will be cleared for a specified approach and landing on the adjacent parallel runway. Example, "Cleared for ILS runway 07 left approach, side-step to runway 07 right." Pilots are expected to commence the side-step maneuver as soon as possible after the runway or runway environment is in sight. (J18) — AIM ¶5-4-17(b)

ALL
4715. How can an IAF be identified on a Standard Instrument Approach Procedure (SIAP) Chart?

A—All fixes that are labeled IAF.
B—Any fix illustrated within the 10 mile ring other than the FAF or stepdown fix.
C—The procedure turn and the fixes on the feeder facility ring.

The instrument approach commences at the Initial Approach Fix (IAP). Initial Approach Fixes are specifically identified with "IAF." (J33) — Instrument Approach Procedure Legend

ALL
4746. Which fixes on the IAP Charts are initial approach fixes?

A—Any fix on the en route facilities ring, the feeder facilities ring, and those at the start of arc approaches.
B—Only the fixes at the start of arc approaches and those on either the feeder facilities ring or en route facilities ring that have a transition course shown to the approach procedure.
C—Any fix that is identified by the letters IAF.

Initial Approach Fixes are the fixes depicted on the Instrument Approach Procedure Charts that identify the beginning of the initial approach segment(s). The letters IAF identify Initial Approach Fixes on instrument approach charts. (J18) — Pilot/Controller Glossary

Answers (A) and (B) are incorrect because the instrument approach fixes are specifically identified with "IAF" with the exception of the procedure turn approach pattern.

ALL
4632. When the approach procedure involves a procedure turn, the maximum speed should not be greater than

A—180 knots IAS.
B—200 knots IAS.
C—250 knots IAS.

When the approach procedure involves a procedure turn, a maximum speed no more than 200 KIAS should be observed, and the turn should be executed within the distance specified in the profile view. (J18) — AIM ¶5-4-8(a)(2)

Answers
4740 [B]   4715 [A]   4746 [C]   4632 [B]

8–10   ASA   2002 Instrument Test Prep

Chapter 8 Arrival and Approach

ALL
4636. What does the absence of the procedure turn barb on the plan view on an approach chart indicate?

A—A procedure turn is not authorized.
B—Teardrop-type procedure turn is authorized.
C—Racetrack-type procedure turn is authorized.

The absence of the procedure turn barb in the Plan View indicates that a procedure turn is not authorized for that procedure. (J42) — Instrument Approach Procedures

ALL
4767. Where a holding pattern is specified in lieu of a procedure turn, the holding maneuver must be executed within

A—the 1 minute time limitation or DME distance as specified in the profile view.
B—a radius of 5 miles from the holding fix.
C—10 knots of the specified holding speed.

Where a holding pattern is specified in lieu of a procedure turn, the holding maneuver must be executed within the 1 minute time limitation or DME distance as specified in the profile view. The maneuver is completed when the aircraft is established on the inbound course after executing the appropriate holding pattern entry. (J17) — AIM ¶5-4-8(b)(3)

ALL
4771. Assume this clearance is received:

"CLEARED FOR ILS RUNWAY 07 LEFT APPROACH, SIDE-STEP TO RUNWAY 07 RIGHT."

When would the pilot be expected to commence the side-step maneuver?

A—As soon as possible after the runway environment is in sight.
B—Any time after becoming aligned with the final approach course of Runway 07 left, and after passing the final approach fix.
C—After reaching the circling minimums for Runway 07 right.

Aircraft that execute a side-step maneuver will be cleared for a specified approach and landing on the adjacent parallel runway. Pilots are expected to commence the side-step maneuver as soon as possible after the runway or runway environment is in sight. (J18) — AIM ¶5-4-17(b)

Answer (B) is incorrect because the pilot has only been cleared for runway 07 right, and should be commenced as soon as the runway environment is in sight. Answer (C) is incorrect because the side-step should commence as soon as the runway environment is in sight.

ALL
4670. When simultaneous approaches are in progress, how does each pilot receive radar advisories?

A—On tower frequency.
B—On approach control frequency.
C—One pilot on tower frequency and the other on approach control frequency.

Pilots will monitor tower frequency for advisories and instructions when simultaneous approaches are in progress. (J18) — AIM ¶5-4-14(b)

Answers

4636 [A]   4767 [A]   4771 [A]   4670 [A]

Chapter 8 **Arrival and Approach**

Instrument Approach Procedures Chart (IAP)

A pilot adhering to the altitudes, flight paths, and weather minimums depicted on the IAP chart, or vectors and altitudes issued by the radar controller, is assured of terrain and obstruction clearance and runway or airport alignment during approach for landing.

The IAP chart may be divided into four distinct areas: the Plan View, showing the route to the airport; the Profile View, showing altitude and descent information; the Minimums Section, showing approach categories, minimum altitudes, and visibility requirements; and the Airport diagram, showing runway alignments, runway lights, and approach lighting systems.

1. The **Plan View** is that portion of the IAP chart depicted at "A" in Figure 8-2. Above the IAP chart is the procedure identifications which will depict the A/C equipment necessary to execute the approach, the runway alignment, the name of the airport, the city and state of airport location. *See* Figure 8-2, #1. An ILS approach, for example, requires the aircraft to have an operable localizer, glide slope, and marker beacon receiver. A LOC/DME approach would require the aircraft to be equipped with both a localizer receiver and distance measuring equipment (DME). If the approach is aligned within 30° of the centerline, the runway number listed at the top of the approach chart means straight-in landing minimums are published for that runway. If the approach course is not within 30° of the runway centerline, an alphabetic code will be assigned to tie IAP identification (for example, NDB-A, VOR-C), indicating that only circle-to-land minimums are published. This would not preclude a pilot from landing straight-in, however, if the pilot has the runway in sight in sufficient time to make a normal approach for landing, and has been cleared to land.

   The IAP plan view will list in either upper corner, the approach control, tower, and other communications frequencies a pilot will need. Some listings may include a direction (for example, North 120.2, South 120.8).

   The IAP plan view may contain a Minimum Sector Altitude (MSA) diagram. The diagram shows the altitude that would provide obstacle clearance of at least 1,000 feet in the defined sector while within 25 NM of the primary omnidirectional NAVAID; usually a VOR or NDB. *See* Figure 8-2, #2.

   An IAP may include a procedural track around a DME arc to intercept a radial. An arc-to-radial altitude restriction applies while established on that segment of the IAP.

2. The **Profile View** is that portion of the IAP chart depicted at "B" in Figure 8-2. The profile view shows a side view of the procedures. This view includes the minimum altitude and maximum distance for the procedure turn, altitudes over prescribed fixes, distances between fixes, and the missed approach procedure.

3. The **Minimums Section** is that portion of the IAP chart depicted at "C" in Figure 8-2. The categories listed on instrument approach charts are based on aircraft speed. The speed is 1.3 times V_{SO} at maximum certificated gross landing weight.

4. The **Aerodrome Data** is that portion of the IAP chart which includes an airport diagram, and depicts runway alignments, runway lights, approach lights, and other important information, such as the touchdown zone elevation (TDZE) and airport elevation. *See* area "D" in Figure 8-2.

See FAA Legends 10, 11, 14, 15, and 18 for IAP legends, which will be available for you to use during the FAA Knowledge tests.

Chapter 8 **Arrival and Approach**

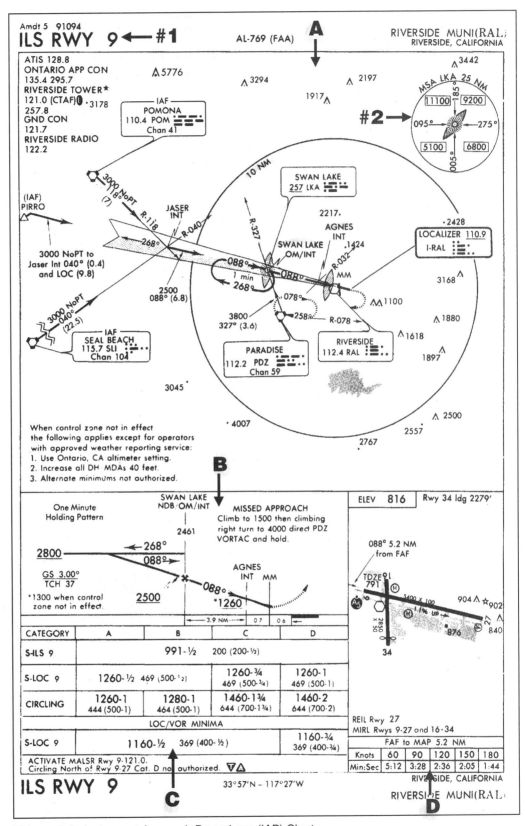

Figure 8-2. Instrument Approach Procedures (IAP) Chart

Chapter 8 **Arrival and Approach**

ALL
4274. (Refer to Figure 29.) What is the TDZ elevation for RWY 16 on Eugene/Mahlon Sweet Field?

A—363 feet MSL.
B—365 feet MSL.
C—396 feet MSL.

The Touchdown Zone Elevation (TDZE) of 363 feet for RWY 16 is shown on the airport sketch of FAA Figure 29. (J42) — Instrument Approach Procedures

ALL
4276. (Refer to Figure 29.) Using a ground speed of 90 knots on the ILS final approach course, what rate of descent should be used as a reference to maintain the ILS glide slope?

A—415 feet per minute.
B—480 feet per minute.
C—555 feet per minute.

On FAA Figure 29, find the glide slope angle of descent on the profile view, in this case, 3°. Enter the rate of descent table (FAA Legend 21) at the 3° point in the angle of descent column. Move directly to the right until intercepting the 90-knot ground speed column and read the rate of descent, in this case, 480 fpm. (J42) — Instrument Approach Procedures

ALL
4282. (Refer to Figure 36A.) Under which condition should the missed approach procedure for the VOR/DME RNAV RWY 33 approach be initiated?

A—Immediately upon reaching the 5.0 DME from the FAF.
B—When passage of the MAP way point is shown on the ambiguity indicator.
C—After the MDA is reached and 1.8 DME fix from the MAP way point.

The missed approach point on the RNAV RWY 33 approach is arrival at the indicated MAP waypoint. Arrival at a waypoint is shown on the TO-FROM ambiguity indicator. (H833) — FAA-H-8083-15, Chapter 7

Answer (A) is incorrect because FAF is identified as 5.0 DME from the MAP, so it can not define when the MAP should be initiated. Answer (C) is incorrect because 1.8 DME from the MAP waypoint is at the MDA.

ALL
4283-1. (Refer to Figure 36A.) What is the MDA and visibility criteria respectively for the S 33 approach procedure?

A—1,240 feet MSL; 1 SM.
B—1,280 feet MSL; 1 and 1/4 SM.
C—1,300 feet MSL; 1 SM.

FAA Figure 36A indicates the MDA and visibility criteria for S33 approach to be 1,240 feet and 1 SM with the Addison altimeter. (H833) — FAA-H-8083-15, Chapter 7

ALL
4283-2. (Refer to Figure 36A.) What is the MDA and visibility criteria respectively for the S 33 approach procedure?

A—1,240 feet MSL; 1/2 SM.
B—1,240 feet MSL; 1 SM.
C—1,280 feet MSL; 1 and 1/4 SM.

FAA Figure 36A indicates the MDA and visibility criteria for the S33 approach to be 1,240 feet and 1 SM with the Addison altimeter. (H833) — FAA-H-8083-15, Chapter 7

ALL
4285. (Refer to Figure 36A.) What is the minimum number of way points required for the complete RNAV RWY 33 approach procedure including the IAF's and missed approach procedure?

A—One way point.
B—Two way points.
C—Three way points.

Two way points are required for the complete RNAV RWY 33 approach: ADDIS and the MAP. (J42) — IAP Chart Legend

Answer (A) is incorrect because the way point is the MAP (as described by the box) even without the way point star. Answer (C) is incorrect because the FAF is not a way point, instead it is a DME fix off the MAP way point.

ALL
4296. (Refer to Figure 42A.) Which navigational information and services would be available to the pilot when using the localizer frequency?

A—Localizer and glide slope, DME, TACAN with no voice capability.
B—Localizer information only, ATIS and DME are available.
C—Localizer and glide slope, DME, and no voice capability.

Answers

| 4274 | [A] | 4276 | [B] | 4282 | [B] | 4283-1 | [A] | 4283-2 | [B] | 4285 | [B] |
| 4296 | [C] | | | | | | | | | | |

ILS-1 RWY 36L provides localizer and glide slope (since an ILS approach requires these components). DME is available, as indicated by the channel numbers. "No voice capability" is indicated by the underline on the I-BXN 111.9 frequency. (J42) — U.S. Terminal Procedures, IAP Chart

Answer (A) is incorrect because TACAN is not available. Answer (B) is incorrect because glide slope information is also available.

ALL
4297. (Refer to Figures 42 and 42A.) What is the difference in elevation (in feet MSL) between the airport elevation and the TDZE for RWY 36L?

A—15 feet.
B—18 feet.
C—22 feet.

The airport elevation is 603 feet MSL (see the upper left corner of the airport diagram). The TDZE (Touchdown Zone Elevation) for RWY 36L is 588 feet MSL (see the left center of the airport diagram). The difference is 603 – 588 = 15 feet. (J42) — U.S. Terminal Procedures, IAP Chart

ALL
4298. (Refer to Figure 42A.) What rate of descent should you plan to use initially to establish the glidepath for the ILS RWY 36L approach? (Use 120 knots ground speed.)

A—425 feet per minute.
B—530 feet per minute.
C—635 feet per minute.

The bottom-right corner of the profile view indicates a 3° glide slope. Enter the Rate-of-Descent Table (FAA Legend 21) in the left-hand column at the given glide slope angle of descent (3°). Move horizontally to the right to the 120-knot Ground speed column and read the recommended rate of descent (635 fpm). (J42) — U.S. Terminal Procedures, IAP Chart

ALL
4299. (Refer to Figures 42A and 43.) What is your position relative to CHAAR intersection? The aircraft is level at 3,000 feet MSL.

A—Right of the localizer course approaching CHAAR intersection and approaching the glide slope.
B—Left of the localizer course approaching CHAAR intersection and below the glide slope.
C—Right of the localizer course, past CHAAR intersection and above the glide slope.

The NAV-1 is on the I-BXN localizer (111.9) indicating right of course (left deviation). The aircraft is approaching CHAAR intersection (DME readout is 7.5, and CHAAR is 7.2). The aircraft is below and approaching the glide slope (the aircraft is shown above the glide slope). The NAV-2 is on the LOVE VOR (114.3) indicating the aircraft is south of CHAAR (it is indicating R-230 and CHAAR is R-233). Therefore, the aircraft is right of the localizer course approaching CHAAR intersection and approaching the glide slope. (J42) — Instrument Approach Procedures

Answer (B) is incorrect because if the aircraft was right of the localizer NAV-1 would have a left deviation. Answer (C) is incorrect because if the aircraft was right of the localizer NAV-1 would have a left deviation; if the aircraft was past CHAAR it the DME would be indicating less than 7.2; and if the aircraft was above the glide slope, the airplane would be pictured below the glide slope.

ALL
4306. (Refer to Figure 49.) What determines the MAP on the LOC/DME RWY 21 approach at Portland International Airport?

A—I-GPO 1.2 DME.
B—5.8 NM from ROBOT FAF.
C—160° radial of BTG VORTAC.

The missed approach point is determined by I-GPO 1.2 DME. (J42) — Instrument Approach Procedures

ALL
4307. (Refer to Figures 44 and 49.) What is the MDA and visibility criteria for a straight-in LOC/DME RWY 21 approach at Portland International?

A—1,100 feet MSL; visibility 1 SM.
B—680 feet MSL; visibility 1 SM.
C—680 feet MSL; visibility 1 NM.

Use FAA Figure 44 to determine what category is needed. Multiply 77 (V_{SO}) by 1.3 to equal 100.1 knots, which is Category B (90 to 120 knots). Look at FAA Figure 49 to find the straight-in minimums for Rwy 21. Category B is 680 feet MSL and 1 SM visibility. (J42) — Instrument Approach Procedures

Answers

4297 [A]   4298 [C]   4299 [A]   4306 [A]   4307 [B]

Chapter 8 **Arrival and Approach**

ALL
4308. (Refer to Figure 49.) When conducting the LOC/DME RWY 21 approach at PDX, what is the Minimum Safe Altitude (MSA) while maneuvering between the BTG VORTAC and CREAK intersection?

A—3,400 feet MSL.
B—5,700 feet MSL.
C—6,100 feet MSL.

The minimum safe altitude (MSA) is found in the small circle in the bottom right corner of the approach chart plan view, and is 6,100 feet MSL between BRG and CREAK intersection. (J42) — Instrument Approach Procedures

ALL
4309. (Refer to Figure 49.) You have been cleared to the CREAK intersection via the BTG 054° radial at 7,000 feet. Approaching CREAK, you are cleared for the LOC/DME RWY 21 approach to PDX. Descent to procedure turn altitude should not begin prior to

A—completion of the procedure turn, and established on the localizer.
B—CREAK outbound.
C—intercepting the glide slope.

If you are above the altitude designated for the course reversal, you may begin descent as soon as you cross the IAF, which is CREAK intersection. (J42) — Instrument Approach Procedures

ALL
4310. (Refer to Figure 49.) With a ground speed of 120 knots, approximately what minimum rate of descent will be required between I-GPO 7 DME fix (ROBOT) and the I-GPO 4 DME fix?

A—1,200 fpm.
B—500 fpm.
C—800 fpm.

Use the profile view of Figure 49 to calculate the rate of descent:

1. Use a flight computer to calculate the leg time given a distance of 3 NM and ground speed of 120 knots (00:01:30 or 1 minute, 30 seconds).
2. Compute the altitude to be lost (2,300 – 1,100 = 1,200 feet).
3. Compute the rate of descent: 1,200 feet in 00:01:30 (1,200/1.5 = 800 fpm).

(J42) — Instrument Approach Procedures

ALL
4311. (Refer to Figure 49.) What is the usable runway length for landing on runway 21 at PDX?

A—5,957 feet.
B—7,000 feet.
C—7,900 feet.

The top right of the airport diagram in Figure 49 indicates the usable runway length for landing on runway 21 is 5,957 feet. (J42) — Instrument Approach Procedures

ALL
4319. (Refer to Figure 55.) Using an average ground speed of 90 knots, what constant rate of descent from 2,400 feet MSL at the 6 DME fix would enable the aircraft to arrive at 2,000 feet MSL at the FAF?

A—200 feet per minute.
B—400 feet per minute.
C—600 feet per minute.

The distance between the 6 DME and the FAF is 3 NM as shown in the profile view. The aircraft will cover that distance in 2 minutes, at 90 knots. Therefore, the aircraft must lose 400 feet in 2 minutes, or 200 feet per minute. Computation:

1. Apply the following formula:

$$\frac{\text{Distance} \times 60}{\text{Ground speed}} = \text{Time}$$

or:

$$\frac{3 \times 60}{90} = 2 \text{ minutes}$$

2. Compute the altitude to be lost:

   2,400 feet
   – 2,000 feet
   ———————
   400 feet (altitude to be lost)

3. Compute the time:

   400 feet in 2 minutes = 200 feet per minute.

(J42) — AC 61-23C, Chapter 8

Answers

4308 [C]   4309 [B]   4310 [C]   4311 [A]   4319 [A]

ALL
4321. (Refer to Figure 55.) Under which condition should a missed approach procedure be initiated if the runway environment (Paso Robles Municipal Airport) is not in sight?

A— After descending to 1,440 feet MSL.
B— After descent to 1,440 feet or reaching the 1 NM DME, whichever occurs first.
C— When you reach the established missed approach point and determine the visibility is less than 1/2 mile.

The established missed approach point is the PRB VOR. If, upon reaching it, the runway environment is not in sight and the visibility is less than 1 mile, the missed approach procedure should be commenced. (J42) — Instrument Approach Procedures

Answers (A) and (B) are incorrect because neither describe the published missed approach point.

ALL
4332. (Refer to Figure 60A.) What is the elevation of the TDZE for RWY 4?

A— 70 feet MSL.
B— 54 feet MSL.
C— 46 feet MSL.

Touchdown Zone Elevation (TDZE) is the highest elevation in the first 3,000 feet of the landing surface. TDZE is indicated on the Instrument Approach Procedure Chart when straight-in landing minimums are authorized. The TDZE is 46 feet MSL, and can be found at the approach end of the landing runway. (J42) — Instrument Approach Procedures

RTC
4340. (Refer to Figure 68.) Upon which maximum airspeed is the COPTER VOR/DME 117° approach category based?

A— 80 knots.
B— 90 knots.
C— 100 knots.

The approach criteria for helicopters are based on airspeeds not exceeding 90 knots, regardless of weight. (J42) — Instrument Approach Procedures

RTC
4341. (Refer to Figure 68.) What minimum visibility must exist to execute the COPTER VOR/DME 117° approach procedure?

A— 3/4 mile.
B— 1/2 mile.
C— 1/4 mile.

The approach minimums for the "helicopter only" approach procedure may differ from the adjusted approach category "A" criteria for the fixed-wing aircraft even though the approach is made to the same general landing area or runway environment. The minimum visibility is 1/2 mile. (J42) — Instrument Approach Procedures

RTC
4342. (Refer to Figure 68.) What is the VASI approach slope angle for RWY 12 at Houma-Terrebonne?

A— 3.0°.
B— 2.8°.
C— 2.5°.

The VASI approach slope angle is found at the top of the A/FD excerpt for "RWY 12: REIL. PVASI (PSIL)— GA 3.0°TCH 42'." (J34) — A/FD Legend

RTC
4343. (Refer to Figure 68.) What would be the approach minimums if you must use the Moisant Field altimeter settings?

A— 440-1.
B— 480 and 1/2.
C— 580 and 1/2.

In the Remarks section of the IAP, it states that when the local altimeter is not available, use the New Orleans International altimeter setting and increase MDA 140 feet:

```
  440   feet and 1/2 mile
+ 140   feet
─────
  580   feet and 1/2 mile
```

(J42) — Instrument Approach Procedures

Chapter 8 **Arrival and Approach**

ALL
4350. (Refer to Figure 72.) How many precision approach procedures are published for Bradley International Airport?

A—One.
B—Three.
C—Four.

The bottom of the A/FD excerpt lists the available precision approaches: ILS/DME for runways 06, 33, and 24. (J34) — A/FD Legend

ALL
4351. (Refer to Figure 73.) What is the minimum altitude at which you should intercept the glide slope on the ILS RWY 6 approach procedure?

A—3,000 feet MSL.
B—1,800 feet MSL.
C—1,690 feet MSL.

The glide slope intercept and final approach fix for a precision approach is identified with a lightning bolt (on the NOS chart) on the profile view of the approach. The minimum altitude to intercept the glide slope on the ILS RWY 6 approach is 1,800 feet MSL. (J42) — U.S. Terminal Procedures, IAP Legend; FAA-H-8083-15

Answer (A) is incorrect because 3,000 feet MSL is the minimum altitude in the holding pattern at PENNA intersection. Answer (C) is incorrect because 1,690 feet MSL is the altitude crossing the LOM on the glide slope.

ALL
4352. (Refer to Figure 73.) At which indication or occurrence should you initiate the published missed approach procedure for the ILS RWY 6 approach provided the runway environment is not in sight?

A—When reaching 374 feet MSL indicated altitude.
B—When 3 minutes (at 90 knots ground speed) have expired or reaching 374 feet MSL, whichever occurs first.
C—Upon reaching 374 feet AGL.

With respect to the operation of aircraft, Decision Height (DH) means the height at which a decision must be made during an ILS, MLS, or PAR instrument approach to either continue the approach or to execute a missed approach. The DH for the ILS RWY 6 approach is 374 feet MSL. (J42) — U.S. Terminal Procedures, IAP Chart Legend

Answer (B) is incorrect because timing is used for localizer approaches, or as a backup to ILS approaches. Answer (C) is incorrect because the DH is 374 feet MSL (not AGL).

ALL
4354. (Refer to Figure 73.) Using an average ground speed of 90 knots on the final approach segment, what rate of descent should be used initially to establish the glidepath for the ILS RWY 6 approach procedure?

A—395 feet per minute.
B—480 feet per minute.
C—555 feet per minute.

Enter the Rate-of-Descent Table (FAA Legend 21) in the left-hand column at the given glide slope angle of descent (3.0°, found on the profile view of the approach). Move horizontally to the right to the 90-knot ground speed column and read the recommended rate of descent (480 fpm). (J42) — U.S. Terminal Procedures, IAP Chart

Answer (A) is incorrect because 395 fpm is the required descent rate for 75 (not 90) knots ground speed. Answer (C) is incorrect because 555 fpm is the required descent rate for 105 (not 90) knots.

ALL
4355. (Refer to Figure 73.) What is the touchdown zone elevation for RWY 6?

A—174 feet MSL.
B—200 feet AGL.
C—270 feet MSL.

The airport sketch shows the Touchdown Zone Elevation (TDZE) for RWY 6 near the approach end of the landing runway (on the NOS IAP) as 174 feet MSL. (J42) — U.S. Terminal Procedures, IAP Chart Legend

Answer (B) is incorrect because 200 feet AGL is the height above touchdown (HAT). Answer (C) is incorrect because 270 feet MSL is the height of an obstruction near the approach end of RWY 6.

ALL
4357. (Refer to Figure 73.) Which runway and landing environment lighting is available for approach and landing on RWY 6 at Bradley International?

A—HIRL, REIL, and VASI.
B—HIRL and VASI.
C—ALSF2 and HIRL.

Looking at the airport sketch in the IAP, there is an "A" in a circle, with a dot above it, near the approach end of RWY 6. FAA Legend 19 indicates this means ALSF-2 approach lighting. Also, at the bottom of the airport sketch, it indicates RWY 6 has HIRL (high intensity runway lights). (J42) — U.S. Terminal Procedures, IAP Legend

Answers

| 4350 | [B] | 4351 | [B] | 4352 | [A] | 4354 | [B] | 4355 | [A] | 4357 | [C] |

ALL
4368. (Refer to Figures 74 and 80.) Which aircraft approach category should be used for a circling approach for a landing on RWY 27?

A—A.
B—B.
C—C.

The approach category is based on 1.3 times V_{SO}. V_{SO} can be found in FAA Figure 74. For this problem, 1.3 x 72 = 93.6 knots, so the aircraft would be Category B (91-120 knots). (J42) — U.S. Terminal Procedures, IAP Legend

ALL
4369. (Refer to Figure 80.) How many initial approach fixes serve the VOR/DME RWY 27R (Billings Logan) approach procedure?

A—Three.
B—Four.
C—Five.

The initial approach fixes (identified by "IAF") are: (1) R-040 and DME arc, (2) R157 and DME arc, (3) MUSTY intersection and (4) BILLINGS VORTAC. (J42) — U.S. Terminal Procedures, IAP Legend

ALL
4371. (Refer to Figure 80.) What is the TDZE for landing on RWY 27R?

A—3,649 feet MSL.
B—3,514 feet MSL.
C—3,450 feet MSL.

The instrument approach plate shows the touchdown zone elevation (TDZE) in the airport diagram (bottom, right-hand quadrant) near the approach end of runway 27R as 3,514 feet MSL. (J42) — Pilot/Controller Glossary

ALL
4470. What does the symbol T within a black triangle in the minimums section of the IAP for a particular airport indicate?

A—Takeoff minimums are 1 mile for aircraft having two engines or less and 1/2 mile for those with more than two engines.
B—Instrument takeoffs are not authorized.
C—Takeoff minimums are not standard and/or departure procedures are published.

A "T" within a black triangle in the minimums section of the instrument approach plate indicates that takeoff minimums are not standard and/or departure procedures are published. (J16) — FAA-H-8083-15

ALL
4635. (Refer to Figure 118.) During the ILS RWY 12L procedure at DSM, what altitude minimum applies if the glide slope becomes inoperative?

A—1,420 feet.
B—1,360 feet.
C—1,121 feet.

When the glide slope becomes inoperative on a precision approach, the approach, if continued, becomes a localizer approach. Looking at the minimums section of the chart for S-LOC, 12L indicates an MDA of 1360 would be applicable for all categories assuming a straight-in landing. (J42) — Instrument Approach Procedures

Answer (A) is incorrect because 1,420 feet is the circling MDA for categories B and C. Answer (C) is incorrect because 1,121 feet is the DH for the ILS, which is not applicable with the glide slope inoperative.

ALL
4642. (Refer to Figure 119.) The final approach fix for the precision approach is located at

A—DENAY intersection.
B—Glide slope intercept (lightning bolt).
C—ROMEN intersection/locator outer marker.

The Final Approach Fix (FAF) is the fix from which the final approach (IFR) to an airport is executed and which identifies the beginning of the final approach segment. On government charts it is designated by the Maltese Cross for non-precision approaches and by the lightning bolt symbol or glide slope intercept for precision approaches. The profile view of FAA Figure 119 depicts the lightning bolt symbol pointing from the glide slope intercept altitude of 2200 to ROMEN LOM which is also the non-precision FAP. (J42) — Pilot/Controller Glossary

Answer (A) is incorrect because Denay intersection is an initial approach fix (not the final approach fix). Answer (C) is incorrect because ROMEN intersection/locator outer marker is the final approach fix for the nonprecision (not precision) approach.

Answers

4368 [B]   4369 [B]   4371 [B]   4470 [C]   4635 [B]   4642 [B]

Chapter 8 **Arrival and Approach**

ALL
4648. (Refer to Figure 120.) Refer to the DEN ILS RWY 35R procedure. The FAF intercept altitude is

A—7,488 feet MSL.
B—7,500 feet MSL.
C—9,000 feet MSL.

The FAF identifies the beginning of the final approach segment. On precision approaches a lightning bolt symbol indicates the FAF. For DEN ILS RWY 35R, the FAF intercept altitude is 7,500 feet. (J42) — U.S. Terminal Procedures, IAP Legend

Answer (A) is incorrect because 7,488 feet is the glide slope altitude over the OM. Answer (C) is incorrect because 9,000 feet is the minimum altitude between Sedal and Engle intersections.

ALL
4649. (Refer to Figure 120.) The symbol on the plan view of the ILS RWY 35R procedure at DEN represents a minimum safe sector altitude within 25 NM of

A—Denver VORTAC.
B—Gandi outer marker.
C—Denver/Stapleton International Airport.

Minimum Safe Altitude (MSA) is defined as altitudes depicted on approach charts which provide at least 1,000 feet of obstacle clearance within a 25-mile radius upon which the procedure is predicated. In this case, it is the Denver VORTAC. MSA DEN 25 NM. (J42) — U.S. Terminal Procedures, IAP Legend

Answers (B) and (C) are incorrect because minimum safe sector altitudes are always based on a VOR or an NDB (not outer markers or airports).

ALL
4650. (Refer to Figure 121.) During the ILS RWY 30R procedure at DSM, the minimum altitude for glide slope interception is

A—2,365 feet MSL.
B—2,500 feet MSL.
C—3,000 feet MSL.

Glide slope intercept altitude is defined as the minimum altitude to intercept the glide slope/path on a precision approach. The intersection of the published intercept altitude with the glide slope/path is designated on government charts by the lightning bolt symbol which is 2,500 feet MSL for the ILS RWY 30R procedure at DSM. (J42) — U.S. Terminal Procedures, IAP Legend

Answer (A) is incorrect because 2,365 feet is the glide slope altitude over FOREM LOM. Answer (C) is incorrect because 3,000 feet is the MSA south of the LOM.

ALL
4651. (Refer to Figure 121.) During the ILS RWY 30R procedure at DSM, what MDA applies should the glide slope become inoperative?

A—1,157 feet.
B—1,320 feet.
C—1,360 feet.

When the ILS glide slope is inoperative or not utilized, the published straight-in localizer minimums apply. In this case, the minimum is 1,320 feet for the S-LOC 30R. (J42) — FAA-H-8083-15

Answer (A) is incorrect because 1,157 feet is DH for the ILS; the Localizer approach must be used when the glide slope is inoperative. Answer (C) is incorrect because 1,360 feet is MDA for a circling approach for Category A.

ALL
4652. (Refer to Figure 122.) The missed approach point of the ATL S-LOC 8L procedure is located how far from the LOM?

A—4.8 NM.
B—5.1 NM.
C—5.2 NM.

The bottom right-hand portion of the IAP, above the time and speed table, states "FAF to MAP 5.2 NM." (J42) — U.S. Terminal Procedures, IAP Legend

Answer (A) is incorrect because 4.8 NM is the distance from the LOM to the MM. Answer (B) is incorrect because 5.1 NM is the distance from the LOM to the IM.

ALL
4654. (Refer to Figure 123.) The symbol on the plan view of the VOR/DME-A procedure at 7D3 represents a minimum safe sector altitude within 25 NM of

A—DEANI intersection.
B—White Cloud VORTAC.
C—Baldwin Municipal Airport.

The Minimum Safe Altitude (MSA) is identified in the bottom right quadrant of the VOR/DME-A procedure at 7D3. The symbol in the middle of the MSA circle indicates the type of navigation aid the altitude is based. The statement above the circle, "MSA HIC 25 NM," indicates White Cloud VORTAC (HIC) as the facility identifier for the MSA. (J42) — Pilot/Controller Glossary

Answers

| 4648 [B] | 4649 [A] | 4650 [B] | 4651 [B] | 4652 [C] | 4654 [B] |

ALL
4655. (Refer to Figure 124.) What options are available concerning the teardrop course reversal for LOC RWY 35 approach to Duncan/Halliburton Field?

A—If a course reversal is required, only the teardrop can be executed.
B—The point where the turn is begun and the type and rate of turn are optional.
C—A normal procedure turn may be made if the 10 DME limit is not exceeded.

When a teardrop procedure turn is depicted and a course reversal is required, this type turn must be executed. (J42) — Instrument Approach Procedures

ALL
4656. (Refer to Figure 124.) The point on the teardrop procedure where the turn inbound (LOC RWY 35) Duncan/Halliburton, is initiated is determined by

A—DME and timing to remain within the 10-NM limit.
B—Timing for a 2 minute maximum.
C—Estimating ground speed and radius of turn.

A procedure turn is the maneuver prescribed when it is necessary to reverse direction to establish the aircraft inbound on an intermediate or final approach course. The maneuver must be completed within the distance specified in the profile view. DME and timing are the only practical methods of determining that you remain within the 10-mile boundary. (J42) — Instrument Approach Procedures

Answer (B) is incorrect because the only limit on the procedure turn is 10 NM from the VOR. Answer (C) is incorrect because estimating ground speed and radius of turn is only part of what is required; the pilot must also remain within 10 NM of the VOR.

ALL
4657. (Refer to Figure 125.) If your aircraft was cleared for the ILS RWY 17R at Lincoln Municipal and crossed the Lincoln VOR at 5,000 feet MSL, at what point in the teardrop could a descent to 3,000 feet commence?

A—As soon as intercepting LOC inbound.
B—Immediately.
C—Only at the point authorized by ATC.

Once the pilot is cleared for the approach and established on part of the published route or procedure, the pilot may descend to the published altitude. LNK is the IAF, and upon crossing LNK the pilot may commence the approach. (J42) — AIM ¶5-4-7

Answer (A) is incorrect because a descent to 3,000 feet may begin once the pilot is LNK VORTAC outbound (not inbound). Answer (C) is incorrect because published altitudes apply when the pilot is on a published route or procedure and has been cleared for the approach.

ALL
4658. (Refer to Figure 125.) If cleared for an S-LOC 17R approach at Lincoln Municipal from over TOUHY, it means the flight should

A—land straight in on runway 17R.
B—comply with straight-in landing minimums.
C—begin final approach without making a procedure turn.

When cleared for an approach, the pilot has authorization to execute the specific approach cleared for. When cleared for the S-LOC 17R approach from over Touhy Intersection, the approach specifies an interception of the localizer with no procedure turn allowed (NoPT). (J18) — AIM ¶5-4-8

Answer (A) is incorrect because the pilot could request to land on a different runway. Answer (B) is incorrect because circling minimums may apply if the pilot requests (or is assigned) a different runway.

ALL
4659. (Refer to Figure 126.) What landing minimums apply for a 14 CFR Part 91 operator at Dothan, AL using a category C aircraft during a circling LOC 31 approach at 120 knots? (DME available).

A—MDA 860 feet MSL and visibility 2 SM.
B—MDA 860 feet MSL and visibility 1 and 1/2 SM.
C—MDA 720 feet MSL and visibility 3/4 SM.

Under the DME minimums for a circling approach for RWY 31, Category C aircraft are required to use the MDA of 860 feet MSL and 1-1/2 statute mile visibility. (J42) — Instrument Approach Procedures

Answer (A) is incorrect because 2 SM is the visibility requirement for Category D aircraft. Answer (C) is incorrect because 720 feet and 3/4 SM are the requirements for Category D aircraft and straight-in localizer approach.

Answers

| 4655 | [A] | 4656 | [A] | 4657 | [B] | 4658 | [C] | 4659 | [B] |

Chapter 8 Arrival and Approach

ALL
4660. (Refer to Figure 126.) If cleared for a straight-in LOC approach from over OALDY, it means the flight should

A—land straight in on runway 31.
B—comply with straight-in landing minimums.
C—begin final approach without making a procedure turn.

A straight-in approach is one begun without first having executed a procedure turn, not necessarily completed with a straight-in landing or made to straight-in landing minimums. (J33) — Pilot/Controller Glossary

Answer (A) is incorrect because the pilot may request a different runway. Answer (B) is incorrect because the circling minimums may be used.

ALL
4661. (Refer to Figure 126.) What is the ability to identify the RRS 2.5 stepdown fix worth in terms of localizer circle-to-land minimums for a category C aircraft?

A—Decreases MDA by 20 feet.
B—Decreases visibility by 1/2 SM.
C—Without the stepdown fix, a circling approach is not available.

The minimum for a circling approach in CAT C aircraft without DME is 880 and 1-1/2. With DME it is 860 and 1-1/2. The only change is the 20-foot decrease in the MDA. (J42) — Instrument Approach Procedures

Answer (B) is incorrect because the visibility minimums are not affected by DME use. Answer (C) is incorrect because circling procedures are authorized without DME.

ALL
4662. (Refer to Figure 127.) If cleared for NDB RWY 28 approach (Lancaster/Fairfield) over ZZV VOR, the flight would be expected to

Category A aircraft
Last assigned altitude 3,000 feet

A—proceed straight in from CRISY, descending to MDA after CASER.
B—proceed to CRISY, then execute the teardrop procedure as depicted on the approach chart.
C—proceed direct to CASER, then straight in to S-28 minimums of 1620-1.

Follow the published feeder route descending to 2,700 outbound on ZZV R-246 "NoPT." From CRISY proceed to CASER LOM via the 277° bearing TO at 2,700. Passing CASER, begin timing and descend to the MDA of 1,620 on the 280° bearing FROM. (J42) — Instrument Approach Procedures

Answer (B) is incorrect because the NoPT symbol under ZZV VOR indicates that the procedure turn is not authorized when approaching from ZZV. Answer (C) is incorrect because the arrow southwest of ZZV VOR indicates the route to CRISY and then straight in.

ALL
4672. During an instrument precision approach, terrain and obstacle clearance depends on adherence to

A—minimum altitude shown on the IAP.
B—terrain contour information.
C—natural and man-made reference point information.

A pilot adhering to the altitudes, flight paths, and weather minimums depicted on the IAP chart is assured terrain and obstruction clearance and runway or airport alignment during approach for landing. (J18) — AIM ¶5-4-5(a)(3)

Answer (B) is incorrect because the IAP takes terrain contour information into account. Answer (C) is incorrect because instrument (not visual) information is used as reference points.

RTC
4673. (Refer to Figure 128.) What is the helicopter landing minimum for the VOR RWY 36 approach at Price/Carbon County Airport?

A—500-foot ceiling and 1/2 mile visibility.
B—1-mile visibility.
C—one-half mile visibility.

Choppers use Category A MDA or DH with half the published visibility (but equal to or not less than 1,200 RVR or 1/4 mile). The visibility minimums for the VOR RWY 36 approach, Category A are one mile, so the helicopter minimum is one-half mile visibility. (B97) — 14 CFR §97.3(d)(1)

RTC
4676. (Refer to Figure 128.) What is the helicopter MDA for a straight-in VOR RWY 36 approach at Price/Carbon County Airport (VOR only)?

A—6,090 feet MSL.
B—500 feet MSL.
C—6,400 feet MSL.

Answers

4660 [C]	4661 [A]	4662 [A]	4672 [A]	4673 [C]	4676 [C]

Choppers use category A MDA or DH with half the published visibility (but equal to or not less than 1,200 RVR or 1/4 mile). The MDA for a straight-in VOR RWY 36 approach, Category A, is 6,400 feet MSL. (B97) — 14 CFR §97.3(d)(1)

ALL
4677. (Refer to Figure 128.) At which points may you initiate a descent to the next lower minimum altitude when cleared for the VOR RWY 36 approach, from the PUC R-095 IAF (DME operative)?

A— Start descent from 8,000 when established on final, from 7,500 when at the 4 DME fix, and from 6,180 when landing requirements are met.
B— Start descent from 8,000 when established on the PUC R-186, from 6,400 at the 4 DME fix, and from 6,180 when landing requirements are met.
C— Start descent from 8,000 at the R-127, from 6,400 at the LR-127, from 6,180 at the 4 DME fix.

The overhead view shows 8,000 while on the arc. Once established on R-186, the pilot can proceed down to 6,400. After the 4 DME fix the pilot can proceed down to 6,180, which is the decision height. With the runway environment in sight, the pilot may continue to descend for the landing. (J42) — Instrument Approach Procedures

Answer (A) is incorrect because 7,500 is the minimum altitude for the procedure turn. Answer (C) is incorrect because the pilot may not descend below 8,000 feet until established on R-186, may not descend below 6,400 until identifying the 4 DME fix, and may not descend below 6,180 until the runway environment is in sight.

ALL
4678. (Refer to Figure 128.) What is the purpose of the 10,300 MSA on the Price/Carbon County Airport Approach Chart?

A— It provides safe clearance above the highest obstacle in the defined sector out to 25 NM.
B— It provides an altitude above which navigational course guidance is assured.
C— It is the minimum vector altitude for radar vectors in the sector southeast of PUC between 020° and 290° magnetic bearing to PUC VOR.

Minimum Safe Altitudes (MSA) are published for emergency use on Approach Procedure Charts, utilizing NDB or VOR type facilities. The altitude shown provides at least 1,000 feet of clearance above the highest obstacle in the defined sector to a distance of 25 NM from the facility upon which the procedure is predicated. As many as four sectors may be depicted with different altitudes for each sector displayed in rectangular boxes in the plan view of the chart. (J18) — Pilot/Controller Glossary

Answer (B) is incorrect because navigational course guidance is not assured at the MSA within these sectors. Answer (C) is incorrect because the MSA provides safe clearance (not vector) altitudes.

RTC
4679. (Refer to Figure 129.) As you approach LABER during a straight-in RNAV RWY 36 approach in a helicopter, Little Rock Approach Control advises that the ceiling is 400 feet and the visibility is 1/4 mile. Do regulations permit you to continue the approach and land?

A— No, you may not reduce the visibility prescribed for Category A airplanes by more than 50 percent.
B— Yes, only a 1/4 mile visibility or an RVR of 1,200 feet is required for any approach, including RNAV.
C— No, neither the ceiling nor the visibility meet regulatory requirements.

Choppers use Category A MDA or DH with half the published visibility (but equal to or not less than 1,200 RVR or 1/4 mile). The visibility minimum for RNAV RWY 36, Category A, is 1 mile. So for helicopters, the minimum is one-half mile. The weather minimums are below this, so the helicopter is not permitted to continue the approach and land. (B97) — 14 CFR §97.3(d)(1)

ALL
4680. (Refer to Figure 129.) What indication should you get when it is time to turn inbound while in the procedure turn at LABER?

A— 4 DME miles from LABER.
B— 10 DME miles from the MAP.
C— 12 DME miles from LIT VORTAC.

DME holding patterns and procedure turn entries follow the usual rules except that distances (nautical miles) are used in lieu of time values. The DME will indicate 4 DME (see the holding pattern in the plan view) from LABER when it is time to turn inbound. (J17) — AIM ¶5-3-7

Answers

| 4677 | [B] | 4678 | [A] | 4679 | [C] | 4680 | [A] |

Chapter 8 **Arrival and Approach**

ALL
4682. (Refer to Figure 129.) How should the missed approach point be identified when executing the RNAV RWY 36 approach at Adams Field?

A—When the TO-FROM indicator changes.
B—Upon arrival at 760 feet on the glidepath.
C—When time has expired for 5 NM past the FAF.

The MAP is a waypoint and the TO/FROM indicator will show passage of the "phantom station." (J42) — Instrument Approach Procedures

Answer (B) is incorrect because the pilot will be at 760 feet on the RWY 36 approach prior to reaching the MAP. Answer (C) is incorrect because on the RNAV, the MAP is designated, so it is to be used (not the time).

ALL
4683. (Refer to Figure 129.) What is the position of LABER relative to the reference facility?

A—316°, 24.3 NM.
B—177°, 10 NM.
C—198°, 8 NM.

The bottom line "113.9 LIT 198.0°-8" of the information box extending from LABER waypoint gives the frequency, the identifier (Little Rock VOR), and the radial and distance from the reference facility (198°, 8 NM). (J42) — Instrument Approach Procedures

ALL
4686. (Refer to Figure 130.) What are the procedure turn restrictions on the LDA RWY 6 approach at Roanoke Regional?

A—Remain within 10 NM of CLAMM INT and on the north side of the approach course.
B—Remain within 10 NM of the airport on the north side of the approach course.
C—Remain within 10 NM of the outer marker on the north side of the approach course.

The profile view for LDA RWY 6 approach has a note in the upper left-hand corner which states "Remain within 10 NM." The 10 NM is in reference to the Clamm INT. The procedure turn is depicted north of the approach with a right turn, so the pilot must remain on the north side of the approach course. (J42) — Instrument Approach Procedures

Answers (B) and (C) are incorrect because the 10 NM is in reference to the FAF (not the airport or the outer marker).

ALL
4687. (Refer to Figure 130.) What are the restrictions regarding circle to land procedures for LDA RWY/GS 6 approach at Roanoke Regional?

A—Circling to runway 24 not authorized.
B—Circling not authorized NW of RWY 6-24.
C—Visibility increased 1/2 mile for circling approach.

See the note under in the remarks section of the instrument approach. Circling not authorized northwest of runway 6/24. (J42) — Instrument Approach Procedures

ALL
4688. (Refer to Figure 130.) At what minimum altitude should you cross CLAMM intersection during the S-LDA 6 approach at Roanoke Regional?

A—4,200 MSL.
B—4,182 MSL.
C—2,800 MSL.

The single asterisk alongside the glide slope intercept altitude in the profile view refers you to the note below specifying 4,200 feet MSL when glide slope is not in use. (J42) — Instrument Approach Procedures

Answer (B) is incorrect because 4,182 is the glide slope altitude at Clamm INT. Answer (C) is incorrect because 2,800 is the minimum altitude between Clamm and Skirt.

ALL
4689. (Refer to Figure 130.) How should the pilot identify the missed approach point for the S-LDA GS 6 approach to Roanoke Regional?

A—Arrival at 1,540 feet on the glide slope.
B—Arrival at 1.0 DME on the LDA course.
C—Time expired for distance from OM to MAP.

The LDA Rwy GS 6 approach is considered a precision approach; therefore, the missed approach point is the DH which is 1,540 feet. (J42) — Instrument Approach Procedures

ALL
4690. (Refer to Figure 131.) The control tower at BOS reports "tall vessels" in the approach area. What are the VOR/DME RNAV RWY 4R straight-in approach minimums for Category A aircraft.

A— 890/24.
B— 840/40
C— 890/40.

The straight-in minimums for a Category A aircraft are 840/24. However, the note in the upper left portion of the approach procedure states "when control tower reports tall vessels in approach area, increase S-4R Cat A visibility to RVR 4000." (J42) — Instrument Approach Procedures

ALL
4691. (Refer to Figure 131.) What determines the MAP for the straight-in VOR/DME RNAV RWY 4R approach at BOS?

A— RULSY way point.
B— .5 NM to RULSY way point.
C— 2.5 NM to RULSY at 840 feet MSL.

The profile view of the chart states the MAP WP is 0.5 NM from RULSY WP. (J42) — Instrument Approach Procedures

ALL
4692. Which of the following statements is true regarding Parallel ILS approaches?

A— Parallel ILS approach runway centerlines are separated by at least 4,300 feet and standard IFR separation is provided on the adjacent runway.
B— Parallel ILS approaches provide aircraft a minimum of 1-1/2 miles radar separation between successive aircraft on the adjacent localizer course.
C— Landing minimums to the adjacent runway will be higher than the minimums to the primary runway, but will normally be lower than the published circling minimums.

Parallel ILS approaches provide aircraft a minimum of 1-1/2 miles separation between successive aircraft on the adjacent localizer course. (J18) — AIM ¶5-4-13

ALL
4693. (Refer to Figure 131.) What is the landing distance available for the VOR/DME RNAV RWY 4R approach at BOS?

A— 7,000 feet.
B— 10,005 feet.
C— 8,850 feet.

The upper-right corner of the airport diagram indicates the landing distance for Rwy 4R is 8,850 feet. (J42) — Instrument Approach Procedures

ALL
4694. (Refer to Figure 131.) During a missed approach from the VOR/DME RNAV RWY 4R approach at BOS, what course should be flown to the missed approach holding way point?

A— 036°.
B— Runway heading.
C— 033°.

The missed approach procedures are in the upper right corner of the approach plate. The missed approach procedure for VOR/DME RNAV RWY 4R states: climb to 3000 via 033° track to WAXEN WP and hold. (J42) — Instrument Approach Procedures

ALL
4671. During an instrument approach, under what conditions, if any, is the holding pattern course reversal not required?

A— When radar vectors are provided.
B— When cleared for the approach.
C— None, since it is always mandatory.

Holding pattern course reversal is not required with radar vectoring or when "NoPT" is shown on the approach course. (J18) — AIM ¶5-4-8(b)(3)

Answer (B) is incorrect because a course reversal may be required when cleared for the approach. Answer (C) is incorrect because a course reversal is not required when being radar vectored or if NoPT is shown on the approach chart.

Answers

4690 [B]   4691 [B]   4692 [B]   4693 [C]   4694 [C]   4671 [A]

Chapter 8 **Arrival and Approach**

ALL
4696. (Refer to Figure 133.) How should a pilot reverse course to get established on the inbound course of the ILS RWY 9, if radar vectoring or the three IAF's are not utilized?

A—Execute a standard 45° procedure turn toward Seal Beach VORTAC or Pomona VORTAC.
B—Make an appropriate entry to the depicted holding pattern at Swan Lake OM/INT.
C—Use any type of procedure turn, but remain within 10 NM of Riverside VOR.

A procedure turn need not be established when an approach can be made from a properly aligned holding pattern. In such cases, the holding pattern is established over an intermediate fix or a final approach fix. The holding pattern maneuver is completed when the aircraft is established on the inbound course after executing the appropriate entry. (J42) — AIM ¶5-4-8(a)(4)

Answer (A) is incorrect because it is not an appropriate procedure. Answer (C) is incorrect because the pilot must follow the depicted holding pattern.

RTC
4697. (Refer to Figure 133.) If the Class D airspace is not effective, what is the LOC/VOR minima for a helicopter if cleared for the S-LOC 9 approach at Riverside Municipal?

A—1,200 and 1/4 mile.
B—991 and RVR 24.
C—1,300 and 1/4 mile.

Helicopters may use airplane approach procedures using the Category A Minimum Descent Altitude (MDA) or Decision Height (DH). The required visibility minimum may be reduced to one-half the published visibility minimum for Category A aircraft, but it must not in any case be reduced to less than one-quarter mile or 1,200 feet RVR.

```
   1,160   MDA
  +   40   from note #2 in plan view
   1,200   New MDA
```
(B97) — U.S. Terminal Procedures, IAP Legend

ALL
4700. (Refer to Figure 133.) Why are two VOR/LOC receivers recommended to obtain an MDA of 1,160 when making an S-LOC 9 approach to Riverside Municipal?

A—To obtain R-327 of PDZ when on the localizer course.
B—In order to identify Riverside VOR.
C—To utilize the published stepdown fix.

A step-down fix may be provided on the final, i.e., between the final approach fix and the airport for the purpose of authorizing a lower MDA after passing an obstruction. This step-down fix may be made by an NDB bearing, fan marker, radar fix, radial from another VOR, or by a DME. The Agnes INT needs to be identified for the step-down fix, in order to proceed to the DH 1160. (J42) — Instrument Approach Procedures

Answer (A) is incorrect because the Swan Lake NDB can be used to determine R-327 of PDZ when on the localizer. Answer (B) is incorrect because the Riverside VOR is not used in this approach.

ALL
4701. (Refer to Figure 133.) What is the minimum altitude descent procedure if cleared for the S-ILS 9 approach from Seal Beach VORTAC?

A—Descend and maintain 3,000 to JASER INT, descend to and maintain 2,500 until crossing SWAN LAKE, descend and maintain 1,260 until crossing AGNES, and to 991 (DH) after passing AGNES.
B—Descend and maintain 3,000 to JASER INT, descend to 2,800 when established on the LOC course, intercept and maintain the GS to 991 (DH).
C—Descend and maintain 3,000 to JASER INT, descend to 2,500 while established on the LOC course inbound, intercept and maintain the GS to 991 (DH).

A pilot shall maintain the last altitude assigned to that pilot until the aircraft is established on a segment of a published route or instrument approach procedure, unless a different altitude is assigned by ATC. Once cleared from Seal Beach VORTAC, the pilot shall descend and maintain 3,000 feet to Jaser INT, then descend to 2,500 feet once established on the localizer course inbound, then intercept and maintain the glide slope to the decision height (991 feet). (J42) — Instrument Approach Procedures

Answer (A) is incorrect because the pilot must maintain 2,500 feet until intercepting the glide slope, and then the glide slope is flown to the decision height at 991 feet. Answer (B) is incorrect because once established on the localizer, the pilot should descend to 2,500 feet (not 2,800) until the glide slope is intercepted.

Answers

4696 [B]   4697 [A]   4700 [C]   4701 [C]

Chapter 8 **Arrival and Approach**

RTC
4713. During a precision instrument approach (using Category A minimums) a helicopter may not be operated below DH unless

A—the ceiling is forecast to be at or above landing minimums prescribed for that procedure.
B—positioned such that a normal approach to the runway of intended landing can be made.
C—the visibility is forecast to be at or above the landing minimums prescribed for that procedure.

Where a DH or MDA is applicable, no pilot may operate an aircraft at any airport below the authorized MDA or continue an approach below the authorized DH unless the aircraft is continuously in a position from which a descent to a landing on the intended runway can be made at a normal rate of descent using normal maneuvers. (B10) — 14 CFR §91.175(c)(i)

ALL
4714. Which procedure should be followed by a pilot who is circling to land in a Category B airplane, but is maintaining a speed 5 knots faster than the maximum specified for that category?

A—Use the approach minimums appropriate for Category C.
B—Use Category B minimums.
C—Use Category D minimums since they apply to all circling approaches.

If it is necessary to maneuver at speeds in excess of the upper limit of the speed range for each category, the minimum for the next higher approach category should be used. (J18) — AIM ¶5-4-1

Answer (B) is incorrect because the pilot should use the minimums in the category appropriate to the approach speed being flown. Answer (C) is incorrect because not all circling approaches have a Category D.

ALL
4717. Aircraft approach categories are based on

A—certificated approach speed at maximum gross weight.
B—1.3 times the stall speed in landing configuration at maximum gross landing weight.
C—1.3 times the stall speed at maximum gross weight.

Aircraft approach category means a grouping of aircraft based on a speed of 1.3 V_{SO} and the maximum certificated landing weight are those values as established for the aircraft by the certificating authority of the country of registry. (J18) — AIM ¶5-4-1

RTC
4722. Upon what maximum airspeed is the instrument approach criteria for a helicopter based?

A—100 knots.
B—90 knots.
C—80 knots.

Helicopters use Category A procedures, which requires a speed less than 91 knots. (J42) — 14 CFR §97.3(b)

RTC
4723. All helicopters are considered to be in which approach category for a helicopter IAP?

A—A.
B—A or B, depending upon weight.
C—B.

When helicopters use instrument flight procedures designed for fixed-wing aircraft, approach Category A approach minima shall apply, regardless of helicopter weight. (J42) — 14 CFR §97.3(d)(1)

RTC
4724. What reduction, if any, to visibility requirements is authorized when using a fixed-wing IAP for a helicopter instrument approach?

A—All visibility requirements may be reduced by one-half.
B—All visibility requirements may be reduced by one-fourth.
C—The visibility requirements may be reduced by one-half, but in no case lower than 1,200 RVR or 1/4 mile.

The required visibility minimum may be reduced to one-half the published visibility minimum for Category A aircraft, but must not in any case be reduced to less than one-quarter mile or 1,200 feet RVR. (B97) — 14 CFR §97.3(d)(1)

Answers

| 4713 | [B] | 4714 | [A] | 4717 | [B] | 4722 | [B] | 4723 | [A] | 4724 | [C] |

Chapter 8 **Arrival and Approach**

ALL

4744. If all ILS components are operating and the required visual references are not established, the missed approach should be initiated upon

A—arrival at the DH on the glide slope.
B—arrival at the middle marker.
C—expiration of the time listed on the approach chart for missed approach.

An ILS approach is a precision approach procedure, because it has an electronic glide slope. The Missed Approach Point (MAP) on a precision approach is always the Decision Height (DH). (J18) — AIM ¶5-5-5

Answer (B) is incorrect because the MM is not the MAP. Answer (C) is incorrect because the time listed on the approach chart for the missed approach is used for a localizer approach when the glide slope has failed.

ALL

4749. When may a pilot make a straight-in landing, if using an IAP having only circling minimums?

A—A straight-in landing may not be made, but the pilot may continue to the runway at MDA and then circle to land on the runway.
B—The pilot may land straight-in if the runway is the active runway and he has been cleared to land.
C—A straight-in landing may be made if the pilot has the runway in sight in sufficient time to make a normal approach for landing, and has been cleared to land.

When either the normal rate of descent or the runway alignment factor of 30° is exceeded, a straight-in-minimum is not published and a circling minimum applies. The fact that a straight-in-minimum is not published does not preclude the pilot from landing straight-in, if the active runway is in sight and there is sufficient time to make a normal approach for landing. (J18) — AIM ¶5-4-18(d)

Answer (A) is incorrect because a straight-in landing can be made if there is sufficient time and the runway is in sight. Answer (B) is incorrect because in addition to having the runway in sight, there must also be sufficient time to make a normal approach for landing.

ALL

4763. If during an ILS approach in IFR conditions, the approach lights are not visible upon arrival at the DH, the pilot is

A—required to immediately execute the missed approach procedure.
B—permitted to continue the approach and descend to the localizer MDA.
C—permitted to continue the approach to the approach threshold of the ILS runway.

No pilot may operate an aircraft at any airport below the authorized MDA, or continue an approach below the authorized DH unless at least one of the following visual references for the intended runway is distinctly visible and identifiable to the pilot. If the approach lights are not visible upon arrival at the DH, then the pilot must execute the missed approach procedure. (B10) — 14 CFR §91.175

Answer (B) is incorrect because the DH is lower than the localizer MDA, and is used as the missed approach point on an ILS approach. Answer (C) is incorrect because the pilot may only continue the approach below the DH if the required visual references are distinctly visible and identifiable.

ALL

4764. Immediately after passing the final approach fix inbound during an ILS approach in IFR conditions, the glide slope warning flag appears. The pilot is

A—permitted to continue the approach and descend to the DH.
B—permitted to continue the approach and descend to the localizer MDA.
C—required to immediately begin the prescribed missed approach procedure.

When the glide slope fails, the ILS reverts to a nonprecision localizer approach. With a glide slope failure, the pilot is permitted to switch to a localizer approach and descend to the localizer minimum descent altitude (MDA) (J01) — 14 CFR §91.175.

Answer (A) is incorrect because without a glide slope, the localizer becomes a nonprecision approach which has a MDA (not a DH). Answer (C) is incorrect because the missed approach procedure should only be executed upon reaching the missed approach point.

Radar Approaches

Where radar is approved for approach control service, it is used not only for Airport Surveillance Radar (ASR) and Precision Approach Radar (PAR) approaches, but is also used to provide course guidance to the final approach course and for the monitoring of nonradar approaches.

Airport Surveillance Radar (ASR) may provide a nonprecision approach to an airport. The controller will vector the aircraft to the final approach course, then provide azimuth guidance and inform the pilot of the aircraft's range from the runway. The pilot will also be advised of the MDA, told when to begin descent, and informed when over the MAP. Upon request, the controller will also provide recommended altitudes each mile on final.

Precision Approach Radar (PAR) provides elevation (glide path) guidance in addition to azimuth and range information.

On occasion, a no-gyro radar approach may be provided. The controller will direct the pilot when to start and stop turns. All turns should be executed at standard rate, except on final approach. On final approach, all turns will be half-standard rate.

Radar service is automatically terminated upon completion of the radar approach. Pilots are authorized to use any radar IAP for which civil minimums are published.

ALL
4711. Where may you use a surveillance approach?

A—At any airport that has an approach control.
B—At any airport which has radar service.
C—At airports for which civil radar instrument approach minimums have been published.

An ASR (Airport Surveillance Radar) approach is conducted by surveillance radar and provides navigational guidance in azimuth only. This type of approach may be made to any airport or heliport having an approved surveillance approach. (J18) — AIM ¶5-4-3

Answers (A) and (B) are incorrect because specific procedures, missed approach points, and minimums must be established before ATC can conduct a surveillance approach.

ALL
4728. How is ATC radar used for instrument approaches when the facility is approved for approach control service?

A—Precision approaches, weather surveillance, and as a substitute for any inoperative component of a navigation aid used for approaches.
B—ASR approaches, weather surveillance, and course guidance by approach control.
C—Course guidance to the final approach course, ASR and PAR approaches, and the monitoring of nonradar approaches.

Where radar is approved for approach control service, it is used not only for radar approaches (ASR and PAR), but is also used to provide vectors in conjunction with published instrument approach procedures predicated on radio NAVAIDs such as ILS, VOR, and NDB. (J18) — AIM ¶5-4-3

Answers (A) and (B) are incorrect because approach control radar is not designed for weather surveillance.

ALL
4741. Which information, in addition to headings, does the radar controller provide without request during an ASR approach?

A—The recommended altitude for each mile from the runway.
B—When reaching the MDA.
C—When to commence descent to MDA, the aircraft's position each mile on final from the runway, and arrival at the MAP.

An ASR approach is one in which a controller provides navigation guidance in azimuth only. The pilot will be advised when to commence descent to the MDA. In addition, the pilot will be advised of the location of the Missed Approach Point (MAP) prescribed for the procedure and the aircraft's position each mile on final from the runway, airport or heliport or MAP, as appropriate. (J18) — AIM ¶5-4-10(c)(2)

Answer (A) is incorrect because altitude recommended for each mile is only available on request. Answer (B) is incorrect because the controller does not have precise enough altitude information.

Answers
4711 [C]   4728 [C]   4741 [C]

Chapter 8 Arrival and Approach

ALL
4822. During a "no-gyro" approach and prior to being handed off to the final approach controller, the pilot should make all turns

A—one-half standard rate unless otherwise advised.
B—any rate not exceeding a 30° bank.
C—standard rate unless otherwise advised.

During a "no-gyro" approach and prior to being handed off to the final approach controller, the pilot should make all turns at standard rate and should execute the turn immediately upon receipt of instructions. (J18) — AIM ¶5-4-10(c)(3)

ALL
4823. After being handed off to the final approach controller during a "no-gyro" surveillance or precision approach, the pilot should make all turns

A—one half standard-rate.
B—based upon the ground speed of the aircraft.
C—standard-rate.

After being handed off to the final approach controller during a "no-gyro" approach, the pilot should make all turns at one-half standard rate and should execute the turn immediately upon receipt of instructions. (J18) — AIM ¶5-4-10(c)(3)

Visual and Contact Approaches

Under certain conditions, a pilot on an IFR flight plan may be authorized to deviate from the published IAP and proceed to the destination airport visually, while remaining on an IFR flight plan and in the IFR system.

A **Visual Approach** may be assigned by ATC or requested by the pilot if:
1. Surface visibility is at least 3 miles and the ceiling is at least 500 feet above the minimum vectoring altitude;
2. The pilot has reported the airport or the preceding aircraft in sight; and
3. The aircraft will remain in VFR conditions.

A **Contact Approach** must be requested by the pilot, provided:
1. The aircraft remains clear of clouds;
2. The pilot has at least 1 mile flight visibility;
3. The pilot can reasonably expect to continue to the airport in the specified conditions.
4. Controllers may honor the pilot's request for a contact approach if reported ground visibility is at least 1 mile, the airport has an IAP, and separation from other aircraft can be maintained.

ALL
4718. What are the main differences between a visual approach and a contact approach?

A—The pilot must request a contact approach; the pilot may be assigned a visual approach and higher weather minimums must exist.
B—The pilot must request a visual approach and report having the field in sight; ATC may assign a contact approach if VFR conditions exist.
C—Any time the pilot reports the field in sight, ATC may clear the pilot for a contact approach; for a visual approach, the pilot must advise that the approach can be made under VFR conditions.

Visual approaches are initiated by ATC to reduce pilot/controller workload and expedite traffic by shortening flight paths to the airport. Contact approaches are initiated by the pilot and are provided if the pilot will remain clear of clouds and have at least 1 mile flight visibility. (J18) — AIM ¶5-4-20 and ¶5-4-22

Answer (B) is incorrect because a visual approach may be assigned and a contact approach must be requested. Answer (C) is incorrect because a contact approach must be requested.

Answers
4822 [C]   4823 [A]   4718 [A]

ALL
4735. What are the requirements for a contact approach to an airport that has an approved IAP, if the pilot is on an instrument flight plan and clear of clouds?

A—The controller must determine that the pilot can see the airport at the altitude flown and can remain clear of clouds.
B—The pilot must agree to the approach when given by ATC and the controller must have determined that the visibility was at least 1 mile and be reasonably sure the pilot can remain clear of clouds.
C—The pilot must request the approach, have at least 1-mile visibility, and be reasonably sure of remaining clear of clouds.

Pilots operating in accordance with an IFR flight plan, provided they are clear of clouds and have at least 1 mile flight visibility and can reasonably expect to continue to the destination airport in those conditions, may request ATC authorization for a contact approach. Controllers may authorize a contact approach if it is specifically requested by the pilot. ATC cannot initiate this approach. (J18) — AIM ¶5-4-22

Answer (A) is incorrect because the pilot (not the controller) must determine whether the flight can continue to the airport. Answer (B) is incorrect because only the pilot can request a contact approach.

ALL
4737. When may you obtain a contact approach?

A—ATC may assign a contact approach if VFR conditions exist or you report the runway in sight and are clear of clouds.
B—ATC may assign a contact approach if you are below the clouds and the visibility is at least 1 mile.
C—ATC will assign a contact approach only upon request if the reported visibility is at least 1 mile.

Pilots operating in accordance with an IFR flight plan, provided they are clear of clouds and have at least 1 mile flight visibility and can reasonably expect to continue to the destination airport in those conditions, may request ATC authorization for a contact approach. (J18) — AIM ¶5-4-22

ALL
4750. A contact approach is an approach procedure that may be used

A—in lieu of conducting a SIAP.
B—if assigned by ATC and will facilitate the approach.
C—in lieu of a visual approach.

A contact approach is an approach wherein an aircraft on an IFR flight plan, having an air traffic control authorization, operating clear of clouds with at least 1 mile flight visibility and a reasonable expectation of continuing to the destination airport on those conditions, may deviate from the standard instrument approach procedure (SIAP) and proceed to the destination airport by visual reference to the surface. This approach will only be authorized when requested by the pilot and the reported ground visibility at the destination airport is at least 1 statute mile. (J18) — AIM ¶5-4-22

Answer (B) is incorrect because a contact approach can only be authorized when requested by the pilot. Answer (C) is incorrect because a contact approach would not be necessary if a visual approach is already in progress.

ALL
4743. What conditions are necessary before ATC can authorize a visual approach?

A—You must have the preceding aircraft in sight, and be able to remain in VFR weather conditions.
B—You must have the airport in sight or the preceding aircraft in sight, and be able to proceed to, and land in IFR conditions.
C—You must have the airport in sight or a preceding aircraft to be followed, and be able to proceed to the airport in VFR conditions.

When it will be operationally beneficial, ATC may authorize an aircraft to conduct a visual approach to an airport or to follow another aircraft when flight to, and landing at, the airport can be accomplished in VFR weather. The pilot must have the airport or the identified aircraft in sight before the clearance is issued. (J18) — AIM ¶5-4-20(a)

Answer (A) is incorrect because the pilot could have either the preceding aircraft in sight or the airport. Answer (B) is incorrect because the landing must be made in VFR conditions.

Answers

4735 [C]   4737 [C]   4750 [A]   4743 [C]

Chapter 8 **Arrival and Approach**

ALL
4712. You arrive at your destination airport on an IFR flight plan. Which is a prerequisite condition for the performance of a contact approach?

A—Clear of clouds and at least 1 SM flight visibility.
B—A ground visibility of at least 2 SM.
C—A flight visibility of at least 1/2 NM.

Pilots operating in accordance with an IFR flight plan, provided they are clear of clouds and have at least 1 mile flight visibility and can reasonably expect to continue to the destination airport in those conditions, may request ATC authorization for a contact approach. Controllers may authorize a contact approach provided the contact approach is specifically requested by the pilot. ATC cannot initiate this approach. (J18) — AIM ¶5-4-22

Answer (B) is incorrect because flight visibility is the requirement for a contact approach. Answer (C) is incorrect because a flight visibility of 1 SM is needed.

Timed Approaches from Holding

Timed approaches may be conducted when the following conditions are met:

1. A control tower is in operation at the airport where the approaches are conducted.
2. Direct communications are maintained between the pilot and the center or approach controller until the pilot is instructed to contact the tower.
3. If more than one missed approach procedure is available, none require a course reversal.
4. If only one missed approach procedure is available, the following conditions are met:
   a. Course reversal is not required; and
   b. Reported ceiling and visibility are equal to or greater than the highest prescribed circling minimums for the IAP.
5. When cleared for the approach, pilots shall not execute a procedure turn.

Although the controller will not specifically state that "timed approaches are in progress," his/her assigning a time to depart the final approach fix inbound is indicative that timed approach procedures are being utilized. In lieu of holding, the controller may use radar vectors to the final approach course to establish a mileage interval between aircraft that will ensure the appropriate time sequence between the final approach fix and the airport.

Each pilot in an approach sequence will be given advance notice as to the time he/she should leave the holding point on approach to the airport. When a time to leave the holding point has been received, the pilot should adjust the flight path to leave the fix as closely as possible to the designated time.

ALL
4627. If only one missed approach procedure is available, which of the following conditions is required when conducting "timed approaches from a holding fix"?

A—The pilot must contact the airport control tower prior to departing the holding fix inbound.
B—The reported ceiling and visibility minimums must be equal to or greater than the highest prescribed circling minimums for the IAP.
C—The reported ceiling and visibility minimums must be equal to or greater than the highest prescribed straight-in MDA minimums for the IAP.

If only one missed approach procedure is available, the reported ceiling and visibility minimums must be equal to or greater than the highest prescribed circling minimums for the instrument approach procedure. (J17) — AIM ¶5-4-9(a)

Answer (A) is incorrect because the pilot is to contact the tower when instructed to do so by approach or center. Answer (C) is incorrect because the ceiling and visibility minimums must exceed the highest circling (not straight-in) minimums.

Answers

4712 [A]   4627 [B]

Chapter 8 Arrival and Approach

ALL
4628. Prior to conducting "timed approaches from a holding fix," which one of the following is required?

A—The time required to fly from the primary facility to the field boundary must be determined by a reliable means.
B—The airport where the approach is to be conducted must have a control tower in operation.
C—The pilot must have established two-way communications with the tower before departing the holding fix.

Prior to conducting timed approaches from a holding fix, a control tower must be in operation at the airport where the approaches are to be conducted. (J18) — AIM ¶5-4-9

Answer (A) is incorrect because the times to fly from the holding fix to the runway at various ground speeds are provided on the approach plates. Answer (C) is incorrect because the pilot must be in contact with approach or center until instructed to contact the tower.

ALL
4629. When making a "timed approach" from a holding fix at the outer marker, the pilot should adjust the

A—holding pattern to start the procedure turn at the assigned time.
B—airspeed at the final approach fix in order to arrive at the missed approach point at the assigned time.
C—holding pattern to leave the final approach fix inbound at the assigned time.

Each pilot in an approach sequence will be given advance notice as to the time they should leave the holding point on approach to the airport. When a time to leave the holding point has been received, the pilot should adjust the flight path to leave the fix as closely as possible to the designated time. (J18) — AIM ¶5-4-9

Answer (A) is incorrect because procedure turns should not be executed during timed approaches. Answer (B) is incorrect because the assigned time is the departure time from the final approach fix (not the arrival time at the MAP).

ALL
4768. Which of the following conditions is required before "timed approaches from a holding fix" may be conducted?

A—If more than one missed approach procedure is available, only one may require a course reversal.
B—If more than one missed approach procedure is available, none may require a course reversal.
C—Direct communication between the pilot and the tower must be established prior to beginning the approach.

Timed approaches may be conducted when the following conditions are met:

1. *A control tower is in operation at the airport where the approaches are conducted.*

2. *Direct communications are maintained between the pilot and the center or approach controller until the pilot is instructed to contact the tower.*

3. *If more than one missed approach procedure is available, none require a course reversal.*

(J17) — AIM ¶5-4-9

Answer (A) is incorrect because if more than one missed approach procedure is available a course reversal is not allowed. Answer (C) is incorrect because the pilot should be in contact with approach prior to beginning the approach.

Answers
4628 [B]   4629 [C]   4768 [B]

Chapter 8 **Arrival and Approach**

Missed Approach

When executing an early missed approach, fly the instrument approach procedure as specified to the missed approach point, at or above the MDA or DH, before executing a turning maneuver.

If visual contact is lost during a circling maneuver, make an initial climbing turn toward the approach runway and continue the turn until established on the missed approach course.

ALL
4667. If an early missed approach is initiated before reaching the MAP, the following procedure should be used unless otherwise cleared by ATC.

A— Proceed to the missed approach point at or above the MDA or DH before executing a turning maneuver.
B— Begin a climbing turn immediately and follow missed approach procedures.
C— Maintain altitude and continue past MAP for 1 minute or 1 mile whichever occurs first.

When an early missed approach is executed, pilots should fly the IAP as specified on the approach plate to the missed approach point at or above the MDA or DH before executing a turning maneuver to ensure terrain/obstacle clearance. (J18) — AIM ¶5-4-19

Answer (B) is incorrect because the pilot should not execute any turns prior to the missed approach point. Answer (C) is incorrect because the pilot need only proceed to (not past) the MAP.

ALL
4631. If the pilot loses visual reference while circling to land from an instrument approach and ATC radar service is not available, the missed approach action should be to

A— execute a climbing turn to parallel the published final approach course and climb to the initial approach altitude.
B— climb to the published circling minimums then proceed direct to the final approach fix.
C— make a climbing turn toward the landing runway and continue the turn until established on the missed approach course.

If visual reference is lost while circling to land from an instrument approach, the missed approach specified for that particular procedure must be followed (unless an alternate missed approach procedure is specified by ATC). To become established on the prescribed missed approach course, the pilot should make an initial climbing turn toward the landing runway and continue the turn until he/she is established on the missed approach course. (J18) — AIM ¶5-4-19

Answers (A) and (B) are incorrect because the climbing turn should be toward the landing runway and then execute the published missed approach procedure.

RVR

Runway Visual Range (RVR) is an instrumentally-derived value that represents the maximum horizontal distance down a specific runway at which a pilot can see and identify standard high intensity runway lights. It is reported in hundreds of feet.

If minimum visibility for an IAP is stated as an RVR value, and RVR is not available, use the comparable values table shown in FAA Legend 11 to get the corresponding value in statute miles.

Answers
4667 [A]   4631 [C]

ALL
4401. What does the Runway Visual Range (RVR) value, depicted on certain straight-in IAP Charts, represent?

A—The slant range distance the pilot can see down the runway while crossing the threshold on glide slope.
B—The horizontal distance a pilot should see when looking down the runway from a moving aircraft.
C—The slant visual range a pilot should see down the final approach and during landing.

Runway Visual Range (RVR) is the maximum horizontal distance down a specified instrument runway at which a pilot can see and identify standard high-intensity runway lights. It is always determined using a transmissiometer and is reported in hundreds of feet. (J33) — AIM ¶7-1-3

Answers (A) and (C) are incorrect because RVR is the horizontal (not slant range) distance the pilot will be able to see down the runway.

ALL
4716. RVR minimums for landing are prescribed in an IAP, but RVR is inoperative and cannot be reported for the intended runway at the time. Which of the following would be an operational consideration?

A—RVR minimums which are specified in the procedures should be converted and applied as ground visibility.
B—RVR minimums may be disregarded, providing the runway has an operative HIRL system.
C—RVR minimums may be disregarded, providing all other components of the ILS system are operative.

If RVR minimums for takeoff or landing are prescribed in an instrument approach procedure, but RVR is not reported for the runway of intended operation, the RVR minimum shall be converted to ground visibility and shall be the visibility minimum for takeoff or landing on that runway. (B10) — 14 CFR §91.175(h)

Answers (B) and (C) are incorrect because the RVR minimums should not be disregarded, instead they should be converted to ground visibility.

ALL
4754. If the RVR is not reported, what meteorological value should you substitute for 2,400 RVR?

A—A ground visibility of 1/2 NM.
B—A slant range visibility of 2,400 feet for the final approach segment of the published approach procedure.
C—A ground visibility of 1/2 SM.

Use the RVR/Meteorological Visibility Comparable Values in FAA Legend 11 to convert RVR (feet) to visibility (statute miles). 2,400 RVR can be substituted for a ground visibility of 1/2 SM. (B10) — 14 CFR §91.175(h)(2)

Answer (A) is incorrect because the RVR value is converted into statute (not nautical) miles. Answer (B) is incorrect because when RVR is not reported, the ground visibility is substituted.

ALL
4759. The RVR minimums for takeoff or landing are published in an IAP, but RVR is inoperative and cannot be reported for the runway at the time. Which of the following would apply?

A—RVR minimums which are specified in the procedure should be converted and applied as ground visibility.
B—RVR minimums may be disregarded, providing the runway has an operative HIRL system.
C—RVR minimums may be disregarded, providing all other components of the ILS system are operative.

If RVR minimums for takeoff or landing are prescribed in an instrument approach procedure, but RVR is not reported for the runway of intended operation, the RVR minimum shall be converted to ground visibility and shall be the visibility minimum for takeoff or landing on that runway. See FAA Legend 11 for the RVR/Meteorological Visibility Comparable Values. (B10) — 14 CFR §91.175(h)(1)

Answers (B) and (C) are incorrect because the RVR minimums must be converted to ground visibility (not disregarded).

ALL
4762. If the RVR equipment is inoperative for an IAP that requires a visibility of 2,400 RVR, how should the pilot expect the visibility requirement to be reported in lieu of the published RVR?

A—As a slant range visibility of 2,400 feet.
B—As an RVR of 2,400 feet.
C—As a ground visibility of 1/2 SM.

If RVR minimums for takeoff or landing are prescribed in an instrument approach procedure, but RVR is not reported for the runway of intended operation, the RVR minimum shall be converted to ground visibility and shall be the visibility minimum for takeoff or landing on that runway. See FAA Legend 11 for the RVR/Meteorological Visibility Comparable Values. (B10) — 14 CFR §91.175(h)(1)

Answers (A) and (B) are incorrect because if the RVR equipment is inoperative, the pilot can expect to receive the visibility requirement reported as ground visibility.

Answers

| 4401 | [B] | 4716 | [A] | 4754 | [C] | 4759 | [A] | 4762 | [C] |

Chapter 8 **Arrival and Approach**

RTC
4328. (Refer to Figures 56 and 60A.) To which value may the visibility criteria be reduced, if any, for the S-ILS 4 approach?

A—RVR 20.
B—RVR 16.
C—RVR 12.

Helicopters may also use the Category A Minimum Descent Altitude (MDA) or Decision Height (DH). The required visibility minimum may be reduced to one-half the published visibility minimum for Category A aircraft, but in no case may it be reduced to less than one-quarter mile or 1,200 feet RVR. Looking at FAA Figure 60A, the RVR is given as 18; the acceptable visibility may be reduced to RVR 12. (J42) — Instrument Approach Procedures

Inoperative Components

If all equipment required to execute an IAP is operational and used by the pilot, approach minimums are as shown on the approach chart. Normal minimums are 1/2-statute mile visibility (RVR 2,400) and a DH with a HAT of 200 feet. If a ground component or visual aid is out of service or not utilized by the pilot, the minimums may be raised. If more than one component is inoperative, apply only the greater adjustment. The Inoperative Components Table, in FAA Legend 22, must be consulted to determine the amount of increase, if any, that must be applied to the minimums. A compass locator or PAR can be substituted for an inoperative middle marker.

ALL
4699. (Refer to Figure 133.) What action should the pilot take if the marker beacon receiver becomes inoperative during the S-ILS 9 approach at Riverside Municipal?

A—Substitute SWAN LAKE INT. for the OM and surveillance radar for the MM.
B—Raise the DH 100 feet (50 feet for the OM and 50 feet for the MM).
C—Substitute SWAN LAKE INT. for the OM and use published minimums.

A compass locator or precision radar may be substituted for the outer or middle marker. DME, VOR, or nondirectional beacon fixes authorized in the standard instrument approach procedure, or surveillance radar may be substituted for the outer marker. With the marker beacon receiver inoperative, the Swan Lake INT (Localizer and R-327 Paradise) can be substituted for the outer marker. (J17) — 14 CFR §91.175(k)

Answer (A) is incorrect because only precision radar (not surveillance) can be substituted for the MM. Answer (B) is incorrect because when two or more components are inoperative, each minimum is raised to the highest minimum required by any single component that is inoperative, but Swan Lake INT can be substituted for the OM and no substitute is necessary for the MM. Therefore, the DH will be as published (not raised 100 feet).

ALL
4356. (Refer to Figure 73.) After passing the OM, Bradley Approach Control advises you that the MM on the ILS RWY 6 approach is inoperative. Under these circumstances, what adjustments, if any, are required to be made to the DH and visibility?

A—DH 424/24.
B—No adjustments are required.
C—DH 374/24.

FAA Legend 22 indicates no adjustments are necessary to DH or RVR for MM out. The Jepp chart is outdated (August 17, 1990). The FAA removed the MM from the Inoperative Components Table on October 15, 1992. Use the NOS chart to answer this question. (J42) — U.S. Terminal Procedures, IAP Chart

ALL
4731. Which pilot action is appropriate if more than one component of an ILS is unusable?

A—Use the highest minimum required by any single component that is unusable.
B—Request another approach appropriate to the equipment that is useable.
C—Raise the minimums a total of that required by each component that is unusable.

Answers

4328 [C]   4699 [C]   4356 [B]   4731 [A]

Chapter 8 Arrival and Approach

If more than one component is inoperative, each minimum is raised to the highest minimum required by any single component that is inoperative. (J42) — Instrument Approach Procedures

Answer (B) is incorrect because although an approach may have inoperative components, it may still be usable. Answer (C) is incorrect because only the highest minimum is required as a result of any component being inoperative.

ALL
4732. Which substitution is permitted when an ILS component is inoperative?

A—A compass locator or precision radar may be substituted for the ILS outer or middle marker.
B—ADF or VOR bearings which cross either the outer or middle marker sites may be substituted for these markers.
C—DME, when located at the localizer antenna site, should be substituted for the outer or middle marker.

The basic ground components of an ILS are the localizer, glide slope, outer marker, middle marker, and when installed for use with Category II or Category III instrument approach procedures, an inner marker. A compass locator or precision radar may be substituted for the outer or middle marker. DME, VOR, or nondirectional beacon fixes authorized in the standard instrument approach procedure or surveillance radar may be substituted for the outer marker. (J01) — AIM ¶1-1-10

Answer (B) is incorrect because an ADF or VOR can only be substituted for the outer (not the middle) marker. Answer (C) is incorrect because DME can only be substituted for the outer (not the middle) marker.

ALL
4733. What facilities, if any, may be substituted for an inoperative middle marker during an ILS approach without affecting the straight-in minimums?

A—ASR.
B—Substitution not necessary, minimums do not change.
C—Compass locator, PAR, and ASR.

Based on the Inoperative Components Table (See FAA Legend 22), loss of the middle marker would not affect minimums. (J42) — 14 CFR §91.175(k)

Answers (A) and (C) are incorrect because an ASR cannot be substituted for the middle marker.

ALL
4742. Which of these facilities may be substituted for an MM during a complete ILS IAP?

A—Surveillance and precision radar.
B—Compass locator and precision radar.
C—A VOR/DME fix.

The basic ground components of an ILS are the localizer, glide slope, outer marker, middle marker, and when installed for use with Category II or Category III instrument approach procedures, an inner marker. A compass locator or precision radar may be substituted for the outer or middle marker. (J01) — AIM ¶1-1-10

Answers (A) and (C) are incorrect because surveillance, precision radar, and a VOR/DME fix may be substituted for the outer marker (not the middle marker).

ALL
4770. Which substitution is appropriate during an ILS approach?

A—A VOR radial crossing the outer marker site may be substituted for the outer marker.
B—LOC minimums should be substituted for ILS minimums whenever the glide slope becomes inoperative.
C—DME, when located at the localizer antenna site, should be substituted for either the outer or middle marker.

Inoperative ILS Components:

Inoperative Localizer—When the localizer fails, an ILS approach is not authorized.
Inoperative Glide Slope—When the glide slope fails, the ILS reverts to a nonprecision localizer approach, and localizer minimums apply.

(J01) — AIM ¶1-1-10

Answer (A) is incorrect because unless the VOR radial is authorized in the standard approach procedure, the VOR radial may not be used. Answer (C) is incorrect because DME can not be substituted for the middle marker.

Answers

4732 [A]   4733 [B]   4742 [B]   4770 [B]

Chapter 8 **Arrival and Approach**

ALL
4706. A pilot is making an ILS approach and is past the OM to a runway which has a VASI. What action should the pilot take if an electronic glide slope malfunction occurs and the pilot has the VASI in sight?

A—The pilot should inform ATC of the malfunction and then descend immediately to the localizer DH and make a localizer approach.
B—The pilot may continue the approach and use the VASI glide slope in place of the electronic glide slope.
C—The pilot must request an LOC approach, and may descend below the VASI at the pilot's discretion.

As long as the flight visibility in not less than the visibility prescribed for the approach, the pilot should continue the approach and use the VASI. (B10) — 14 CFR §91.175(c)(3)(vi)

Answer (A) is incorrect because once the necessary visual references are in sight, the approach may continue visually. Answer (C) is incorrect because once the necessary visual references are in sight, the approach may continue visually, and the pilot may descend below the VASI glidepath only when it is necessary for a safe landing.

RTC
4330. (Refer to Figure 60A.) What is the DA and visibility criteria for a helicopter for the ILS RWY 4 approach if the MM is inoperative?

A—246/24.
B—246/12.
C—460/12.

Since helicopters can decrease RVR by 50% but not less than 1,200, and no adjustment is necessary to the DA (no adjustment is necessary for the MM out), the correct choice is 246/12. (J42) — Instrument Approach Procedures

Hazards on Approach

Hydroplaning will radically reduce braking effectiveness. It is most likely to occur in conditions of standing water, slush, high speed and smooth runway texture. The best technique to use when landing under such conditions is to apply moderate braking after the wheels have had time to "spin up." Anti-skid should be on.

Wake turbulence is caused by high-pressure air circulating outward, upward, and around a wing tip. These wing-tip vortices are at their greatest strength when the generating aircraft is heavy, clean and slow.

When the vortices of a large aircraft are close to ground level, they tend to move laterally. Thus, if a large aircraft is landing or taking off, a light crosswind or a light quartering tailwind would tend to hold the upwind vortex over the runway. If landing behind a large aircraft, plan to land past its touchdown point. If departing behind a large aircraft, lift off prior to its rotation point. Climb above and upwind of its flight path.

ALL
4408. The operation of an airport rotating beacon during daylight hours may indicate that

A—the in-flight visibility is less than 3 miles and the ceiling is less than 1,500 feet within Class E airspace.
B—the ground visibility is less than 3 miles and/or the ceiling is less than 1,000 feet in Class B, C, or D airspace.
C—an IFR clearance is required to operate within the airport traffic area.

In Class B, C, D, or E airspace, operation of the airport beacon during the hours of daylight often indicates that the ground visibility is less than 3 miles and/or the ceiling is less than 1,000 feet. Pilots should not rely solely on the operation of the airport beacon to indicate if weather conditions are IFR or VFR. (J03) — AIM ¶2-1-8(d)

Answers

4706 [B]   4330 [C]   4408 [B]

Chapter 8 Arrival and Approach

ALL

4707. What wind condition prolongs the hazards of wake turbulence on a landing runway for the longest period of time?

A—Direct headwind.
B—Direct tailwind.
C—Light quartering tailwind.

A tailwind condition can move the vortices of the preceding aircraft forward into the touchdown zone. The light quartering tailwind requires maximum caution. Pilots should be alert to large aircraft upwind from their approach and takeoff flight paths. (J27) — AIM ¶7-3-4

Answer (A) is incorrect because a direct headwind will move the vortices to the sides of the runway. Answer (B) is incorrect because a direct tailwind will move the vortices to the sides of the runway.

ALL

4708. Wake turbulence is near maximum behind a jet transport just after takeoff because

A—the engines are at maximum thrust output at slow airspeed.
B—the gear and flap configuration increases the turbulence to maximum.
C—of the high angle of attack and high gross weight.

The strength of the vortex is governed by the weight, speed, and shape of the wing of the generating aircraft. The greatest vortex strength occurs when the generating aircraft is heavy, clean, and slow. (J27) — AIM ¶7-3-3

Answer (A) is incorrect because vortices are related to airflow around the wing tips (not engines). Answer (B) is incorrect because when the gear and flap configuration changes, the characteristics of the vortex change, and the basic factor increasing vortex strength is weight.

ALL

4709. What effect would a light crosswind of approximately 7 knots have on vortex behavior?

A—The light crosswind would rapidly dissipate vortex strength.
B—The upwind vortex would tend to remain over the runway.
C—The downwind vortex would tend to remain over the runway.

A crosswind will decrease the lateral movement of the upwind vortex and increase the movement of the downwind vortex. Thus, a light wind of 3 to 7 knots could result in the upwind vortex remaining in the touchdown zone for a period of time and hasten the drift of the downwind vortex toward another runway. (J27) — AIM ¶7-3-4

Answer (A) is incorrect because a strong wind would help dissipate vortex strength quicker than a light wind. Answer (C) is incorrect because the upwind vortex would remain over the runway in a light crosswind.

ALL

4710. When landing behind a large jet aircraft, at which point on the runway should you plan to land?

A—If any crosswind, land on the windward side of the runway and prior to the jet's touchdown point.
B—At least 1,000 feet beyond the jet's touchdown point.
C—Beyond the jet's touchdown point.

When landing behind a large aircraft on the same runway, stay at or above the large aircraft's final approach flight path. Note the touchdown point and land beyond it. (J27) — AIM ¶7-3-6

Answer (A) is incorrect because the pilot should land beyond the jet's touchdown point (not prior). Answer (B) is incorrect because there is no specific minimum distance behind a jet's touchdown point that the pilot should land.

ALL

4738. Under which conditions is hydroplaning most likely to occur?

A—When rudder is used for directional control instead of allowing the nosewheel to contact the surface early in the landing roll on a wet runway.
B—During conditions of standing water, slush, high speed, and smooth runway texture.
C—During a landing on any wet runway when brake application is delayed until a wedge of water begins to build ahead of the tires.

Hydroplaning may occur as a result of aircraft speed and water on the runway. As the speed of the aircraft and the depth of the water increase, the water layer builds up an increasing resistance to displacement, resulting in the formation of a wedge of water beneath the tire. Dynamic hydroplaning occurs when there is standing water on the runway surface. Water about one-tenth of an inch deep acts to lift the tire off the runway. (V14) — AC 91-6A, page 5

Answer (A) is incorrect because the nosewheel cannot hydroplane if it is not on the ground. Answer (C) is incorrect because brake timing does not affect hydroplaning.

Answers

| 4707 | [C] | 4708 | [C] | 4709 | [B] | 4710 | [C] | 4738 | [B] |

Chapter 8 **Arrival and Approach**

Airport Lighting and Marking Aids

The Visual Approach Slope Indicator (VASI) is a system of lights arranged to provide visual descent guidance information during the approach to a runway. These lights are visible from 3-5 miles during the day and up to 20 miles or more at night. The visual glide path of the VASI provides safe obstruction clearance within ±10 degrees of the extended runway centerline and to 4 NM from the runway threshold. Descent, using the VASI, should not be initiated until the aircraft is visually aligned with the runway. Lateral course guidance is provided by the runway or runway lights. *See* Figure 8-3.

The Precision Approach Path Indicator (PAPI) uses light units similar to the VASI, but they are installed in a single row of either two- or four-light units. These systems have an effective visual range of about 5 miles during the day and up to 20 miles at night. The row of light units is usually installed on the left side of the runway, and the glide path indications are as depicted. *See* Figure 8-4.

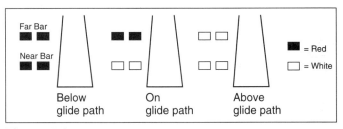

Figure 8-3

Runway lighting is described in FAA Legends 19 and 20.

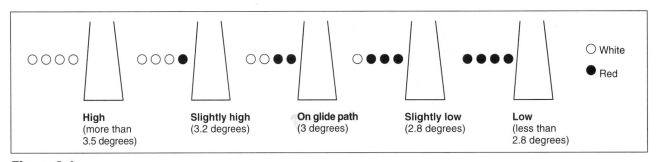

Figure 8-4

ALL
4774. (Refer to Figure 134.) Unless a higher angle is necessary for obstacle clearance, what is the normal glidepath angle for a 2-bar VASI?

A—2.75°.
B—3.00°.
C—3.25°.

Two-bar VASI installations provide one visual glidepath which is normally set at 3°. (J03) — AIM ¶2-1-2(a)(3)

ALL
4775. Which of the following indications would a pilot see while approaching to land on a runway served by a 2-bar VASI?

A—If on the glidepath, the near bars will appear red, and the far bars will appear white.
B—If departing to the high side of the glidepath, the far bars will change from red to white.
C—If on the glidepath, both near bars and far bars will appear white.

When on the glidepath with a two-bar VASI (four light units shown), the pilot will see red-over-white. As the airplane levels off and deviates above the visual glidepath, it flies through the transition zone for the far bars. The far bars will turn from red to white, and the airplane will be above the glidepath. See the figure on the next page. (J03) — AIM ¶2-1-2

Answer (A) is incorrect because this type of indication is not possible. Answer (C) is incorrect because both bars must be white to indicate the aircraft above (not on) the glidepath.

Answers
4774 [B]   4775 [B]

Chapter 8 **Arrival and Approach**

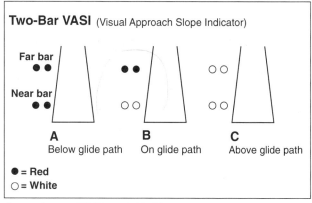

Questions 4775 and 4778

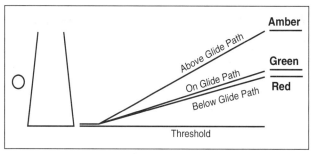

Question 4777

ALL
4776. The middle and far bars of a 3-bar VASI will

A—both appear white to the pilot when on the upper glidepath.
B—constitute a 2-bar VASI for using the lower glidepath.
C—constitute a 2-bar VASI for using the upper glidepath.

Three-bar VASI installations provide two visual glidepaths. The lower path is provided by the near and middle bars and is normally set at 3° higher, while the upper path, provided by the middle and far bars, is normally 1/4 degree higher. (J03) — AIM ¶2-1-2

Answer (A) is incorrect because if a pilot was on the upper glidepath, the middle bar would be white and the far bar would be red. Answer (B) is incorrect because the middle and far bars are for the upper glidepath.

ALL
4777. Tricolor Visual Approach Indicators normally consist of

A—a single unit, projecting a three-color visual approach path.
B—three separate light units, each projecting a different color approach path.
C—three separate light projecting units of very high candle power with a daytime range of approximately 5 miles.

Tri-color visual approach slope indicators normally consist of a single light unit projecting a three-color visual approach path into the final approach area of the runway upon which the indicator is installed. The below glidepath indication is red, the above glidepath indication is amber, and the on glidepath indication is green. See the following figure. (J03) — AIM ¶2-1-2

Answers (B) and (C) are incorrect because a tricolor visual approach indicator consists of a single unit (not three separate light units).

ALL
4778. When on the proper glidepath of a 2-bar VASI, the pilot will see the near bar as

A—white and the far bar as red.
B—red and the far bar as white.
C—white and the far bar as white.

When on the glidepath with a two-bar VASI, the pilot will see red-over-white. See the figure to the left. (J03) — AIM ¶2-1-2

ALL
4779. If an approach is being made to a runway that has an operating 3-bar VASI and all the VASI lights appear red as the airplane reaches the MDA, the pilot should

A—start a climb to reach the proper glidepath.
B—continue at the same rate of descent if the runway is in sight.
C—level off momentarily to intercept the proper approach path.

See the figure below. An airplane approaching to land on a runway served by a visual approach slope indicator, shall maintain an altitude at or above the glide slope until a lower altitude is necessary for a safe landing. If the VASI is indicating all red lights, the aircraft is below the glidepaths, and the pilot should level off momentarily to intercept the proper approach path. (J03) — AIM ¶2-1-2

Answer (A) is incorrect because climbing is not necessary to acquire the proper glidepath. Answer (B) is incorrect because if you continue your present rate of descent, you will remain below the glidepath.

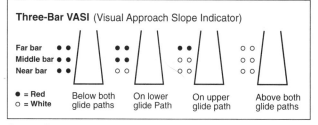

Question 4779

Answers

| 4776 | [C] | 4777 | [A] | 4778 | [A] | 4779 | [C] |

Chapter 8 Arrival and Approach

ALL
4780. Which is a feature of the tricolor VASI?

A—One light projector with three colors: red, green, and amber.
B—Two visual glidepaths for the runway.
C—Three glidepaths, with the center path indicated by a white light.

Tri-color visual approach slope indicators normally consist of a single light unit projecting a three-color visual approach path into the final approach area of the runway upon which the indicator is installed. The below-glidepath indication is red, the above-glidepath indication is amber, and the on-glidepath indication is green. See the figure below. (J03) — AIM ¶2-1-2

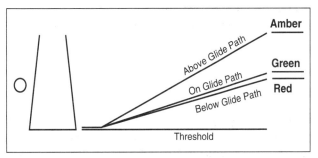

Question 4780

ALL
4781. Which approach and landing objective is assured when the pilot remains on the proper glidepath of the VASI?

A—Continuation of course guidance after transition to VFR.
B—Safe obstruction clearance in the approach area.
C—Course guidance from the visual descent point to touchdown.

The VASI is a system of lights arranged to provide visual descent guidance information during the approach to a runway. These lights are visible from 3 to 5 miles during the day, and up to 20 miles or more at night. The visual glidepath of the VASI provides safe obstruction clearance within plus or minus 10° of the extended runway centerline, and to 4 NM from the runway threshold. (J03) — AIM ¶2-1-2

Answers (A) and (C) are incorrect because a VASI does not provide course guidance.

ALL
4782. (Refer to Figure 135.) Unless a higher angle is required for obstacle clearance, what is the normal glidepath for a 3-bar VASI?

A—2.3°.
B—2.75°.
C—3.0°.

Three-bar VASI installations provide two visual glidepaths. The lower glidepath is provided by the near and middle bars and is normally set at 3° while the upper glidepath, provided by the middle and far bars, is normally 1/4-degree higher. This higher glidepath is intended for use only by high cockpit aircraft to provide a sufficient threshold-crossing height. Although normal glidepath angles are 3°, angles at some locations may be as high as 4.5° to give proper obstacle clearance. (J03) — AIM ¶2-1-2

ALL
4783. (Refer to Figure 135.) Which illustration would a pilot observe when on the lower glidepath?

A—4.
B—5.
C—6.

On a lower glidepath of a 3-bar VASI, the near lights will be white and the middle and far bars will be red, which is illustration 5. See the figure below. (J03) — AIM ¶2-1-2

ALL
4784. (Refer to Figure 135.) Which illustration would a pilot observe if the aircraft is above both glidepaths?

A—5.
B—6.
C—7.

Above a glidepath of a 3-bar VASI, all of the lights will be white, which is illustration 7. See the figure below. (J03) — AIM ¶2-1-2

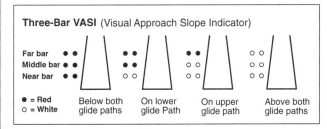

Questions 4783, 4784 and 4785

Answers

4780 [A]   4781 [B]   4782 [C]   4783 [B]   4784 [C]

ALL
4785. (Refer to Figure 135.) Which illustration would a pilot observe if the aircraft is below both glidepaths?

A—4.
B—5.
C—6.

Below a glidepath of a 3-bar VASI, all of the lights will be red, which is illustration 4. See the previous figure. (J03) — AIM ¶2-1-2

ALL
4786. (Refer to Figure 136.) Which illustration depicts an "on glidepath" indication?

A—8.
B—10.
C—11.

The precision approach path indicator (PAPI) uses light units similar to the VASI, but are installed in a single row of either two- or four-light units. The on-glidepath indication will have 2 red lights and 2 white lights, which is illustration 10. (J03) — AIM ¶2-1-2

ALL
4787. (Refer to Figure 136.) Which illustration depicts a "slightly low" (2.8°) indication?

A—9.
B—10.
C—11.

The precision approach path indicator (PAPI) uses light units similar to the VASI, but are installed in a single row of either two- or four-light units. The slightly-low position will be indicated with 3 red lights and 1 white light, which is illustration 11. (J03) — AIM ¶2-1-2

ALL
4788. (Refer to Figure 136.) Which illustration would a pilot observe if the aircraft is on a glidepath higher than 3.5°?

A—8.
B—9.
C—11.

The precision approach path indicator (PAPI) uses light units similar to the VASI, but are installed in a single row of either two- or four-light units. The high-glidepath indication will have all white lights, which is illustration 8. (J03) — AIM ¶2-1-2

ALL
4789. (Refer to Figure 136.) Which illustration would a pilot observe if the aircraft is "slightly high" (3.2°) on the glidepath?

A—8.
B—9.
C—11.

The precision approach path indicator (PAPI) uses light units similar to the VASI, but are installed in a single row of either two- or four-light units. The slightly-high position will be indicated with 3 white lights, and 1 red light, which is illustration 9. (J03) — AIM ¶2-1-2

ALL
4790. (Refer to Figure 136.) Which illustration would a pilot observe if the aircraft is less than 2.5°?

A—10.
B—11.
C—12.

The precision approach path indicator (PAPI) uses light units similar to the VASI, but are installed in a single row of either two- or four-light units. The low-glidepath indication will have all red lights, which is illustration 12. (J03) — AIM ¶2-1-2

Answers

4785 [A]   4786 [B]   4787 [C]   4788 [A]   4789 [B]   4790 [C]

Chapter 8 Arrival and Approach

ALL
4795. Which type of runway lighting consists of a pair of synchronized flashing lights, one on each side of the runway threshold?

A—RAIL.
B—HIRL.
C—REIL.

Runway End Identifier Lights (REIL) are installed for rapid and positive identification of the approach end of a runway. The system consists of a pair of synchronized flashing lights, located laterally one on each side of the runway threshold facing the approach area. See the figure below. (J03) — AIM ¶2-1-3

Answer (A) is incorrect because RAIL are runway alignment indicator lights, which are sequenced flashing lights that are used with other approach lighting systems. Answer (B) is incorrect because HIRL are high-intensity runway lights, which is a runway edge lighting system.

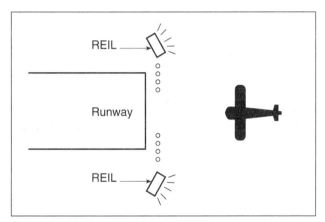

Question 4795

ALL
4796. The primary purpose of runway end identifier lights, installed at many airfields, is to provide

A—rapid identification of the approach end of the runway during reduced visibility.
B—a warning of the final 3,000 feet of runway remaining as viewed from the takeoff or approach position.
C—rapid identification of the primary runway during reduced visibility.

Runway End Identifier Lights (REIL) are installed for rapid and positive identification of the approach end of a runway. The system consists of a pair of synchronized flashing lights, located laterally one on each side of the runway threshold facing the approach area. (J03) — AIM ¶2-1-3

ALL
4791. (Refer to Figure 137.) What is the distance (A) from the beginning of the runway to the fixed distance marker?

A—500 feet.
B—1,000 feet.
C—1,500 feet.

The aiming point marking (or fixed distance marker) serves as a visual aiming point for a landing aircraft. These two rectangular markings consist of a broad white stripe located on each side of the runway centerline and approximately 1,000 feet from the landing threshold. (J05) — AIM ¶2-3-3

ALL
4792. (Refer to Figure 137.) What is the distance (B) from the beginning of the runway to the touchdown zone marker?

A—250 feet.
B—500 feet.
C—750 feet.

The touchdown zone markings identify the touchdown zone for landing operations and are coded to provide distance information in 500-foot increments. These markings consist of groups of one, two, and three rectangular bars symmetrically arranged in pairs about the runway centerline. (J05) — AIM ¶2-3-3

ALL
4793. (Refer to Figure 137.) What is the distance (C) from the beginning of the touchdown zone marker to the beginning of the fixed distance marker?

A—1,000 feet.
B—500 feet.
C—250 feet.

The touchdown zone marker is 500 feet from the runway threshold, and the fixed distance marker is 1,000 feet from the runway threshold. Therefore, the distance from the beginning of the touchdown zone marker to the beginning of the fixed distance marker is 500 feet (1,000 – 500). (J05) — AIM ¶2-3-3

Answers
4795 [C]   4796 [A]   4791 [B]   4792 [B]   4793 [B]

ALL
4794. Which runway marking indicates a displaced threshold on an instrument runway?

A—Arrows leading to the threshold mark.
B—Centerline dashes starting at the threshold.
C—Red chevron marks in the nonlanding portion of the runway.

A displaced threshold is a threshold that is not at the beginning of the full strength runway pavement. The paved area behind the displaced runway threshold is available for taxiing, the landing rollout, and the takeoff of aircraft. White arrows are located along the centerline in the area between the beginning of the runway and the displaced threshold. (J05) — AIM ¶2-3-3(h)

ALL
4797. (Refer to Figure 138.) What night operations, if any, are authorized between the approach end of the runway and the threshold lights?

A—No aircraft operations are permitted short of the threshold lights.
B—Only taxi operations are permitted in the area short of the threshold lights.
C—Taxi and takeoff operations are permitted, providing the takeoff operations are toward the visible green threshold lights.

On displaced thresholds, runway edge lights emit red light toward the runway to indicate the end of the runway to a departing aircraft, and green light outward from the runway end to indicate the threshold to landing aircraft. The area behind the displaced threshold is available for taxiing, landing rollout, and takeoff. (J05) — AIM ¶2-1-4(c)

Closing the Flight Plan

If landing at an airport with an operating control tower, the tower will automatically close the IFR flight plan. When landing at any other airport, it is the pilot's responsibility to close the flight plan with the last controller or the nearest flight service station (FSS). Flight service stations will provide airport advisories at non-tower airports.

ALL
4076. When may a pilot cancel the IFR flight plan prior to completing the flight?

A—Any time.
B—Only if an emergency occurs.
C—Only in VFR conditions when not in Class A airspace.

An IFR flight plan may be canceled at any time the flight is operating in VFR conditions outside Class A airspace by the pilot stating, "Cancel my IFR flight plan" to the controller or air/ground station with which they are communicating. (J15) — AIM ¶5-1-13

Answer (A) is incorrect because the pilot must be in VFR conditions and outside Class A airspace before canceling an IFR flight plan. Answer (B) is incorrect because ATC will provide assistance in an emergency while under an IFR flight plan.

ALL
4058. How is your flight plan closed when your destination airport has IFR conditions and there is no control tower or flight service station (FSS) on the field?

A—The ARTCC controller will close your flight plan when you report the runway in sight.
B—You may close your flight plan any time after starting the approach by contacting any FSS or ATC facility.
C—Upon landing, you must close your flight plan by radio or by telephone to any FSS or ATC facility.

If operating on an IFR flight plan to an airport where there is no functioning control tower, the pilot must initiate cancellation of the IFR flight plan. This can be done after landing if there is a functioning FSS or other means of direct communications (radio or telephone) with ATC. (J15) — AIM ¶5-1-13

Answer (A) is incorrect because the pilots (not ARTCC) is responsible for closing the flight plan at uncontrolled airports. Answer (B) is incorrect because while under an IFR clearance (required for IFR conditions) the flight plan can not be closed until after landing.

Answers

4794 [A]   4797 [C]   4076 [C]   4058 [C]

Cross-Reference A:
Answer, Subject Matter Knowledge Code, Category & Page Number

The FAA does not publish a category list for any exam; therefore, the category assigned below is based on historical data and the best judgement of our researchers.

ALL All aircraft
AIR Airplane
RTC Rotorcraft (applies to both helicopter and gyroplane)
LTA Lighter-Than-Air (applies to hot air balloon, gas balloon and airship)

It is important to answer every question assigned on your FAA Knowledge Test. If in their ongoing review, the FAA test authors decide a question has no correct answer, is no longer applicable, or is otherwise defective, your answer will be marked correct no matter which one you choose. You will not be given the automatic credit unless you have marked an answer. For those questions for which none of the answer choices provide an accurate response, we have noted [X] as the Answer.

Question	Answer	SMK Code	Category	Page
4001	[C]	(A20)	ALL	5-8
4002	[C]	(A24)	AIR	5-4
4003	[C]	(B08)	ALL	5-19
4004	[A]	(B07)	ALL	5-14
4005	[B]	(B10)	ALL	6-6
4006	[C]	(B10)	ALL	5-33
4007	[B]	(B11)	ALL	5-12
4008	[A]	(A20)	ALL	5-18
4009	[B]	(A20)	ALL	5-18
4010	[B]	(A20)	ALL	5-19
4011	[A]	(B08)	ALL	5-19
4012	[A]	(A20)	ALL	5-6
4013	[A]	(A20)	ALL	5-6
4014	[B]	(A20)	ALL	5-6
4015	[A]	(A20)	ALL	5-7
4016	[C]	(A20)	RTC	5-9
4017	[A]	(A20)	ALL	5-7
4018	[C]	(A20)	RTC	5-5
4019	[A]	(A20)	RTC	5-9
4020	[B]	(A20)	ALL	5-7
4021	[B]	(A20)	ALL	5-7
4022	[C]	(A20)	RTC	5-9
4023	[A]	(A20)	AIR	5-8
4024	[C]	(A20)	ALL	5-3
4025	[B]	(A20)	ALL	5-3
4026	[C]	(A20)	AIR	5-8
4027	[A]	(A20)	ALL	5-8
4028	[C]	(A20)	ALL	5-3
4029	[B]	(A20)	AIR	5-4
4030-1	[B]	(A20)	RTC	5-5
4030-2	[B]	(A20)	AIR	5-5
4031	[B]	(A20)	ALL	5-4
4032	[B]	(B10)	ALL	6-6
4033	[C]	(B08)	ALL	5-20
4034	[C]	(A20)	AIR	5-4
4035	[C]	(A20)	AIR	5-5
4036	[C]	(B10)	AIR	5-15
4037	[B]	(B11)	ALL	5-10
4038	[A]	(B11)	ALL	5-10
4039	[C]	(B07)	ALL	5-16
4040	[A]	(B10)	RTC	6-15
4041	[A]	(B10)	RTC	6-15
4042	[C]	(B11)	ALL	5-16
4043	[C]	(B11)	ALL	5-12
4044	[B]	(B10)	ALL	4-9
4045	[C]	(B11)	ALL	5-16
4046	[A]	(B10)	ALL	4-9
4047	[C]	(B13)	ALL	5-15
4048	[A]	(B10)	ALL	5-15
4049	[C]	(B13)	ALL	5-16
4050	[A]	(B11)	ALL	5-11
4051	[C]	(B11)	ALL	5-10
4052	[C]	(B11)	ALL	5-17
4053	[C]	(B11)	ALL	5-17
4054	[A]	(J01)	ALL	4-9
4055	[C]	(B11)	ALL	5-11
4056	[B]	(H814)	ALL	3-12
4057	[A]	(B11)	ALL	5-12
4058	[C]	(J15)	ALL	8-45
4059	[B]	(J15)	ALL	6-7
4060	[A]	(J15)	ALL	6-7
4061	[A]	(J15)	ALL	6-7
4062	[C]	(B10)	ALL	6-4
4063	[B]	(B10)	ALL	6-5
4064	[A]	(B10)	ALL	6-5
4065	[C]	(B10)	ALL	6-5
4066	[B]	(B10)	ALL	6-5
4067	[B]	(B10)	ALL	6-5
4068	[C]	(B10)	ALL	6-6
4069	[A]	(H862)	ALL	4-32
4070	[B]	(J34)	ALL	6-14
4071	[B]	(J15)	ALL	5-35
4072	[C]	(J15)	ALL	6-8
4073	[B]	(J15)	ALL	6-10
4074	[A]	(J15)	ALL	6-10
4075	[C]	(J15)	ALL	6-8
4076	[C]	(J15)	ALL	8-45
4077	[A]	(J15)	ALL	7-7
4078	[A]	(J15)	ALL	5-35
4079	[C]	(J34)	ALL	5-26
4080	[C]	(J06)	ALL	5-27
4081	[B]	(B10)	ALL	6-10
4082-1	[C]	(B10)	RTC	6-12
4082-2	[A]	(B10)	AIR	6-12
4083-1	[C]	(B10)	AIR	6-12
4083-2	[C]	(B10)	RTC	6-12

Cross-Reference A: Answer, SMK Code, Category & Page Number

Question	Answer	SMK Code	Category	Page
4083-3	[B]	(B10)	AIR	6-13
4084-1	[A]	(B10)	RTC	6-16
4084-2	[C]	(B10)	RTC	6-16
4085-1	[A]	(B10)	AIR	6-10
4085-2	[A]	(B10)	RTC	6-10
4086	[C]	(B10)	ALL	6-13
4087-1	[A]	(B10)	AIR	6-11
4087-2	[C]	(B10)	RTC	6-11
4088	[C]	(G10)	ALL	5-26
4089	[A]	(I22)	ALL	3-6
4090	[B]	(I22)	ALL	3-7
4091	[C]	(I22)	ALL	3-7
4092	[A]	(I31)	ALL	2-7
4093	[B]	(I22)	ALL	3-9
4094	[C]	(I21)	ALL	1-5
4095	[C]	(I21)	ALL	1-4
4096	[A]	(I21)	ALL	1-4
4097	[C]	(I20)	ALL	1-3
4098	[C]	(I27)	ALL	1-15
4099	[C]	(I24)	ALL	1-9
4100	[C]	(I29)	ALL	1-22
4101	[A]	(I24)	ALL	1-8
4102	[A]	(I24)	ALL	1-8
4103	[B]	(I24)	ALL	1-8
4104	[A]	(I24)	ALL	1-8
4105	[B]	(I23)	ALL	1-6
4106	[B]	(I23)	ALL	1-7
4107	[B]	(I23)	ALL	1-7
4108	[C]	(I23)	ALL	1-7
4109	[B]	(I22)	ALL	3-7
4110	[B]	(I22)	ALL	3-10
4111	[A]	(H808)	ALL	3-7
4112	[A]	(I21)	ALL	1-5
4113	[B]	(I21)	ALL	1-5
4114	[A]	(I21)	ALL	1-5
4115	[C]	(I25)	ALL	1-10
4116	[A]	(I27)	ALL	1-14
4117	[B]	(I27)	ALL	1-15
4118	[C]	(I25)	ALL	1-10
4119	[C]	(I27)	ALL	1-15
4120	[C]	(I27)	ALL	1-14
4121	[B]	(I25)	ALL	1-9
4122	[B]	(I25)	ALL	1-10
4123	[C]	(I25)	ALL	1-10
4124	[A]	(I25)	ALL	1-10
4125	[A]	(I21)	ALL	1-5
4126	[A]	(I30)	ALL	1-18
4127	[A]	(I27)	ALL	1-15
4128	[B]	(I27)	ALL	1-15
4129	[B]	(I28)	ALL	1-12
4130	[C]	(I28)	ALL	1-12
4131	[B]	(I26)	ALL	1-11
4132	[B]	(I28)	ALL	1-12
4133	[B]	(I29)	ALL	1-11
4134	[C]	(I26)	ALL	1-11
4135	[C]	(I30)	ALL	1-24
4136	[A]	(I27)	ALL	1-15
4137	[C]	(I30)	ALL	1-18
4138	[C]	(I28)	ALL	1-23
4139	[C]	(I28)	ALL	1-23
4140	[A]	(I28)	ALL	1-24
4141	[C]	(I27)	ALL	1-19
4142	[A]	(I57)	ALL	1-18
4143	[B]	(I30)	ALL	1-18
4144	[A]	(I30)	ALL	1-19
4145	[B]	(I30)	ALL	1-18
4146	[C]	(I30)	ALL	1-19
4147	[B]	(I30)	ALL	1-18
4148	[C]	(I30)	ALL	1-17
4149	[A]	(I28)	ALL	1-12
4150	[C]	(I28)	ALL	1-23
4151	[C]	(I29)	ALL	1-22
4152	[C]	(I29)	ALL	1-22
4153	[B]	(I29)	ALL	1-22
4154	[C]	(I20)	ALL	1-3
4155	[C]	(I32)	ALL	1-4
4156	[A]	(I31)	ALL	1-13
4157	[C]	(I26)	ALL	1-11
4158	[C]	(I27)	ALL	1-14
4159	[B]	(I24)	ALL	1-8
4160	[B]	(I30)	ALL	1-16
4161	[C]	(I24)	ALL	1-9
4162	[C]	(I32)	ALL	1-14
4163	[B]	(I31)	ALL	1-13
4164	[A]	(I31)	ALL	1-13
4165	[A]	(I31)	ALL	1-13
4166	[C]	(I31)	ALL	1-13
4167	[A]	(I31)	ALL	1-13
4168	[B]	(I32)	ALL	1-4
4169	[C]	(I31)	ALL	1-13
4170	[B]	(I57)	ALL	2-5
4171	[C]	(I29)	ALL	1-22
4172	[A]	(I63)	ALL	2-9
4173	[B]	(I61)	ALL	2-21
4174	[A]	(I60)	ALL	2-18
4175	[A]	(I64)	ALL	2-15
4176	[C]	(I57)	ALL	2-5
4177	[A]	(I57)	ALL	2-5
4178	[B]	(I57)	ALL	2-5
4179	[C]	(I57)	ALL	2-6
4180	[A]	(I57)	ALL	2-5
4181	[B]	(I57)	ALL	2-12
4182	[B]	(I56)	ALL	2-4
4183	[A]	(I57)	ALL	2-13
4184	[C]	(I56)	ALL	2-13
4185	[C]	(I57)	ALL	2-14
4186	[C]	(I57)	ALL	2-12
4187	[C]	(I57)	ALL	2-13
4188	[B]	(I63)	ALL	2-10
4189	[C]	(I63)	ALL	2-10
4190	[B]	(I63)	ALL	2-10
4191	[C]	(I63)	ALL	2-9
4192	[C]	(I63)	ALL	2-10
4193	[C]	(I63)	ALL	2-10
4194	[A]	(I63)	ALL	2-11
4195	[C]	(I61)	ALL	2-21
4196	[A]	(I55)	ALL	2-4
4197	[A]	(I65)	ALL	2-27
4198	[A]	(I56)	ALL	2-8
4199	[B]	(I63)	ALL	2-11
4200	[A]	(I21)	ALL	1-6
4201	[C]	(I57)	ALL	2-6
4202	[C]	(I55)	ALL	2-3
4203	[A]	(I57)	ALL	2-4
4204	[C]	(I57)	ALL	2-6
4205	[C]	(I55)	ALL	2-4
4206	[B]	(I59)	ALL	2-16
4207	[B]	(I59)	ALL	2-17
4208	[C]	(I59)	ALL	2-17
4209	[C]	(I58)	ALL	2-16

Cross-Reference A: Answer, SMK Code, Category & Page Number

Question	Answer	SMK Code	Category	Page
4210	[A]	(I67)	ALL	1-17
4211	[A]	(I64)	ALL	2-24
4212	[C]	(I64)	ALL	2-24
4213	[A]	(I64)	ALL	2-25
4214	[C]	(I64)	ALL	2-24
4215	[B]	(I65)	ALL	2-15
4216	[B]	(I64)	ALL	2-25
4217	[A]	(I64)	ALL	2-25
4218	[A]	(I64)	ALL	2-25
4219	[A]	(I64)	ALL	2-25
4220	[C]	(I56)	ALL	2-8
4221	[A]	(I64)	ALL	2-26
4222	[C]	(I64)	ALL	2-26
4223	[B]	(I64)	ALL	2-26
4224	[C]	(I64)	ALL	2-26
4225	[C]	(I64)	ALL	2-27
4226	[B]	(I57)	ALL	2-15
4227	[C]	(I20)	ALL	1-3
4228	[C]	(I43)	ALL	2-5
4229	[B]	(I64)	ALL	2-27
4230	[C]	(I60)	ALL	2-18
4231	[B]	(I60)	ALL	2-18
4232	[C]	(I60)	ALL	2-19
4233	[B]	(I60)	ALL	2-20
4234	[A]	(I60)	ALL	2-20
4235	[B]	(I60)	ALL	2-21
4236	[A]	(I60)	ALL	2-20
4237	[A]	(I60)	ALL	2-20
4238	[C]	(I23)	ALL	1-24
4239	[C]	(I65)	ALL	2-27
4240	[B]	(I65)	ALL	2-28
4241	[A]	(I54)	ALL	2-13
4242	[B]	(I64)	ALL	2-22
4243	[C]	(I64)	ALL	2-22
4244	[C]	(I64)	ALL	2-22
4245	[C]	(I64)	ALL	2-22
4246	[A]	(I63)	ALL	2-11
4247	[C]	(I63)	ALL	2-11
4248	[A]	(I65)	ALL	2-28
4249	[C]	(I63)	ALL	2-11
4250	[B]	(I63)	ALL	2-12
4251	[C]	(J25)	ALL	1-20
4252	[C]	(J25)	ALL	1-20
4253	[C]	(J25)	ALL	1-20
4254	[C]	(J25)	ALL	1-20
4255	[C]	(J25)	ALL	1-20
4256	[C]	(J25)	ALL	1-21
4257	[A]	(J25)	ALL	1-21
4258	[B]	(J25)	ALL	1-21
4259	[A]	(H342)	ALL	7-27
4260	[A]	(H342)	ALL	7-27
4261	[A]	(J17)	ALL	7-10
4262	[B]	(J35)	ALL	7-11
4263	[B]	(J35)	ALL	7-7
4264	[C]	(J35)	ALL	7-11
4265	[A]	(H342)	ALL	7-28
4266	[C]	(J15)	ALL	6-8
4267	[B]	(H342)	ALL	3-3
4268	[C]	(H342)	ALL	7-22
4269	[A]	(H832)	ALL	4-32
4270	[C]	(H342)	ALL	6-21
4271	[A]	(J40)	ALL	4-3
4272	[B]	(J40)	ALL	6-22
4273	[C]	(J01)	ALL	4-4
4274	[A]	(J42)	ALL	8-14
4275	[B]	(J34)	ALL	6-14
4276	[B]	(J42)	ALL	8-14
4277	[C]	(J15)	ALL	6-8
4278	[B]	(H342)	ALL	3-3
4279	[A]	(H342)	ALL	7-23
4280	[A]	(J35)	ALL	7-9
4281	[A]	(J41)	ALL	8-4
4282	[B]	(H833)	ALL	8-14
4283-1	[A]	(H833)	ALL	8-14
4283-2	[B]	(H833)	ALL	8-14
4284	[A]	(K26)	ALL	5-14
4285	[B]	(J42)	ALL	8-14
4286	[C]	(J41)	ALL	8-3
4287	[A]	(J35)	ALL	7-7
4288	[B]	(J15)	ALL	6-8
4289	[A]	(H342)	ALL	3-4
4290	[C]	(H342)	ALL	7-23
4291	[A]	(J35)	ALL	7-7
4292	[B]	(J41)	ALL	8-4
4293	[C]	(J41)	ALL	8-6
4294	[C]	(J41)	ALL	8-4
4295	[C]	(J18)	ALL	8-6
4296	[C]	(J42)	ALL	8-14
4297	[A]	(J42)	ALL	8-15
4298	[C]	(J42)	ALL	8-15
4299	[A]	(J42)	ALL	8-15
4300	[C]	(J15)	ALL	6-9
4301	[A]	(H342)	ALL	3-4
4302	[B]	(H342)	ALL	7-24
4303	[C]	(J40)	ALL	6-22
4304	[A]	(J40)	ALL	6-22
4305	[B]	(J34)	ALL	6-14
4306	[A]	(J42)	ALL	8-15
4307	[B]	(J42)	ALL	8-15
4308	[C]	(J42)	ALL	8-16
4309	[B]	(J42)	ALL	8-16
4310	[C]	(J42)	ALL	8-16
4311	[A]	(J42)	ALL	8-16
4312	[A]	(J15)	ALL	6-9
4313	[B]	(H342)	ALL	3-4
4314	[C]	(H342)	ALL	7-24
4315	[B]	(J40)	ALL	6-22
4316	[C]	(J40)	ALL	6-23
4317	[A]	(J35)	ALL	7-7
4318	[B]	(J35)	ALL	7-8
4319	[A]	(J42)	ALL	8-16
4320	[C]	(H832)	ALL	4-6
4321	[C]	(J42)	ALL	8-17
4322	[C]	(J15)	RTC	6-9
4323	[A]	(H342)	RTC	3-4
4324	[C]	(H342)	RTC	7-25
4325	[B]	(J34)	ALL	7-11
4326	[A]	(J01)	ALL	4-12
4327	[A]	(J34)	ALL	7-5
4328	[C]	(J42)	RTC	8-36
4329	[A]	(B11)	RTC	5-14
4330	[C]	(J42)	RTC	8-38
4331	[C]	(J42)	ALL	4-31
4332	[C]	(J42)	ALL	8-17
4333	[C]	(J15)	RTC	6-8
4334	[B]	(H342)	RTC	3-5
4335	[B]	(H342)	RTC	7-26
4336	[C]	(J35)	ALL	7-8
4337	[A]	(J34)	ALL	4-12
4338	[A]	(J35)	ALL	4-15

Cross-Reference A: Answer, SMK Code, Category & Page Number

Question	Answer	SMK Code	Category	Page
4339	[B]	(J35)	ALL	7-6
4340	[B]	(J42)	RTC	8-17
4341	[B]	(J42)	RTC	8-17
4342	[A]	(J34)	RTC	8-17
4343	[C]	(J42)	RTC	8-17
4344	[A]	(J15)	ALL	6-9
4345	[A]	(H342)	ALL	3-5
4346	[B]	(H342)	ALL	7-26
4347	[A]	(J35)	ALL	4-16
4348	[C]	(J35)	ALL	7-8
4349	[B]	(J41)	ALL	8-4
4350	[B]	(J34)	ALL	8-18
4351	[B]	(J42)	ALL	8-18
4352	[A]	(J42)	ALL	8-18
4353	[C]	(J01)	ALL	4-35
4354	[B]	(J42)	ALL	8-18
4355	[A]	(J42)	ALL	8-18
4356	[B]	(J42)	ALL	8-36
4357	[C]	(J42)	ALL	8-18
4358	[C]	(J15)	ALL	6-10
4359	[B]	(H342)	ALL	3-5
4360	[C]	(H342)	ALL	7-25
4361	[B]	(J40)	ALL	6-23
4362	[C]	(J01)	ALL	4-9
4363	[C]	(J40)	ALL	6-23
4364	[C]	(J40)	ALL	6-23
4365	[C]	(J34)	ALL	7-6
4366	[B]	(J35)	ALL	7-8
4367	[C]	(J35)	ALL	4-32
4368	[B]	(J42)	ALL	8-19
4369	[B]	(J42)	ALL	8-19
4370-1	[B]	(J35)	ALL	7-9
4370-2	[A]	(H842)	ALL	7-9
4371	[B]	(J42)	ALL	8-19
4372	[A]	(J01)	ALL	4-11
4373	[B]	(J02)	ALL	6-28
4374	[A]	(J24)	ALL	5-38
4375	[B]	(J08)	ALL	5-13
4376	[C]	(J01)	ALL	4-12
4377	[B]	(J01)	ALL	4-10
4378	[B]	(J01)	ALL	4-12
4379	[C]	(J12)	ALL	5-35
4380	[B]	(J14)	ALL	5-35
4381	[A]	(J21)	ALL	5-36
4382	[A]	(J01)	ALL	4-10
4383	[B]	(J01)	ALL	4-10
4384	[B]	(J01)	ALL	4-11
4385	[C]	(J01)	ALL	4-10
4386	[C]	(J01)	ALL	4-10
4387	[C]	(J01)	ALL	4-11
4388	[C]	(J01)	ALL	4-11
4389	[A]	(J01)	ALL	4-11
4390	[B]	(J11)	ALL	6-26
4391	[A]	(J01)	ALL	4-11
4392	[B]	(J14)	ALL	6-19
4393	[C]	(J14)	ALL	6-20
4394	[B]	(J16)	ALL	6-19
4395	[B]	(J14)	ALL	6-18
4396	[B]	(J14)	ALL	6-18
4397	[A]	(J01)	ALL	4-4
4398	[C]	(J14)	ALL	6-19
4399	[B]	(J01)	ALL	4-4
4400	[C]	(J01)	ALL	4-4
4401	[B]	(J33)	ALL	8-35
4402	[C]	(J26)	ALL	3-11
4403	[B]	(J11)	ALL	8-5
4404	[A]	(J11)	ALL	8-6
4405	[B]	(J15)	ALL	6-15
4406	[B]	(J15)	ALL	5-27
4407	[A]	(J14)	ALL	6-17
4408	[B]	(J03)	ALL	8-38
4409	[C]	(J08)	ALL	6-26
4410	[A]	(J01)	ALL	4-12
4411	[C]	(J01)	ALL	4-13
4412	[C]	(J01)	ALL	4-6
4413	[B]	(J01)	ALL	4-5
4414	[B]	(J16)	ALL	6-19
4415	[C]	(J11)	ALL	6-26
4416	[B]	(J11)	ALL	6-26
4417	[C]	(J15)	ALL	6-23
4418	[C]	(J40)	ALL	6-24
4419	[C]	(J16)	ALL	6-24
4420	[B]	(J16)	ALL	6-27
4421	[A]	(J11)	ALL	6-26
4422	[A]	(J33)	ALL	6-26
4423	[B]	(J33)	ALL	6-27
4424	[C]	(J33)	ALL	6-27
4425	[A]	(J07)	ALL	5-32
4426	[C]	(J08)	ALL	5-13
4427	[C]	(J08)	ALL	6-6
4428	[C]	(J07)	ALL	5-32
4429	[A]	(J33)	ALL	7-3
4430	[B]	(J19)	ALL	6-30
4431	[C]	(J19)	ALL	6-30
4432	[A]	(J33)	ALL	7-3
4433	[C]	(J06)	ALL	6-30
4434	[B]	(J09)	ALL	5-22
4435	[A]	(J33)	ALL	7-3
4436	[C]	(J06)	ALL	7-4
4437	[A]	(J33)	ALL	7-4
4438	[A]	(J08)	ALL	5-12
4439	[C]	(J08)	ALL	5-14
4440	[A]	(J08)	ALL	5-14
4441	[A]	(B08)	ALL	5-33
4442	[C]	(J14)	ALL	6-21
4443	[B]	(J14)	ALL	6-20
4444	[C]	(J26)	ALL	3-10
4445	[C]	(J26)	ALL	3-9
4446	[B]	(J26)	ALL	3-10
4447	[A]	(J08)	ALL	6-30
4448	[C]	(B08)	ALL	5-11
4449	[B]	(J14)	ALL	6-31
4450	[A]	(J14)	ALL	6-31
4451	[A]	(J14)	ALL	6-31
4452	[B]	(J14)	ALL	6-31
4453	[C]	(J14)	ALL	6-31
4454	[B]	(J14)	ALL	6-31
4455	[A]	(J14)	ALL	6-32
4456	[B]	(J17)	ALL	5-34
4457	[C]	(J14)	ALL	6-32
4458	[A]	(J14)	ALL	6-20
4459	[C]	(B08)	ALL	5-11
4460	[A]	(J17)	ALL	5-35
4461	[A]	(J14)	ALL	6-17
4462	[B]	(J21)	ALL	5-37
4463	[B]	(J21)	ALL	5-37
4464	[A]	(J24)	ALL	5-37
4465	[A]	(J24)	ALL	5-37
4466	[A]	(J24)	ALL	5-38
4467	[C]	(J25)	ALL	2-13

Cross-Reference A: Answer, SMK Code, Category & Page Number

Question	Answer	SMK Code	Category	Page
4468	[A]	(J25)	ALL	2-8
4469	[C]	(J08)	ALL	8-6
4470	[C]	(J16)	ALL	8-19
4471	[C]	(J19)	ALL	6-28
4472	[B]	(H832)	ALL	4-5
4473	[A]	(J08)	ALL	5-22
4474	[B]	(J08)	ALL	5-23
4475	[A]	(J07)	ALL	5-23
4476	[C]	(J08)	ALL	5-23
4477	[A]	(H808)	ALL	3-7
4478	[A]	(H808)	ALL	3-7
4479	[B]	(H808)	ALL	3-8
4480	[C]	(J26)	ALL	3-10
4481	[C]	(J26)	ALL	3-10
4482	[C]	(J26)	ALL	3-11
4483	[C]	(H808)	ALL	3-12
4484	[B]	(H808)	ALL	3-12
4485	[A]	(J17)	ALL	5-23
4486	[B]	(J16)	ALL	6-19
4487	[B]	(I07)	ALL	4-5
4488	[C]	(J40)	ALL	6-24
4489	[A]	(J40)	ALL	6-24
4490	[B]	(J40)	ALL	6-24
4491	[A]	(J40)	ALL	6-25
4492	[C]	(J40)	ALL	6-25
4493	[C]	(J35)	ALL	7-9
4494	[A]	(J17)	ALL	7-13
4495	[A]	(J17)	ALL	4-16
4496	[C]	(J35)	ALL	7-9
4497	[C]	(J35)	ALL	7-9
4498	[C]	(J35)	ALL	7-10
4499	[A]	(J35)	ALL	7-10
4500	[C]	(J24)	ALL	5-38
4501	[B]	(J35)	ALL	7-7
4502	[B]	(J35)	ALL	7-10
4503	[C]	(B11)	ALL	5-17
4504	[C]	(J35)	ALL	2-14
4505	[B]	(B10)	ALL	5-38
4506	[C]	(J35)	ALL	7-6
4507	[A]	(J35)	ALL	4-16
4508	[C]	(J35)	ALL	5-23
4509	[B]	(J35)	ALL	7-10
4510	[C]	(J35)	ALL	6-32
4511	[C]	(H342)	ALL	7-27
4512	[C]	(J03)	ALL	7-11
4513	[C]	(B11)	ALL	5-17
4514	[B]	(H342)	ALL	7-28
4515	[B]	(J35)	ALL	7-11
4516	[B]	(J35)	ALL	7-8
4517	[B]	(J35)	ALL	7-10
4518	[B]	(B09)	ALL	5-24
4519	[A]	(B09)	ALL	5-25
4520	[C]	(B09)	ALL	5-25
4521	[B]	(B09)	ALL	5-25
4522	[C]	(B09)	ALL	5-25
4523	[A]	(B09)	ALL	5-25
4524	[B]	(B09)	ALL	5-25
4525	[C]	(B09)	ALL	5-26
4526	[B]	(J08)	ALL	5-21
4527	[B]	(J08)	ALL	5-23
4528	[B]	(J08)	ALL	5-21
4529	[C]	(J08)	ALL	5-22
4530	[B]	(J06)	ALL	5-22
4531	[B]	(J08)	ALL	5-22
4532	[C]	(J08)	ALL	5-22
4533	[C]	(J08)	ALL	5-22
4534	[B]	(J05)	ALL	6-29
4535	[B]	(J05)	ALL	6-29
4536	[A]	(J05)	ALL	6-29
4537	[B]	(J05)	ALL	6-29
4538	[C]	(J13)	ALL	6-28
4539	[A]	(J08)	ALL	5-24
4540	[B]	(J18)	ALL	7-5
4541	[C]	(B08)	ALL	5-33
4542	[C]	(J33)	ALL	7-4
4543	[C]	(B08)	ALL	5-33
4544	[B]	(J33)	ALL	7-4
4545	[A]	(J33)	ALL	7-4
4546	[B]	(J10)	ALL	7-5
4547	[B]	(J33)	ALL	7-5
4548	[C]	(H831)	ALL	4-13
4549	[C]	(H831)	ALL	4-13
4550	[C]	(H831)	ALL	4-13
4551	[C]	(H831)	ALL	4-13
4552	[B]	(H576)	ALL	4-13
4553	[B]	(I08)	ALL	4-14
4554	[C]	(H831)	ALL	4-14
4555	[C]	(J14)	ALL	6-21
4556	[C]	(H831)	ALL	4-14
4557	[A]	(H576)	ALL	4-14
4558	[C]	(H831)	ALL	4-14
4559	[B]	(H831)	ALL	4-14
4560	[C]	(H831)	ALL	4-15
4561	[A]	(H831)	ALL	4-15
4562	[C]	(H831)	ALL	4-15
4563	[A]	(H831)	ALL	4-19
4564	[B]	(H831)	ALL	4-19
4565	[C]	(H831)	ALL	4-19
4566	[C]	(H831)	ALL	4-19
4567	[C]	(H831)	ALL	4-20
4568	[A]	(H831)	ALL	4-20
4569	[B]	(H831)	ALL	4-20
4570	[B]	(H831)	ALL	4-20
4571	[C]	(H831)	ALL	4-21
4572	[C]	(H831)	ALL	4-21
4573	[B]	(H831)	ALL	4-21
4574	[C]	(H831)	ALL	4-22
4575	[A]	(H831)	ALL	4-22
4576	[C]	(H831)	ALL	4-23
4577	[C]	(H831)	ALL	4-23
4578	[B]	(H830)	ALL	4-24
4579	[B]	(H831)	ALL	4-29
4580	[B]	(H831)	ALL	4-30
4581	[A]	(H831)	ALL	4-30
4582	[B]	(H831)	ALL	4-30
4583	[C]	(H830)	ALL	4-24
4584	[A]	(H830)	ALL	4-24
4585	[B]	(H830)	ALL	4-25
4586	[B]	(H830)	ALL	4-25
4587	[A]	(H831)	ALL	4-30
4588	[C]	(H831)	ALL	4-30
4589	[A]	(H831)	ALL	4-31
4590	[C]	(H831)	ALL	4-31
4591	[C]	(H830)	ALL	4-25
4592	[C]	(H830)	ALL	4-25
4593	[B]	(H830)	ALL	4-26
4594	[A]	(H830)	ALL	4-26
4595	[B]	(H830)	ALL	4-26
4596	[A]	(H830)	ALL	4-27
4597	[B]	(H830)	ALL	4-27

Cross-Reference A: Answer, SMK Code, Category & Page Number

Question	Answer	SMK Code	Category	Page
4598	[C]	(H830)	ALL	4-27
4599	[A]	(H830)	ALL	4-28
4600	[C]	(H830)	ALL	4-28
4601	[B]	(H831)	ALL	4-15
4602	[B]	(H831)	ALL	4-31
4603	[A]	(H831)	ALL	4-31
4604	[Removed by the FAA]			
4605	[B]	(J17)	ALL	5-36
4606	[A]	(H831)	ALL	4-16
4607	[C]	(H831)	ALL	4-17
4608	[C]	(H831)	ALL	4-17
4609	[B]	(J17)	ALL	7-13
4610	[A]	(J17)	ALL	7-14
4611	[B]	(J17)	ALL	7-14
4612	[C]	(J17)	ALL	7-15
4613	[B]	(J17)	ALL	7-15
4614	[C]	(J17)	ALL	7-16
4615	[C]	(J17)	AIR	7-16
4616	[A]	(J17)	ALL	7-17
4617	[B]	(J17)	ALL	7-20
4618	[C]	(J17)	ALL	7-17
4619	[B]	(J17)	ALL	7-18
4620	[C]	(J17)	ALL	7-20
4621	[C]	(J17)	ALL	7-18
4622	[B]	(J17)	ALL	7-19
4623	[A]	(J17)	ALL	7-19
4624	[A]	(J17)	ALL	7-20
4625	[C]	(J17)	ALL	7-20
4626	[B]	(J17)	ALL	7-20
4627	[B]	(J17)	ALL	8-32
4628	[B]	(J18)	ALL	8-33
4629	[C]	(J18)	ALL	8-33
4630	[A]	(B10)	ALL	6-13
4631	[C]	(J18)	ALL	8-34
4632	[B]	(J18)	ALL	8-10
4633	[C]	(J19)	ALL	6-28
4634	[C]	(J14)	ALL	5-32
4635	[B]	(J42)	ALL	8-19
4636	[A]	(J42)	ALL	8-11
4637	[C]	(B10)	ALL	6-14
4638	[B]	(J16)	ALL	6-25
4639	[C]	(J15)	ALL	7-11
4640	[B]	(J18)	ALL	8-3
4641	[C]	(J18)	ALL	8-7
4642	[B]	(J42)	ALL	8-19
4643	[C]	(J14)	ALL	6-32
4644	[C]	(J14)	RTC	6-20
4645	[C]	(J35)	ALL	7-6
4646	[A]	(J35)	ALL	7-6
4647	[C]	(J35)	ALL	7-6
4648	[B]	(J42)	ALL	8-20
4649	[A]	(J42)	ALL	8-20
4650	[B]	(J42)	ALL	8-20
4651	[B]	(J42)	ALL	8-20
4652	[C]	(J42)	ALL	8-20
4653	[B]	(J42)	ALL	5-13
4654	[B]	(J42)	ALL	8-20
4655	[A]	(J42)	ALL	8-21
4656	[A]	(J42)	ALL	8-21
4657	[B]	(J42)	ALL	8-21
4658	[C]	(J18)	ALL	8-21
4659	[B]	(J42)	ALL	8-21
4660	[C]	(J33)	ALL	8-22
4661	[A]	(J42)	ALL	8-22
4662	[A]	(J42)	ALL	8-22
4663	[B]	(J01)	ALL	4-5
4664	[A]	(J01)	ALL	4-6
4665	[C]	(J01)	ALL	4-33
4666	[B]	(H831)	ALL	4-15
4667	[A]	(J18)	ALL	8-34
4668	[C]	(J18)	ALL	7-21
4669	[A]	(J42)	ALL	4-6
4670	[A]	(J18)	ALL	8-11
4671	[A]	(J18)	ALL	8-25
4672	[A]	(J18)	ALL	8-22
4673	[C]	(B97)	RTC	8-22
4674	[C]	(J01)	ALL	4-5
4675	[A]	(J17)	ALL	7-21
4676	[C]	(B97)	RTC	8-22
4677	[B]	(J42)	ALL	8-23
4678	[A]	(J18)	ALL	8-23
4679	[C]	(B97)	RTC	8-23
4680	[A]	(J17)	ALL	8-23
4681	[C]	(J17)	ALL	7-21
4682	[A]	(J42)	ALL	8-24
4683	[C]	(J42)	ALL	8-24
4684	[C]	(J42)	ALL	4-33
4685	[B]	(H837)	ALL	4-38
4686	[A]	(J42)	ALL	8-24
4687	[B]	(J42)	ALL	8-24
4688	[A]	(J42)	ALL	8-24
4689	[A]	(J42)	ALL	8-24
4690	[B]	(J42)	ALL	8-25
4691	[B]	(J42)	ALL	8-25
4692	[B]	(J18)	ALL	8-25
4693	[C]	(J42)	ALL	8-25
4694	[C]	(J42)	ALL	8-25
4695	[A]	(J42)	ALL	5-13
4696	[B]	(J42)	ALL	8-26
4697	[A]	(B97)	RTC	8-26
4698	[A]	(J17)	ALL	7-21
4699	[C]	(J17)	ALL	8-36
4700	[C]	(J42)	ALL	8-26
4701	[C]	(J42)	ALL	8-26
4702	[A]	(J01)	ALL	4-35
4703	[A]	(J01)	ALL	4-36
4704	[B]	(J01)	ALL	4-36
4705	[B]	(J01)	ALL	4-36
4706	[B]	(B10)	ALL	8-38
4707	[C]	(J27)	ALL	8-39
4708	[C]	(J27)	ALL	8-39
4709	[B]	(J27)	ALL	8-39
4710	[C]	(J27)	ALL	8-39
4711	[C]	(J18)	ALL	8-29
4712	[A]	(J18)	ALL	8-32
4713	[B]	(B10)	RTC	8-27
4714	[A]	(J18)	ALL	8-27
4715	[A]	(J33)	ALL	8-10
4716	[A]	(B10)	ALL	8-35
4717	[B]	(J18)	ALL	8-27
4718	[A]	(J18)	ALL	8-30
4719	[C]	(B10)	ALL	6-13
4720	[B]	(K04)	ALL	1-25
4721	[B]	(I10)	ALL	3-34
4722	[B]	(J42)	RTC	8-27
4723	[A]	(J42)	RTC	8-27
4724	[C]	(B97)	RTC	8-27
4725	[C]	(J19)	ALL	6-29
4726	[B]	(J18)	ALL	8-6
4727	[C]	(K04)	ALL	1-25

Cross-Reference A: Answer, SMK Code, Category & Page Number

Question	Answer	SMK Code	Category	Page
4728	[C]	(J18)	ALL	8-29
4729	[C]	(J01)	ALL	4-36
4730	[C]	(J01)	ALL	4-36
4731	[A]	(J42)	ALL	8-36
4732	[A]	(J01)	ALL	8-37
4733	[B]	(J42)	ALL	8-37
4734	[A]	(J18)	ALL	8-7
4735	[C]	(J18)	ALL	8-31
4736	[A]	(J19)	ALL	6-27
4737	[C]	(J18)	ALL	8-31
4738	[B]	(V14)	ALL	8-39
4739	[B]	(K04)	ALL	1-25
4740	[B]	(J18)	ALL	8-10
4741	[C]	(J18)	ALL	8-29
4742	[B]	(J01)	ALL	8-37
4743	[C]	(J18)	ALL	8-31
4744	[A]	(J18)	ALL	8-28
4745	[C]	(K04)	ALL	3-34
4746	[C]	(J18)	ALL	8-10
4747	[B]	(J01)	ALL	4-36
4748	[C]	(K04)	ALL	3-34
4749	[C]	(J18)	ALL	8-28
4750	[A]	(J18)	ALL	8-31
4751	[C]	(J18)	ALL	8-4
4752	[C]	(K04)	ALL	3-34
4753	[B]	(J01)	ALL	4-37
4754	[C]	(B10)	ALL	8-35
4755	[A]	(K04)	ALL	1-26
4756	[B]	(H815)	ALL	3-35
4757	[C]	(J18)	ALL	8-7
4758	[B]	(J11)	ALL	6-17
4759	[A]	(B10)	ALL	8-35
4760-1	[B]	(B10)	ALL	6-11
4760-2	[B]	(B10)	ALL	6-11
4761	[C]	(B10)	ALL	6-15
4762	[C]	(B10)	ALL	8-35
4763	[A]	(B10)	ALL	8-28
4764	[B]	(J01)	ALL	8-28
4765	[C]	(B10)	ALL	5-33
4766	[B]	(J17)	ALL	7-20
4767	[A]	(J17)	ALL	8-11
4768	[B]	(J17)	ALL	8-33
4769	[C]	(B10)	ALL	6-14
4770	[B]	(J01)	ALL	8-37
4771	[A]	(J18)	ALL	8-11
4772	[B]	(K04)	ALL	3-35
4773	[A]	(H837)	ALL	4-37
4774	[B]	(J03)	ALL	8-40
4775	[B]	(J03)	ALL	8-40
4776	[C]	(J03)	ALL	8-41
4777	[A]	(J03)	ALL	8-41
4778	[A]	(J03)	ALL	8-41
4779	[C]	(J03)	ALL	8-41
4780	[A]	(J03)	ALL	8-42
4781	[B]	(J03)	ALL	8-42
4782	[C]	(J03)	ALL	8-42
4783	[B]	(J03)	ALL	8-42
4784	[C]	(J03)	ALL	8-42
4785	[A]	(J03)	ALL	8-43
4786	[B]	(J03)	ALL	8-43
4787	[C]	(J03)	ALL	8-43
4788	[A]	(J03)	ALL	8-43
4789	[B]	(J03)	ALL	8-43
4790	[C]	(J03)	ALL	8-43
4791	[B]	(J05)	ALL	8-44
4792	[B]	(J05)	ALL	8-44
4793	[B]	(J05)	ALL	8-44
4794	[A]	(J05)	ALL	8-45
4795	[C]	(J03)	ALL	8-44
4796	[A]	(J03)	ALL	8-44
4797	[C]	(J05)	ALL	8-45
4798	[B]	(J01)	ALL	4-39
4799	[A]	(J01)	ALL	4-39
4800	[B]	(J01)	RTC	4-39
4801	[C]	(J01)	ALL	4-39
4802	[A]	(H800)	ALL	5-28
4803	[A]	(J31)	ALL	5-30
4804	[A]	(J31)	ALL	5-30
4805	[A]	(J31)	ALL	5-29
4806	[C]	(J31)	ALL	5-30
4807	[A]	(J31)	ALL	5-29
4808	[B]	(J31)	ALL	5-29
4809	[B]	(J31)	ALL	5-18
4810	[C]	(J31)	ALL	5-28
4811	[C]	(J31)	ALL	5-28
4812	[B]	(J31)	ALL	5-31
4813	[B]	(J31)	ALL	5-28
4814	[C]	(J31)	ALL	5-28
4815	[C]	(J31)	ALL	5-28
4816	[B]	(J31)	ALL	5-18
4817	[A]	(J31)	ALL	5-31
4818	[A]	(J31)	ALL	5-31
4819	[C]	(J31)	ALL	5-30
4820	[B]	(H814)	ALL	3-14
4821	[A]	(L57)	ALL	3-30
4822	[C]	(J18)	ALL	8-30
4823	[A]	(J18)	ALL	8-30
4824	[B]	(H837)	ALL	4-37
4825	[C]	(H837)	ALL	4-37
4826	[A]	(H837)	ALL	4-37
4827	[C]	(H809)	ALL	3-28
4828	[A]	(H809)	ALL	3-28
4829	[C]	(H809)	ALL	3-29
4830	[C]	(L57)	ALL	3-30
4831	[B]	(H812)	ALL	3-17
4832	[A]	(H814)	ALL	3-36
4833-1	[A]	(H816)	ALL	3-19
4833-2	[C]	(H825)	ALL	3-19
4834	[C]	(H812)	ALL	3-25
4835	[A]	(H812)	ALL	3-14
4836	[A]	(H815)	ALL	3-36
4837	[A]	(H815)	ALL	3-36
4838	[B]	(H816)	ALL	3-37
4839	[A]	(H810)	ALL	3-18
4840	[C]	(H813)	ALL	3-33
4841	[C]	(H822)	RTC	3-16
4842	[C]	(H812)	ALL	3-14
4843	[A]	(H807)	ALL	3-19
4844	[B]	(H807)	ALL	3-19
4845-1	[A]	(H815)	ALL	3-37
4845-2	[C]	(H823)	RTC	3-37
4846	[B]	(H828)	RTC	3-16
4847	[A]	(H810)	ALL	3-18
4848	[C]	(H815)	ALL	3-37
4849	[B]	(H827)	RTC	3-42
4850-1	[A]	(H816)	ALL	3-37
4850-2	[C]	(H822)	ALL	3-37
4851	[C]	(H816)	ALL	3-38
4852	[C]	(H827)	RTC	3-16
4853	[B]	(H815)	ALL	3-38

Cross-Reference A: Answer, SMK Code, Category & Page Number

Question	Answer	SMK Code	Category	Page
4854	[A]	(L57)	ALL	3-30
4855	[B]	(H813)	ALL	3-33
4856	[A]	(H816)	ALL	3-18
4857	[C]	(H810)	ALL	3-14
4858	[A]	(H816)	ALL	3-38
4859	[C]	(H813)	ALL	3-33
4860	[B]	(H810)	ALL	3-15
4861	[B]	(H825)	RTC	3-13
4862	[B]	(H813)	ALL	3-34
4863	[C]	(H814)	ALL	3-38
4864	[A]	(H810)	ALL	3-5
4865	[C]	(H814)	ALL	3-38
4866	[B]	(H816)	ALL	3-39
4867	[A]	(H818)	ALL	3-43
4868	[B]	(H807)	ALL	3-20
4869	[C]	(H813)	ALL	3-39
4870-1	[C]	(H807)	ALL	3-18
4870-2	[C]	(H703)	RTC	3-18
4871	[C]	(H814)	ALL	3-40
4872	[C]	(H816)	ALL	3-40
4873-1	[B]	(H818)	AIR	3-43
4873-2	[B]	(H826)	RTC	3-43
4874	[A]	(H816)	ALL	3-41
4875-1	[B]	(H818)	AIR	3-43
4875-2	[B]	(H826)	RTC	3-44
4876	[C]	(H813)	ALL	3-41
4877	[C]	(H809)	ALL	3-25
4878-1	[B]	(H816)	ALL	3-20
4878-2	[B]	(H816)	ALL	3-20
4878-3	[B]	(H825)	RTC	3-21
4879	[C]	(H808)	ALL	3-31
4880	[C]	(H812)	ALL	3-11
4881	[C]	(L59)	ALL	3-29
4882	[B]	(L59)	ALL	3-17
4883	[A]	(L59)	ALL	3-17
4884	[C]	(H807)	ALL	3-18
4885	[A]	(H812)	ALL	3-24
4886	[C]	(H314)	ALL	3-26
4887	[A]	(H314)	ALL	3-26
4888	[B]	(H314)	ALL	3-26
4889	[C]	(H809)	ALL	3-26
4890	[A]	(H314)	ALL	3-26
4891	[C]	(H314)	ALL	3-27
4892	[B]	(H314)	ALL	3-27
4893	[C]	(H314)	ALL	3-27
4894	[B]	(H314)	ALL	3-27
4895	[C]	(H810)	ALL	3-21
4896	[A]	(H810)	ALL	3-21
4897	[B]	(H810)	ALL	3-21
4898	[C]	(H807)	ALL	3-21
4899	[A]	(H807)	ALL	3-41
4900	[A]	(H810)	ALL	3-15
4901	[C]	(H810)	ALL	3-14
4902	[B]	(H810)	ALL	3-13
4903	[A]	(H810)	ALL	3-21
4904	[C]	(H807)	ALL	3-22
4905	[C]	(H807)	ALL	3-22
4906	[A]	(H815)	ALL	3-41
4907	[A]	(H815)	ALL	3-41
4908	[C]	(H859)	ALL	3-31
4909	[A]	(H808)	ALL	3-31
4910	[B]	(H312)	ALL	3-11
4911	[B]	(H808)	ALL	3-8
4912	[B]	(H808)	ALL	3-8
4913	[C]	(H808)	ALL	3-8
4914	[B]	(H807)	ALL	3-22
4915	[B]	(H807)	ALL	3-22
4916	[B]	(H303)	ALL	1-17
4917	[B]	(K04)	ALL	1-26
4918	[A]	(H810)	ALL	3-15
4919	[C]	(H810)	ALL	3-15
4920	[C]	(H814)	ALL	3-41
4921	[C]	(H816)	ALL	3-22
4922	[B]	(I04)	ALL	3-8
4923	[B]	(H808)	ALL	3-8
4924	[C]	(H815)	ALL	3-42
4925	[B]	(H815)	ALL	3-42
4926	[B]	(H815)	ALL	3-42
4927	[C]	(H818)	ALL	3-44
4928	[C]	(H815)	ALL	3-15
4929	[A]	(H815)	ALL	3-15
4930	[B]	(H312)	ALL	3-31
4931	[A]	(H814)	ALL	3-23
4932	[A]	(H814)	ALL	3-23
4933	[B]	(H814)	ALL	3-23
4934	[B]	(H814)	ALL	3-23
4935	[C]	(H814)	ALL	3-23
4936	[B]	(H818)	ALL	3-44
4937	[A]	(H818)	ALL	3-32
4938	[B]	(H818)	ALL	3-44
4939	[B]	(H818)	ALL	3-32
4940	[C]	(H818)	ALL	3-32
4941	[A]	(H818)	ALL	3-32
4942	[B]	(H826)	RTC	3-32

Cross-Reference B:
Subject Matter Knowledge Code & Question Number

The subject matter knowledge codes establish the specific reference for the knowledge standard. When reviewing results of your knowledge test, you should compare the subject matter knowledge code(s) on your test report to the ones found below. All the questions on the Instrument test have been broken down into their subject matter knowledge codes and listed under the appropriate reference. This will be helpful for both review and preparation for the practical test. This list of Subject Matter Knowledge Codes is taken from Advisory Circular 60-25E.

14 CFR Part 61: Certification: Pilots, Flight Instructors, and Ground Instructors

A20......... General
4001, 4008, 4009, 4010, 4012, 4013, 4014, 4015, 4016, 4017, 4018, 4019, 4020, 4021, 4022, 4023, 4024, 4025, 4026, 4027, 4028, 4029, 4030-1, 4030-2, 4031, 4034, 4035

A24......... Commercial Pilots
4002

14 CFR Part 91: General Operating and Flight Rules

B07......... General
4004, 4039

B08......... Flight Rules: General
4003, 4011, 4033, 4441, 4448, 4459, 4541, 4543

B09......... Visual Flight Rules
4518, 4519, 4520, 4521, 4522, 4523, 4524, 4525

B10......... Instrument Flight Rules
4005, 4006, 4032, 4036, 4040, 4041, 4044, 4046, 4048, 4062, 4063, 4064, 4065, 4066, 4067, 4068, 4081, 4082-1, 4082-2, 4083-1, 4083-2, 4083-3, 4084-1, 4084-2, 4085-1, 4085-2, 4086, 4087-1, 4087-2, 4505, 4630, 4637, 4706, 4713, 4716, 4719, 4754, 4759, 4760-1, 4760-2, 4761, 4762, 4763, 4765, 4769

B11......... Equipment, Instrument, and Certificate Requirements
4007, 4037, 4038, 4042, 4043, 4045, 4050, 4051, 4052, 4053, 4055, 4057, 4329, 4503, 4513

B13......... Maintenance, Preventive Maintenance, and Alterations
4047, 4049

14 CFR Part 97: Standard Instrument Approach Procedures

B97......... General
4673, 4676, 4679, 4697, 4724

Cross-Reference B: SMK Code & Question Number

NTSB Part 830: Rules Pertaining to the Notification and Reporting of Aircraft Accidents or Incidents and Overdue Aircraft, and Preservation of Aircraft Wreckage, Mail, Cargo, and Records

G10 .... General
    4088

AC 61-23: Pilot's Handbook of Aeronautical Knowledge

H303 .... Loads and Load Factors
    4916

H312 .... The Pitot-Static System and Associated Instruments
    4910, 4930

H314 .... Magnetic Compass
    4886, 4887, 4888, 4890, 4891, 4892, 4893, 4894

H342 .... Basic Calculations
    4259, 4260, 4265, 4267, 4268, 4270, 4278, 4279, 4289, 4290, 4301, 4302, 4313, 4314, 4323, 4324, 4334, 4335, 4345, 4346, 4359, 4360, 4511, 4514

FAA-H-8083-3: Airplane Flying Handbook

H576 .... VOR Navigation
    4552, 4557

FAA-H-8083-21: Rotorcraft Flying Handbook

H703 .... Aerodynamics of Flight
    4870-2

FAA-H-8083-15: Instrument Flying Handbook

Human Factors

H800 .... Sensory Systems
    4802

Aerodynamics

H807 .... Basic Aerodynamics
    4843, 4844, 4868, 4870-1, 4884, 4898, 4899, 4904, 4905, 4914, 4915

Flight Instruments

H808 .... Pitot Static
    4111, 4477, 4478, 4479, 4483, 4484, 4879, 4909, 4911, 4912, 4913, 4923

H809 .... Compass
    4827, 4828, 4829, 4877, 4889

H810 .... Gyroscopic
    4839, 4847, 4857, 4860, 4864, 4895, 4896, 4897, 4900, 4901, 4902, 4903, 4918, 4919

Cross-Reference B: SMK Code & Question Number

H812 .... Systems Preflight
    4831, 4834, 4835, 4842, 4880, 4885

Airplane Attitude Instrument Flying

H813 .... Fundamental Skills
    4840, 4855, 4859, 4862, 4869, 4876

Airplane Basic Flight Maneuvers

H814 .... Straight-and-Level Flight
    4056, 4820, 4832, 4863, 4865, 4871, 4920, 4931, 4932, 4933, 4934, 4935

H815 .... Straight Climbs and Descents
    4756, 4836, 4837, 4845-1, 4848, 4853, 4906, 4907, 4924, 4925, 4926, 4928, 4929

H816 .... Turns
    4833-1, 4838, 4850-1, 4851, 4856, 4858, 4866, 4872, 4874, 4878-1, 4878-2, 4921

H818 .... Unusual Attitude Recoveries
    4867, 4873-1, 4875-1, 4927, 4936, 4937, 4938, 4939, 4940, 4941

Helicopter Attitude Instrument Flying

H822 .... Straight-and-Level
    4841, 4850-2

H823 .... Straight Climbs
    4845-2

H825 .... Turns
    4833-2, 4861, 4878-3

H826 .... Unusual Attitude Recoveries
    4873-2, 4875-2, 4942

H827 .... Emergencies
    4849, 4852

H828 .... Instrument Takeoff
    4846

Navigation Systems

H830 .... Nondirectional Beacon (NDB)
    4578, 4583, 4584, 4585, 4586, 4591, 4592, 4593, 4594, 4595, 4596, 4597, 4598, 4599, 4600

H831 .... Very High Frequency Omnidirectional Range (VOR)
    4548, 4549, 4550, 4551, 4554, 4556, 4558, 4559, 4560, 4561, 4562, 4563, 4564, 4565, 4566, 4567, 4568, 4569, 4570, 4571, 4572, 4573, 4574, 4575, 4576, 4577, 4579, 4580, 4581, 4582, 4587, 4588, 4589, 4590, 4601, 4602, 4603, 4606, 4607, 4608, 4666

H832 .... Distance Measuring Equipment (DME)
    4269, 4320, 4472

H833 .... Area Navigation (RNAV)
    4282, 4283-1, 4283-2

Cross-Reference B: SMK Code & Question Number

H837 .... Instrument Landing System (ILS)
    4685, 4773, 4824, 4825, 4826

National Airspace System
H842 .... IFR Enroute Charts
    4370-2

Emergency Operations
H859 .... Aircraft System Malfunction
    4908

Glossary
H862 .... Glossary
    4069

AC 61-27: Instrument Flying Handbook

I04 .... Basic Flight Instruments
    4922

I07 .... Electronic Aids to Instrument Flying
    4487

I08 .... Using the Navigation Instruments
    4553

I10 .... The Federal Airways System and Controlled Airspace
    4721

AC 00-6: Aviation Weather

I20 .... The Earth's Atmosphere
    4097, 4154, 4227

I21 .... Temperature
    4094, 4095, 4096, 4112, 4113, 4114, 4125, 4200

I22 .... Atmospheric Pressure and Altimetry
    4089, 4090, 4091, 4093, 4109, 4110

I23 .... Wind
    4105, 4106, 4107, 4108, 4238

I24 .... Moisture, Cloud Formation, and Precipitation
    4099, 4101, 4102, 4103, 4104, 4159, 4161

I25 .... Stable and Unstable Air
    4115, 4118, 4121, 4122, 4123, 4124

I26 .... Clouds
    4131, 4134, 4157

I27 .... Air Masses and Fronts
4098, 4116, 4117, 4119, 4120, 4127, 4128, 4136, 4141, 4158

I28 .... Turbulence
4129, 4130, 4132, 4138, 4139, 4140, 4149, 4150

I29 .... Icing
4100, 4133, 4151, 4152, 4153, 4171

I30 .... Thunderstorms
4126, 4135, 4137, 4143, 4144, 4145, 4146, 4147, 4148, 4160

I31 .... Common IFR Producers
4092, 4156, 4163, 4164, 4165, 4166, 4167, 4169

I32 .... High Altitude Weather
4155, 4162, 4168

AC 00-45: Aviation Weather Services

I43 .... Aviation Weather Forecasts
4228

I54 .... The Aviation Weather Service Program
4241

I55 .... Aviation Routine Weather Report (METAR)
4196, 4202, 4205

I56 .... Pilot and Radar Reports, Satellite Pictures, and Radiosonde Additional Data (RADATs)
4182, 4184, 4198, 4220

I57 .... Aviation Weather Forecasts
4142, 4170, 4176, 4177, 4178, 4179, 4180, 4181, 4183, 4185, 4186, 4187, 4201, 4203, 4204, 4226

I58 .... Surface Analysis Chart
4209

I59 .... Weather Depiction Chart
4206, 4207, 4208

I60 .... Radar Summary Chart
4174, 4230, 4231, 4232, 4233, 4234, 4235, 4236, 4237

I61 .... Constant Pressure Analysis Charts
4173, 4195

I63 .... Winds and Temperatures Aloft
4172, 4188, 4189, 4190, 4191, 4192, 4193, 4194, 4199, 4246, 4247, 4249, 4250

I64 .... Significant Weather Prognostics
4175, 4211, 4212, 4213, 4214, 4216, 4217, 4218, 4219, 4221, 4222, 4223, 4224, 4225, 4229, 4242, 4243, 4244, 4245

Cross-Reference B: SMK Code & Question Number

I65 .... Convective Outlook Chart
4197, 4215, 4239, 4240, 4248

I67 .... Turbulence Locations, Conversion and Density Altitude Tables, Contractions and Acronyms, Station Identifiers, WSR-88D Sites, and Internet Addresses
4210

AIM: Aeronautical Information Manual

J01 .... Air Navigation Radio Aids
4054, 4273, 4326, 4353, 4362, 4372, 4376, 4377, 4378, 4382, 4383, 4384, 4385, 4386, 4387, 4388, 4389, 4391, 4397, 4399, 4400, 4410, 4411, 4412, 4413, 4663, 4664, 4665, 4674, 4702, 4703, 4704, 4705, 4729, 4730, 4732, 4742, 4747, 4753, 4764, 4770, 4798, 4799, 4800, 4801

J02 .... Radar Services and Procedures
4373

J03 .... Airport Lighting Aids
4408, 4512, 4774, 4775, 4776, 4777, 4778, 4779, 4780, 4781, 4782, 4783, 4784, 4785, 4786, 4787, 4788, 4789, 4790, 4795, 4796

J05 .... Airport Marking Aids and Signs
4534, 4535, 4536, 4537, 4791, 4792, 4793, 4794, 4797

J06 .... Airspace: General
4080, 4433, 4436, 4530

J07 .... Class G Airspace
4425, 4428, 4475

J08 .... Controlled Airspace
4375, 4409, 4426, 4427, 4438, 4439, 4440, 4447, 4469, 4473, 4474, 4476, 4526, 4527, 4528, 4529, 4531, 4532, 4533, 4539

J09 .... Special Use Airspace
4434

J10 .... Other Airspace Areas
4546

J11 .... Service Available to Pilots
4390, 4403, 4404, 4415, 4416, 4421, 4758

J12 .... Radio Communications Phraseology and Techniques
4379

J13 .... Airport Operations
4538

J14 .... ATC Clearance/Separations
4380, 4392, 4393, 4395, 4396, 4398, 4407, 4442, 4443, 4449, 4450, 4451, 4452, 4453, 4454, 4455, 4457, 4458, 4461, 4555, 4634, 4643, 4644

J15 .... Preflight
4058, 4059, 4060, 4061, 4071, 4072, 4073, 4074, 4075, 4076, 4077, 4078, 4266, 4277, 4288, 4300, 4312, 4322, 4333, 4344, 4358, 4405, 4406, 4417, 4639

J16 .... Departure Procedures
4394, 4414, 4419, 4420, 4470, 4486, 4638

J17 .... En Route Procedures
4261, 4456, 4460, 4485, 4494, 4495, 4605, 4609, 4610, 4611, 4612, 4613, 4614, 4615, 4616, 4617, 4618, 4619, 4620, 4621, 4622, 4623, 4624, 4625, 4626, 4627, 4675, 4680, 4681, 4698, 4699, 4766, 4767, 4768

J18 .... Arrival Procedures
4295, 4540, 4628, 4629, 4631, 4632, 4640, 4641, 4658, 4667, 4668, 4670, 4671, 4672, 4678, 4692, 4711, 4712, 4714, 4717, 4718, 4726, 4728, 4734, 4735, 4737, 4740, 4741, 4743, 4744, 4746, 4749, 4750, 4751, 4757, 4771, 4822, 4823

J19 .... Pilot/Controller Roles and Responsibilities
4430, 4431, 4471, 4633, 4725, 4736

J21 .... Emergency Procedures: General
4381, 4462, 4463

J24 .... Two-Way Radio Communications Failure
4374, 4464, 4465, 4466, 4500

J25 .... Meteorology
4251, 4252, 4253, 4254, 4255, 4256, 4257, 4258, 4467, 4468

J26 .... Altimeter Setting Procedures
4402, 4444, 4445, 4446, 4480, 4481, 4482

J27 .... Wake Turbulence
4707, 4708, 4709, 4710

J31 .... Fitness for Flight
4803, 4804, 4805, 4806, 4807, 4808, 4809, 4810, 4811, 4812, 4813, 4814, 4815, 4816, 4817, 4818, 4819

J33 .... Pilot Controller Glossary
4401, 4422, 4423, 4424, 4429, 4432, 4435, 4437, 4542, 4544, 4545, 4547, 4660, 4715

J34 .... Airport/Facility Directory
4070, 4079, 4275, 4305, 4325, 4327, 4337, 4342, 4350, 4365

J35 .... En Route Low Altitude Chart
4262, 4263, 4264, 4280, 4287, 4291, 4317, 4318, 4336, 4338, 4339, 4347, 4348, 4366, 4367, 4370-1, 4493, 4496, 4497, 4498, 4499, 4501, 4502, 4504, 4506, 4507, 4508, 4509, 4510, 4515, 4516, 4517, 4645, 4646, 4647

J40 .... Instrument Departure Procedure (DP) Chart
4271, 4272, 4303, 4304, 4315, 4316, 4361, 4363, 4364, 4418, 4488, 4489, 4490, 4491, 4492

J41 .... Standard Terminal Arrival (STAR) Chart
4281, 4286, 4292, 4293, 4294, 4349

Cross-Reference B: SMK Code & Question Number

J42 .... Instrument Approach Procedures (IAP)
   4274, 4276, 4285, 4296, 4297, 4298, 4299, 4306, 4307, 4308, 4309, 4310, 4311, 4319, 4321, 4328, 4330, 4331, 4332, 4340, 4341, 4343, 4351, 4352, 4354, 4355, 4356, 4357, 4368, 4369, 4371, 4635, 4636, 4642, 4648, 4649, 4650, 4651, 4652, 4653, 4654, 4655, 4656, 4657, 4659, 4661, 4662, 4669, 4677, 4682, 4683, 4684, 4686, 4687, 4688, 4689, 4690, 4691, 4693, 4694, 4695, 4696, 4700, 4701, 4722, 4723, 4731, 4733

Additional Advisory Circulars

K04 .... AC 00-54, Pilot Wind Shear Guide
   4720, 4727, 4739, 4745, 4748, 4752, 4755, 4772, 4917

K26 .... AC 20-138, Airworthiness Approval of Global Positioning System (GPS) Navigation
   4284

L57 .... AC 91-43, Unreliable Airspeed Indications
   4821, 4830, 4854

L59 .... AC 91-46, Gyroscopic Instruments: Good Operating Practices
   4881, 4882, 4883

FAA Accident Prevention Program Bulletins

V14 .... FAA-P-8740-50, On Landings, Part III
   4738

Note: AC 00-2, Advisory Circular Checklist, transmits the status of all FAA advisory circulars (ACs), as well as FAA internal publications and miscellaneous flight information such as Aeronautical Information Manual (AIM), Airport/Facility Directory, practical test standards, knowledge test guides, and other material directly related to airman certificates and ratings. To obtain a free copy of AC 00-2, send your request to:

   U.S. Department of Transportation
   General Services Section, M-45.3
   Washington, DC 20590

Notes

Notes

Notes

Notes

More Instrument Products from ASA

ASA has many other books and supplies for the Instrument Pilot. They are listed below and available from an aviation retailer in your area. Need help locating a retailer? Call ASA at 1-800-272-2359 or visit our website (www.asa2fly.com). We can also send you the latest ASA Catalog which includes our *complete* line of publications and pilot supplies…for all types of pilots and aviation technicians.

Instrument Books and Software

	Product Code	Suggested Price
The Pilot's Manual Series: Instrument Flying	ASA-PM-3	$34.95
Instrument Rating Syllabus (for Pilot's Manual Series)	ASA-PM-S-I2	10.95
Instrument Oral Exam Guide by Michael Hayes	ASA-OEG-I4	9.95
Instrument Pilot Practical Test Standards	ASA-8081-4C	4.95
Flight Instructor Instrument Practical Test Standards **New!**	ASA-8081-9B	4.95
Instrument Test Prep	ASA-TP-I	19.95
Instrument Flying Handbook **New!**	ASA-8083-15	24.95
Prepware™ for Instrument Rating	ASA-TW-I	49.95
IP Trainer Courseware and Simulator Software	ASA-IP-6.0	195.00

Pilot Supplies

		Product Code	Price
IFR Kneeboard		ASA-KB-2	$14.95
Folding Lapboard		ASA-KB-LAP	26.95
Holding Pattern Visualizer		ASA-HPC-2	12.95
Instrument Plotter		ASA-CP-IFR	6.95
Jiffyhood IFR Training Hood		ASA-H2G	12.95
Instrument Covers for Partial Panel Practice		ASA-STICKY	4.95

Instrument Approach Binder Products

		Product Code	Price
Kits (includes Binder, Dividers, 10 Sheet Protectors)	*NOS style*	ASA-AP-KT-NOS	19.95
	7-ring style	ASA-AP-KT-7RNG	19.95
Sheet Protectors (10)	*4-ring NOS style*	ASA-AP-SP-NOS	4.95
	7-ring style	ASA-AP-SP-7RNG	4.95
Binders	*4-ring NOS style*	ASA-AP-BD-NOS	9.95
	Flip-style NOS binder	ASA-AP-BD-FLIP	9.95
	7-ring style	ASA-AP-BD-7RNG	9.95
5 Color Dividers	*4-ring NOS style*	ASA-AP-DIV-NOS5	4.95
	7-ring style	ASA-AP-DIV-7RNG	4.95

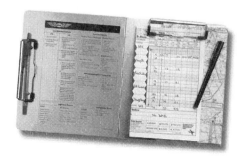

All prices based on U.S. currency.
Prices subject to change without notice.

1.800.ASA.2.FLY   **www.asa2fly.com**

IP Trainer™
Instrument Pilot Procedures Course

You studied for your written exam at home...
Now make room for your instructor and an airplane.

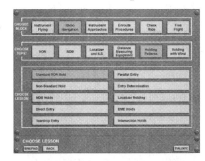

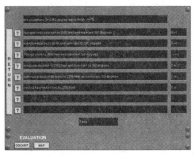

ASA's **IP Trainer**™ Instrument Pilot Procedures Course is more than just another IFR simulator. It's the only program of its kind that includes 133 interactive lessons, taking the instrument student through each element of the IFR Practical Test Standards.

From basic attitude instrument flying through intersection holds and partial-panel work, **IP Trainer's** built-in CFII demonstrates, explains, and tests you on each maneuver, using the same standards that you'll face on your checkride. Each lesson builds on the previous one, until you're executing even complex maneuvers with skill and confidence.

IP Trainer's built-in CFII talks you through complex, real-world IFR procedures, with explanations every step of the way.

- 133 Interactive Lessons that cover every facet of instrument flight training.
- Digitized voice for your CFII and ATC, preparing you for the world of Instrument Flying.
- Fly each lesson in an aerodynamically correct C172 simulator.
- Includes charts, plates and legends necessary to fly lessons.
- Includes the comprehensive textbook *Instrument Flying*, for an integrated flight and ground training program.
- Fly real approaches into real airports, with a virtual CFII monitoring your performance.
- Understand concepts before you fly, saving money on your aircraft rental, your dual time, and your entire rating program.
- Visit the ASA website to download a free demo: www.asa2fly.com

Aviation Supplies & Academics, Inc.
7005 132nd Place SE
Newcastle, Washington 98059-3153
1-800-ASA-2-FLY